D1288196

Theoretical and Computational Methods in Genome Research

Theoretical and Computational Methods in Genome Research

Edited by

Sándor Suhai
German Cancer Research Center
Heidelberg, Germany

Plenum Press • New York and London

Library of Congress Cataloging-in-Publication Data

Theoretical and computational methods in genome research / edited by
Sándor Suhai.
p. cm.
Includes bibliographical references and index.
ISBN 0-306-45503-X
1. Genomes--Data processing--Congresses. 2. Gene libraries--Data
processing--Congresses. 3. Nucleotide sequence--Data processing-
-Congresses. I. Suhai, Sándor.
QH442.T45 1997
572.8'6'0285--dc21 97-1548
CIP

Proceedings of the International Symposium on Theoretical and Computational Genome Research,
held March 24–27, 1996, in Heidelberg, Germany

ISBN 0-306-45503-X

A Division of Plenum Publishing Corporation
233 Spring Street, New York, N.Y. 10013

http://www.plenum.com

10 9 8 7 6 5 4 3 2 1

Printed in the United States of America

PREFACE

The application of computational methods to solve scientific and practical problems in genome research created a new interdisciplinary area that transcends boundaries traditionally separating genetics, biology, mathematics, physics, and computer science. Computers have, of course, been intensively used in the field of life sciences for many years, even before genome research started, to store and analyze DNA or protein sequences; to explore and model the three-dimensional structure, the dynamics, and the function of biopolymers; to compute genetic linkage or evolutionary processes; and more. The rapid development of new molecular and genetic technologies, combined with ambitious goals to explore the structure and function of genomes of higher organisms, has generated, however, not only a huge and exponentially increasing body of data but also a new class of scientific questions. The nature and complexity of these questions will also require, beyond establishing a new kind of alliance between experimental and theoretical disciplines, the development of new generations both in computer software and hardware technologies.

New theoretical procedures, combined with powerful computational facilities, will substantially extend the horizon of problems that genome research can attack with success. Many of us still feel that computational models rationalizing experimental findings in genome research fulfill their promises more slowly than desired. There is also an uncertainty concerning the real position of a "theoretical genome research" in the network of established disciplines integrating their efforts in this field. There seems to be an obvious parallel between the present situation of genome research and that of atomic physics at the end of the first quarter of our century. Advanced experimental techniques made it possible at that time to fill huge data catalogues with the frequencies of spectral lines, yet all attempts at an empirical systematization or classically founded explanation remained unsuccessful until a completely new conceptual framework, quantum mechanics, emerged that made sense of all the data.

The present situation of the life sciences is certainly more intricate due to the more amorphous nature of the field. One has to ask whether it is a fair demand at all that genome research should reach the internal coherence of a unifying theoretical framework like theoretical physics or (to a smaller extent) chemistry. The fear seems to be, however, not unjustified that genomic data accumulation will get so far ahead of its assimilation into an appropriate theoretical framework that the data themselves might eventually prove an encumbrance in developing such new concepts. The aim of most of the computational methods presented in this volume is to improve upon this situation by trying to provide a bridge between experimental databases (information) on the one hand and theoretical concepts (biological and genetic knowledge) on the other.

The content of this volume was presented as plenary lectures at the International Symposium on Theoretical and Computational Genome Research held March 24–27, 1996, at the Deutsches Krebsforschungszentrum (DKFZ) in Heidelberg. It is a great pleasure to thank here Professor Harald zur Hausen and the coworkers of DKFZ for their help and hospitality extended to the lecturers and participants during the meeting and the Commission of the European Communities for the funding of the symposium. The organizers profited much from the help of the scientific committee of the symposium: Martin Bishop, Philippe Dessen, Reinhold Haux, Ralf Hofestädt, Willi Jäger, Jens G. Reich, Otto Ritter, Petre Tautu, and Martin Vingron. Furthermore, the editor is deeply indebted to Michaela Knapp-Mohammady and Anke Retzmann for their help in organizing the meeting and preparing this volume.

Sándor Suhai
Heidelberg, July 1996

CONTENTS

Theoretical and Computational Methods in Genome Research

1

EVALUATING THE STATISTICAL SIGNIFICANCE OF MULTIPLE DISTINCT LOCAL ALIGNMENTS

Stephen F. Altschul

National Center for Biotechnology Information
National Library of Medicine
National Institutes of Health
Bethesda, Maryland 20894

ABSTRACT

A comparison of two sequences may uncover multiple regions of local similarity. While the significance of each local alignment may be evaluated independently, sometimes a combined assessment is appropriate. This paper discusses a variety of statistical and algorithmic issues that such an assessment presents.

1. INTRODUCTION

The most widely used techniques for comparing DNA and protein sequences are local alignment algorithms, which seek similar regions of the sequences under consideration [1–3]. Given a particular measure of similarity, an important question is how strong a local alignment must be to be considered statistically significant. For a broad range of measures, including most of those in common use, this question has been addressed both analytically and experimentally [4–17].

Occasionally a sequence comparison program will uncover not one but multiple local alignments, representing several distinct regions of sequence similarity [17–22]. This is particularly common with the BLAST algorithms [3], which disallow gaps within the alignments found, but is true also of local alignment algorithms that permit gaps [17–21]. Can one arrive at a joint statistical assessment of these several alignments? We will discuss below the various issues that arise in reaching such an assessment. We will confine attention first to local alignments lacking gaps, for which analytic results are available, and then discuss the generalization of these results to alignments with gaps.

Theoretical and Computational Methods in Genome Research, edited by Suhai
Plenum Press, New York, 1997

2. LOCAL ALIGNMENT STATISTICS

The simple measure of local similarity with which we start requires a substitution matrix, or set of scores, s_{ij} for aligning the various pairs of amino acids. Assume we are given two protein sequences to compare. Choosing any pair of equal-length segments, one from each sequence, we may construct an ungapped subalignment. The score for such a segment pair may be taken as the aggregate score of its aligned pairs of amino acids. The segment pair with greatest score is called the maximal segment pair (or MSP), and its score the maximal segment pair score (or MSP score). The MSP score may be calculated using a simplification, that disallows gaps, of the Smith-Waterman dynamic programming algorithm [1].

To analyze the statistical behavior of MSP scores, one may model proteins as random sequences of independently selected amino acids; the amino acids, labeled 1 to 20, occur with respective probabilities p_i. For our theory to progress, we need to assume that the expected score $\Sigma_{i,j=1}^{20} p_i p_j s_{ij}$ for aligning two random amino acids is negative. This is, in fact, a desirable condition: were the expected score positive, long segment pairs would have high score independent of any biological relationship. Given this constrain on the scores, as well as the existence of at least one positive score, it is always possible to find a unique positive solution λ to the equation

$$\sum_{i,j=1}^{20} p_i p_j e^{s_{ij} x} = 1. \tag{1}$$

The parameter λ may be thought of as a natural scale for the scores. A second positive parameter K important for the statistical theory, depends upon the p_i and s_{ij}, and is given by a geometrically convergent infinite series [9,10,14]. A program in C for calculating λ and K is available from the author.

When the lengths m and n of the two sequences being compared are sufficiently large, asymptotic results concerning the distribution of the MSP score S come into play [9,10,14]. The probability that S is at least x is then well approximated by the formula

$$Prob(S \geq x) = 1 - \exp(-Kmn\, e^{-\lambda x}). \tag{2}$$

S is said to follow an extreme value distribution [23]. If only the highest scoring segment pair were ever of interest, we would need progress no further.

3. LOCAL SIMILARITY AND MULTIPLE LOCAL ALIGNMENTS

Because two sequences may share more than a single region of similarity, it is desirable to consider not only the maximal segment pair, but other segment pairs that also achieve high scores. The immediate problem is that the second, third and fourth highest scoring segment pairs are likely to be slight variations of the maximal segment pair. We therefore need a definition of when two subalignments are distinct.

The first such definition, in the context of subalignments with gaps, is due to Sellers [18], who defined the "neighborhood" of a subalignment to be the set of all subalignments it touches within a path graph. A subalignment is then considered "locally optimal" if it has score at least as large as any subalignment within its neighborhood. Sellers provided an algorithm for locating all locally optimal subalignments [8], which had the disadvantage or requiring an undetermined number of sweeps of a path graph. An alternative algorithm, whose time complexity is provably $O(mn)$, was described by Altschul & Erickson [19].

While Sellers' definition is mathematically the most natural one, and allows the simplest algorithms, it has a major practical problem. The existence of a very high-scoring subalignment can suppress, through an intermediate, the discovery of a nearby subalignment it never actually touches. Specifically, suppose subalignments A and C involve completely different sequence segments, but the score of A is very large. Then it is possible that an extension B of subalignment A will in fact touch C, but have intermediate score. By Sellers' definition, C can not be locally optimal because B is within its neighborhood. In order to overcome this problem, Altschul & Erickson [5,19] introduced the definition that a subalignment is "weakly locally optimal" if it intersects no higher scoring subalignment that is weakly locally optimal. This definition is recursive rather than circular: it is anchored by the optimal subalignment. Waterman & Eggert [20], and Huang *et al.* [21] describe efficient algorithms for locating all weakly locally optimal subalignments.

Although the precise definition of local optimality employed makes a difference for the discovery of biologically meaningful similarities, it has essentially no effect on the asymptotic statistics for multiple high-scoring subalignments. The reason is that for the comparison of random sequences, locally optimal subalignments rarely have extremely high scores, and furthermore are unlikely to occur very close to one another. Thus the two definitions above are in most cases congruent, and effectively so in the asymptotic limit.

For subalignments without gaps, we will refer to locally optimal subalignments as high-scoring segment pairs (or HSPs). For the comparison of a pair of sequences, the scores of the r highest-scoring HSPs can be thought of as random variables $S_1, S_2, \ldots, S_r$, which have taken on the values $x_1, x_2, \ldots, x_r$. Frequently, in order to describe a given result, a p-value is reported, which is the probability that a result "as good as" the one observed would have occurred by chance. If we consider just the highest score, the p-value is the probability that $S_1 \geq x_1$, which is given by equation (2). If we instead consider the two highest scores, it is tempting to ask for the probability that $S_1 \geq x_1$ and $S_2 \geq x_2$. This, however, would be invalid, because it encompasses no general notion of "better".

The reason can be understood by considering two possible results: outcome A in which $S_1 = 65$ and $S_2 = 40$, and outcome B in which $S_1 = 52$ and $S_2 = 45$ (see Figure 1). Implicitly, the approach we have just described would not consider outcome A better than outcome B, because its second highest score x_2 is worse than that of B. However, neither would outcome B be considered better than outcome A, because its highest score x_1 is worse than that of A. Basically, outcomes A and B are incomparable, and thus the probability we have defined is not a valid p-value. To construct a joint statistical assessment of multiple HSPs, a linear ordering of all possible outcomes is required. This can be accomplished in several ways.

4. POISSON STATISTICS

One way to define an ordering on outcomes is to consider only the rth highest score x_r, and ask for the probability that $S_r \geq x_r$. In Figure 1, with $r = 2$, this is equivalent to drawing a horizontal line through the point (x_1, x_2), and calculating the probability that a

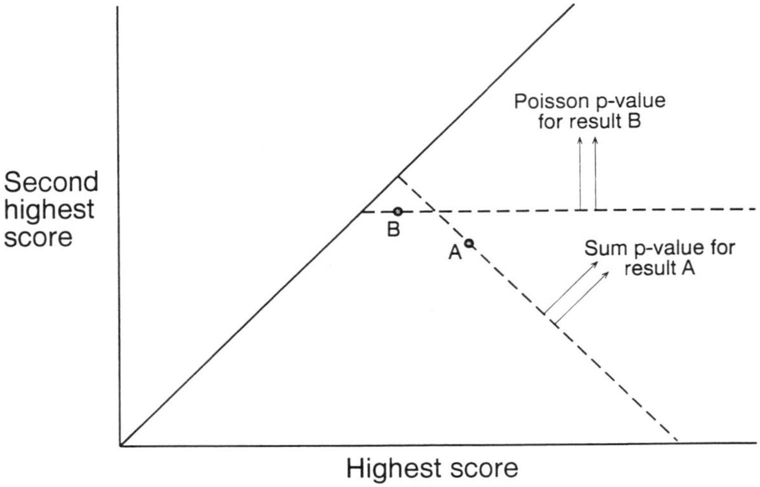

Figure 1. The calculation of Poisson and sum p-values. The region on or below the main diagonal represents feasible joint values of the 1st and 2nd highest locally optimal subalignment scores. A probability density may be imagined to cover this space. The probability that the 2nd highest score is greater than or equal to 52 is that lying on or above the horizontal line through point B. The probability that the sum of the two highest scores is greater than or equal to 105 is that lying on or above the anti-diagonal line through point A. Which result is considered the more significant depends upon which criterion is employed.

random outcome lies on or above this line. By this criterion, outcome B discussed above is considered better than outcome A.

For sequences of sufficient length, the number of HSPs with score exceeding x can be shown to approach a Poisson distribution, with parameter $a = Kmn\, e^{-\lambda x}$ [9,10,15]. Equivalently, the probability that S_r is at least x_r is given approximately by the formula

$$Prob(S_r \geq x_r) = 1 - e^{-a} \sum_{i=0}^{r-1} \frac{a^i}{i!}. \quad (3)$$

For $r = 1$, this is equivalent to equation (2).

For several years, the BLAST server maintained by the NCBI [24] used these Poisson statistics to assess the significance of results involving multiple HSPs. One obvious problem with this approach is that it takes account only of the rth highest score. Intuitively, it would seem that a more appropriate measure of a result's overall quality would depend upon the scores of all the HSPs involved.

5. SUM STATISTICS

An alternative, and perhaps more natural, method for ordering outcomes is by the sum of their scores [22]. In Figure 1, this is equivalent to drawing an anti-diagonal line through the point (x_1, x_2), and calculating the probability that a random outcome lies on, or above and to the right, of this line. By this criterion, outcome A above is considered better than outcome B.

The distribution of the sum of the r highest scoring HSPs can be most easily described in terms of normalized scores [22]. If S_i is the random variable describing the ith highest HSP score, let $S'_i = \lambda S_i - \ln Kmn$. For sufficiently long sequences, the joint probability density function (p.d.f.) for $S'_1,\ldots,S'_r$ is then well approximated by the function

$$f(x_1,\ldots,x_r) = \exp(-e^{-x_r} - \sum_{i=1}^{r} x_i) \tag{4}$$

where $x_1 \geq x_2 \geq \ldots \geq x_r$. The distribution for any function of the S'_i may be calculated from this joint density by appropriate integration. Most simply, the marginal distribution of S'_r is easily calculated, yielding equation (3) above. Our interest here, however, is in the sum of the r highest normalized scores $T_r = S'_1 + \ldots + S'_r$. In general, to calculate the distribution of more complicated functions of the S'_i, a full r-dimensional integration is required. However, because the density of equation (4) depends only upon x_r and the sum of the x_i, some algebraic manipulation allows us to reduce the p.d.f. for T_r to the one dimensional integral

$$f(t) = \frac{e^{-t}}{r!(r-2)!} \int_0^\infty y^{r-2} \exp(-e^{(y-t)/r})\, dy, \tag{5}$$

when $r \geq 2$. The p.d.f. for T_1 is of course given directly by equation (4), with r=1, and is simply that of the extreme value distribution. The densities for T_1, T_3, and T_5 are shown superposed in Figure 2. It should of course be remembered that these are the distributions for sums of normalized scores; the center of the distributions for unnormalized scores would move progressively to the right.

All moments of the distribution for T_r may be calculated using Laplace transforms [22]. The mean is given by $r(1 + \gamma - \Sigma_{i=1}^{r} 1/i)$, where $\gamma \approx 0.577$ is Euler's constant. This quantity is well approximated by $r(1 - \ln r) - 1/2$. The variance is $r^2(\pi^2/6 - \Sigma_{i=1}^{r} 1/i^2) + r$, which is approximately $2r - 1/2$.

To calculate p-values, one needs to obtain the tail probability that $T_r \geq x$, which requires integrating equation (5) for t from x to ∞. This double integral may easily be calculated numerically, and a program for the purpose in C is available from the author. In the limit of large x, the value obtained from the double integral approaches the simple expression

$$Prob(T_r \geq x) \approx \frac{e^{-x}x^{r-1}}{r!(r-1)!}. \tag{6}$$

This is an evocative formula. The probability that $T_r \geq x$ is concentrated near $T_r = x$. Imagine that, for any particular division of the normalized score x into r pieces of decreasing size, the joint probability is proportional to e^{-x}. In how many ways can one so divide the score? Think of the score as a line segment of length x, with r–1 dividers placed somewhere along it. If the dividers are indistinguishable, the number of ways to place them will be proportional to $x^{r-1}/(r-1)!$. Since we are interested only in the lengths of the pieces that result and not in their order, we need to divide again by $r!$, yielding $x^{r-1}/r!(r-1)!$. Thus may formula (6) be intuitively understood.

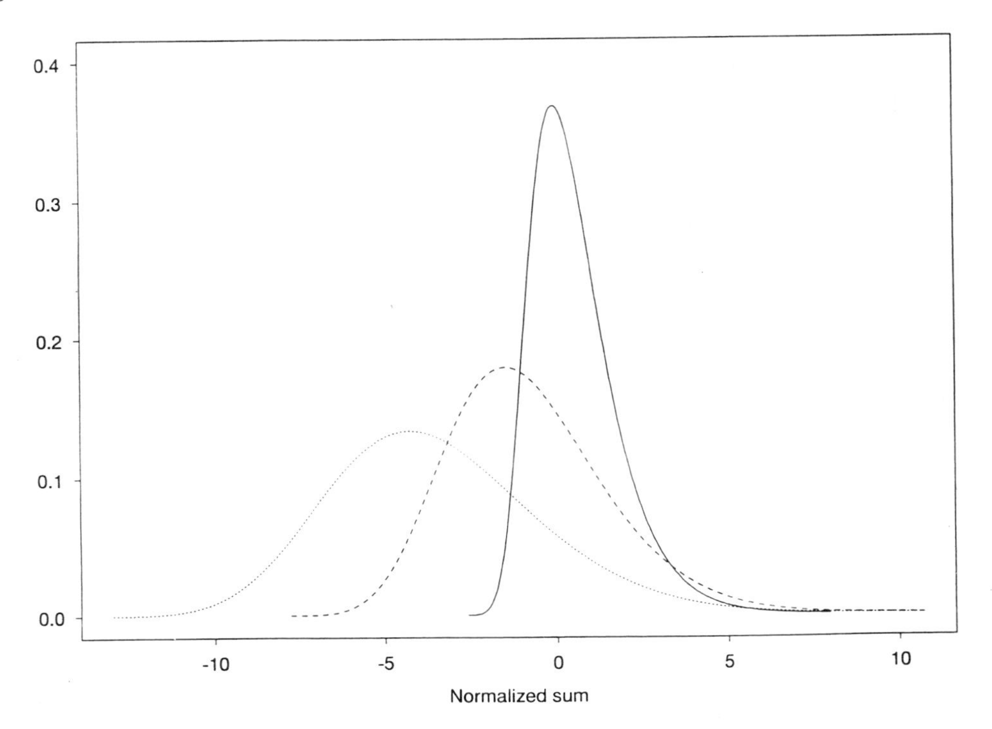

Figure 2. The probability density functions for the normalized sums of the one (solid line), three (dashed line), and five (dotted line) best local alignment scores.

6. MULTIPLE TESTS

In the previous section we described how to calculate a p-value for sum of the r highest HSP scores. A problem is that when we are comparing a particular pair of sequences, we do not in general know *a priori* how many HSPs we should seek. One idea is to find all HSPs that exceed some low cutoff score, and then consider the highest scoring HSP in isolation, the collection of the two highest scoring HSPs, and so forth. For each set we may calculate a sum-statistic p-value, and then choose the "most significant" set, i.e. the one whose p-value is smallest. The problem with this procedure is that, because we have performed multiple tests by considering many different cases, the original p-value for the collection of HSPs ultimately selected is no longer valid.

A way around this difficulty has been suggested by Dr. Phil Green (personal communication). If one performs t separate tests, a conservative p-value may always be calculated by multiplying the p-value for any single test by t. Alternatively, using a Bayesian-like approach, one can assign a "prior" probability α_i to each case one considers and divide the p-value for that particular case by α_i. The lowest corrected p-value will then be conservative.

For multiple HSPs there is, in principle, no fixed limit to the number that might be considered. An attractive procedure, therefore, is to let the α_r assigned to the case of r HSPs be a geometric series in r that sums to one [25]. Specifically, set $\alpha_r = \beta^{r-1}(1-\beta)$ for any β between 0 and 1. If β is chosen to be 0.5, then the original p-value calculated for a single HSP is multiplied by 2, for two HSPs by 4, for three by 8, and so forth. This procedure can be employed for p-values calculated from Poisson statistics as well as from sum statistics.

7. CONSISTENT ORDERING

Related proteins frequently share strongly conserved regions, generally corresponding to secondary structural elements, separated by poorly conserved regions generally corresponding to loops. The conserved regions often may be aligned without gaps, because insertion or deletion of residues would destroy the structure, while the distance between these regions may vary significantly from one protein to another [26,27]. As a result, if the comparison of two protein sequences A and B yields multiple HSPs representing an actual biological relationship, these HSPs generally are "consistently ordered". In other words, the order in which the first segments from the various HSPs are arranged within sequence A is the same as the order in which the second segments from the HSPs are arranged within sequence B. In contrast, there is no reason for multiple HSPs from the comparison of unrelated sequences to be consistently ordered. We can use this fact to improve our separation of biologically meaningful relationships from background noise.

It is possible to insist that a set of HSPs be consistently ordered before we permit a combined assessment of its statistical significance. How does this affect the random distribution of the sum of HSP scores? The random variable T_r discussed above can be written as $\lambda\ (\Sigma_{i=1}^{r} S_i) - \ln K^r - \ln (mn)^r$. The last term may be understood as correcting for the $(mn)^r$ different possible sets of starting positions for the r segment pairs whose scores are added. If we allow ourselves to consider only consistently ordered sets of HSPs, the number of possible sets of starting positions is divided by $r!$. Accordingly, the formula for the normalized sum T_r must be adjusted; the new value is simply the old one plus $\ln r!$ [22]. The net effect, of course, is that the p-value for any given score is decreased. To a first approximation, by formula (6), the p-value is divided by the factor $r!$, as might be expected. The adjustment that consistent ordering imposes upon the calculation of Poisson statistics is somewhat more complicated [22], and will not be described here.

8. GAP LENGTH CONSTRAINTS

While the requirement of consistent ordering excludes many sets of HSPs that are unlikely to represent biological relationships, it is possible to sharpen our statistical assessment even further. One may observe, for example, that the number of residues between conserved blocks is almost always smaller than some maximum length g. Imposing this as a rigid constraint again changes the way the normalized score T_r should be calculated [22]. First, one may imagine arranging the HSPs by the order in which the segments they comprise appear within the sequences being compared. Disregarding edge effects (to be discussed below), there are then mn possible starting positions for the segments of the first HSP. For each subsequent HSP, there are no more than g^2 possible starting positions. Thus, the number of sets of starting positions for an ordered set of r HSPs that satisfy the "small gap" constraint is bounded by $r!mng^{2r-2}$. When long sequences are being compared and r is not too large, this number may be very much smaller than $(mn)^r$, or even $(mn)^r/r!$. The normalized sum score T_r is then correspondingly larger, yielding a greatly reduced p-value.

Clearly, the smaller the length g of the gap that is allowed, within one or both sequences, between the segments of adjacent HSPs, the sharper the statistics become. The danger, of course, is that as g is reduced, more instances of biologically meaningful collections of HSPs are excluded from a combined statistical assessment. There are several possible ways around this dilemma. One is to use a number of different values for g, and discount the resulting p-values to allow for the multiple tests, as described above. A second is to introduce explicit gap costs, creating single gapped subalignments [1,2,18] rather than collections of ungapped HSPs. The statistics of such gapped subalignments will be discussed below.

It may indeed appear that trying to assess the significance of a collection of HSPs is simply a poor man's way of dealing with gapped subalignments. This is not necessarily the case. If two proteins share two or more well conserved regions, separated by weakly conserved or unrelated regions, a traditional subalignment will need to force many badly matching amino acids into alignment in order to connect the related regions. It may frequently be better to pay a single, implicit penalty for excluding these intervening regions from the alignment. A rigorous comparison of the relative sensitivities of this measure and traditional gap-cost measures awaits study.

9. SOME ALGORITHMIC ISSUES

We are concerned in this paper primarily with how to assess the statistical significance of a collection of distinct subalignments, but the problem raises a few algorithmic issues that are worth attention. First, given a collection C of HSPs with associated scores, how do we determine which subset A yields the lowest sum-statistic p-value? This is not trivial because the p-value depends non-linearly upon the sum of the scores of the HSPs in A, and their number. In general, one can not decide whether a particular HSP belongs in A without knowing the number and scores of the other HSPs in A.

To simplify matters, for any fixed number r of HSPs it is always better to have a higher aggregate score. Therefore, one can order the HSPs in C by decreasing score and calculate p-values for just the first HSP, the first two HSPs, etc. One of these collections will be the subset with lowest p-value.

Finding the best subset A becomes more involved if one imposes the condition of consistent ordering or small gaps. A dynamic programming algorithm needs to keep track, at each node, of the best score for subsets of various sizes. In the worst case, this requires time proportional to the cube of the number of HSPs in C. Alternatively, one may take a heuristic approach. For any particular p-value, analysis of formula (6) shows that x and r have an approximately linear relationship. Furthermore, one is typically most interested in results within a certain range of p-values. One may use these facts to calculate a cutoff score for including additional HSPs. This reduces the time required for dynamic programming to the square of the number of HSPs in C. The result produced, however, is not guaranteed to be optimal.

Once one has determined and reported the collection of HSPs with lowest p-value, how does one assess any remaining HSPs? One possible approach is that taken, in a slightly different context, for weakly locally optimal subalignments [19,20]. Remove from the set C the HSPs in the optimal subset A. Find the optimal subset of this remaining set, and if it is statistically significant, report it. Repeat this procedure until no remaining collection of HSPs is statistically significant.

10. EDGE EFFECTS

The statistical results upon which we have been relying are valid in the asymptotic limit of large sequence lengths m and n. One may ask how large m and n must be for the results to apply, and whether there are any corrections that can be made for finite sequence length. An essential point is that m and n enter the formulas above to describe the size of the search space. The segments that constitute an HSP may begin anywhere within the two sequences being compared, i.e. at any of m positions within the first sequence and n positions within the second. However, HSPs do not occur at single points, but must have a certain length before they can attain an appreciable score. When two random sequences are compared, how long should one expect HSPs to be?

It may be shown [9,10] that within HSPs from the comparison of random sequences, the "target frequency" q_{ij} with which amino acid i in the first sequence is aligned with amino acid j in the second converges to

$$q_{ij} = p_i p_j e^{\lambda s_{ij}}. \tag{7}$$

One can use these target frequencies to calculate the expected score per position within HSPs:

$$\sum_{i,j} q_{ij} s_{ij} = \frac{1}{\lambda} \sum_{i,j} q_{ij} \ln \frac{q_{ij}}{p_i p_j} = \frac{H}{\lambda} \tag{8}$$

where H is the relative entropy of the scoring system, expressed in nats [28]. The expected length l of an HSP is just its expected score divided by the expected score per alignment position. By equation (2), the expected score of the highest scoring segment pair is approximately $\ln Kmn/\lambda$, so its expected length is approximately

$$l = \frac{\ln Kmn}{H}. \tag{9}$$

Because it is expected to have length l, an optimal segment pair is unlikely to start within l residues of the end of either sequence being compared. Thus one may correct for "edge effects" by subtracting l from the sequence lengths m and n to obtain effective lengths $m' = m-l$ and $n' = n-l$ [17]. These effective lengths may be used in place of m and n in the formulas above. If l is a large proportion of either m or n, it suggests that edge effects become important, and that the asymptotic results discussed here lose validity.

11. ALIGNMENTS WITH GAP COSTS

The most traditional method for assigning scores to local alignments involves adding scores for gaps (insertions and deletions) as well as for aligned pairs of amino acids [1,2]. It is generally advantageous to employ "affine" gap costs, which charge one penalty for opening a gap and another one for extending it [29–32]. The asymptotic distribution of "Smith-Waterman" (SW) scores, calculated using a substitution matrix and affine gap costs, is not known, but it can be shown that in typical scoring regimes the mean score grows asymptotically as $\ln mn$ [14]. Various computational experiments suggest that the distribution of SW scores, like that of MSP scores, approaches an extreme value distribution [4,7,12,16,17]. Furthermore, it appears that the asymptotic statistical theory for locally optimal SW scores is fully analogous to that described above for HSPs, and can be completely described with the two parameters λ and K [16,17]. The only problem is that no formulas to calculate these parameters are available, and they must therefore be estimated by random simulation or the comparison of real but unrelated sequences [4,7,12,16,17]. Estimates of λ and K have been published for a range of affine gap costs used in conjunction with several popular substitution matrices [17].

It is instructive to compare the statistical assessment of collections of HSPs using sum statistics with that of single SW alignments. A collection of HSPs has implicit gaps with implicit costs arising from three sources. First, normalizing the score for the rth HSP involves subtracting $\ln Kmn$ if the HSP is allowed to appear anywhere, or a smaller number if some spatial relationship among the HSPs is enforced. Second, adding an rth HSP to the collection involves using the distribution for r HSPs rather than that for $r-1$. Equation (6) implies that even the sum of normalized scores must generally increase to retain the same significance level. Finally, increasing r should trigger a p-value adjustment for multiple tests, as described above. In contrast, SW alignments charge for gaps explicitly with gaps costs, and implicitly by lowering the value of λ used in statistical assessment [17]. The definition of a gap is different in the two cases. Collections of HSPs allow gaps of arbitrary size in either or both sequences, while a gap in an SW alignment is restricted to a single sequence. So long as a gap between HSPs does not violate any constraints imposed, there is no preference for short as opposed to long gaps, while gap scores automatically favor short gaps. As mentioned above, it is not immediately clear that one of these alternative definitions of gap is superior.

12. DATABASE SEARCHES

When one quotes a p-value it is always in the context of a particular search. We have so far described how to calculate the probability that a result as good as a given one will arise by chance within the context of a particular pairwise sequence comparison. A database search, however, involves many pairwise comparisons, and any "pairwise p-value" must accordingly be corrected to yield a "database p-value."

Perhaps the most straightforward approach is to treat p-values from all pairwise comparisons as equivalent. Given a database of D sequences, one expects $E = Dp$ of them to attain a pairwise p-value p in comparison with the query. The applicable Poisson distribution then implies that the probability that at least one database sequence yields such a pairwise p-value is $1 - e^{-E}$. This is then the database p-value.

The calculation above makes the implicit assumption that all database sequences are *a priori* equally likely to share an actual relationship with the query. An alternative view, based upon the idea that proteins contain multiple domains, and that a longer protein is therefore more likely to share a biologically meaningful similarity with the query than is a shorter one, implies a different correction [25]. Assume the alignment of interest involves a database sequence of n residues, and that the database has a total of N residues. If the *a priori* likelihood that the query is similar to a given protein is proportional to that protein's length, then D in the above equation for E should be replaced by N/n. The resulting database p-values for average length proteins will be unaffected, but database p-values for short sequences will increase and for long sequences will decrease.

For results involving a single local alignment, the length n enters the formulas for the pairwise and database p-values in reciprocal manners. For example, an HSP with nominal score 75 might achieve a pairwise p-value of 10^{-7} if it involves a database protein of length 200, but a pairwise p-value of 10^{-6} if it involves a database protein of length 2000. If the database itself contains 50,000 proteins and 20,000,000 total residues, the length-proportional correction described above would yield a database p-value of about 0.01 for the alignment, independent of the database sequence it involved. The alternative view, that corrects with reference only to the total number of database sequences, would yield a database p-value of 0.005 for the alignment within the length 200 protein, and a database p-value of 0.05 for the alignment within the length 2000 protein.

It may be argued that proteins are discrete units, and that a query sequence is *a priori* equally likely to share similarity with any one. This view is no longer tenable once one is dealing with DNA sequences, where the database entries are of essentially arbitrary length. Thus for DNA sequence comparison, the only sensible p-value correction is the length-proportional one described above.

13. EXAMPLE

In order to illustrate how the various ideas discussed in this paper are combined, we consider a specific example. First, using the BLOSUM-62 substitution matrix [33], and a standard set of amino acid frequencies, the key parameters may be calculated [9] to be $\lambda \approx 0.319$ and $K \approx 0.133$. Searching the SWISS-PROT database [34], release 32 (49,824 sequences, 17,390,026 residues) with the human neurofibromatosis 1 protein [35] (SWISS-PROT accession number P21359; 2839 residues), we find that when compared with a yeast inhibitory regulator protein [36] (SWISS-PROT accession number P33314; 1104 residues) the two local alignments shown in Figure 3 appear.

The best local alignment has nominal score 70. If we choose not to apply an edge-effect correction, equation (2) implies that the probability of finding an alignment this good within the particular pair of sequences compared is $1 - \exp(-0.133 * 2839 * 1104\, e^{-0.319*70}) \approx 8.4 * 10^{-5}$. As discussed above, since we will consider results involving an arbitrary number of HSPs, we should multiply p-values for results involving r HSPs by the factor $1/\alpha_r$, here taken to be 2^r.

```
                              Score = 70

P21359:   1263 KEVELADSMQTLFRGNSLASKIMTFCFKVYGATYLQKLLDPLLRIVITSS 1312
               K ++      +LFRGNS+ +K +   F   G  YL K L  +L+ +I S+
P33314:    548 KNLDSKHVFNSLFRGNSILTKSIEQYFFRVGNEYLSKALSAILKEIIESN  597


                              Score = 66

P21359:   1402 IGAVGSAMFLRFINPAIVSPYEAGILDKKPPPRIERGLKLMSKILQSIANHVLFTKEE 1459
               +  +   +FLRF  P I++P     + +       R L L+SK+L +++    F  +E
P33314:    672 LNGISGLLFLRFFCPVILNPKLFKYVSQNLNETARRNLTLISKVLLNLSTLTQFANKE  729
```

Figure 3. Two HSPs from the comparison of the human neurofibromatosis 1 protein [35] (SWISS-PROT accession number P21359) with a yeast inhibitory regulator protein [36] (SWISS-PROT accession number P33314). Alignments are scored using the BLOSUM-62 amino acid substitution matrix [33]. Central lines of alignments echo identities, and use "+" to indicate positions with positive substitution score.

Thus our pairwise p-value becomes $1.7 * 10^{-4}$. Using the length-proportional database correction discussed above, we expect $17{,}390{,}026/1104 \approx 15{,}752$ times as many results as good to appear by chance in a database search. This number is $E \approx 2.6$, so our database p-value is $1 - e^{-2.6} \approx 0.93$. In other words, in the context of a database search, this alignment in isolation is by no means statistically significant.

We next consider the two best alignments, with scores 70 and 66 (Figure 3), in conjunction. If we require HSPs to be consistently ordered to qualify for a combined statistical assessment, then these pass the test. The normalized sum score is then $0.319*(70+66) - 2*\ln(0.133*2839*1104) + \ln 2! \approx 18.2$; the $\ln 2!$ term is included because of the consistent ordering requirement. The double integral required to calculate the corresponding p-value for $r = 2$ yields $1.1 * 10^{-7}$; the asymptotic equation (6) gives virtually the same result. Discounting for multiple tests, this time by a factor of 4, yields a pairwise p-value of $4.6 * 10^{-7}$. Correcting, as above, for the database size gives a database p-value of 0.0072, this time quite significant. Further HSPs, if they are consistently ordered, may be added to this pair, and the combined result with the lowest p-value reported.

14. CONCLUSION

Considering that they are based upon a very simple random protein model, the statistics discussed here do remarkably well at estimating the frequency of high scoring alignments among real but unrelated proteins. One set of alignments for which the statistics break down, however, are those involving regions of extremely restricted amino acid composition [25,37]. Such regions most likely arise in evolution by gene slippage, and it is questionable whether alignments are even a reasonable construct for their comparison. Therefore, when two proteins are compared, it is best to filter the sequences to remove such "low complexity" regions beforehand [25,37]. It is, of course, worth noting the existence of these regions, but traditional alignment algorithms and scoring systems are poorly adapted to their analysis.

The tools provided above allow the statistical significance of a set of gapped or ungapped alignments to be assessed. To employ them properly, one needs to decide beforehand what constraints will be imposed, and these will depend to some extent upon the problem at hand. Is it reasonable to expect that, when multiple local alignments are found, they will be consistently ordered? If there are gaps between local alignments, can they be of arbitrary or of only restricted size? To what extent should results involving fewer local alignments be favored? Are there any *a priori* reasons for expecting that true relationships are more likely to be found among certain classes of proteins? These essentially Bayesian considerations are built into our calculations, and the user who is aware of them will employ the statistical tests to best effect.

REFERENCES

1. Smith, T.F. & Waterman, M.S. (1981). Identification of common molecular subsequences. *J. Mol. Biol.* **147**:195–197.
2. Pearson, W.R. & Lipman, D.J. (1988). Improved tools for biological sequence comparison. *Proc. Natl. Acad. Sci. U.S.A.* **85**:2444–2448.
3. Altschul, S.F., Gish, W., Miller, W., Myers, E.W. & Lipman, D.J. (1990). Basic local alignment search tool. *J. Mol. Biol.* **215**:403–410.

4. Smith, T.F., Waterman, M.S. & Burks, C. (1985). The statistical distribution of nucleic acid similarities. *Nucl. Acids Res.* **13**:645–656.
5. Altschul, S.F. & Erickson, B.W. (1986). A nonlinear measure of subalignment similarity and its significance levels. *Bull. Math. Biol.* **48**:617–632.
6. Arratia, R., Gordon, L. & Waterman, M.S. (1986). An extreme value theory for sequence matching. *Ann. Stat.* **14**:971–993.
7. Collins, J.F., Coulson, A.F.W. & Lyall, A. (1988). The significance of protein sequence similarities. *CABIOS* **4**:67–71.
8. Arratia, R. & Waterman, M.S. (1989). The Erdos-Renyi strong law for pattern matching with a given proportion of mismatches. *Ann. Prob.* **17**:1152–1169.
9. Karlin, S. & Altschul, S.F. (1990). Methods for assessing the statistical significance of molecular sequence features by using general scoring schemes. *Proc. Natl. Acad. Sci. U.S.A.* **87**:2264–2268
10. Dembo, A. & Karlin, S. (1991). Strong limit theorems of empirical functionals for large exceedances of partial sums of i.i.d. variables. *Ann. Prob.* **19**:1737–1755.
11. Dembo, A. & Karlin, S. (1991). Strong limit theorems of empirical distributions for large segmental exceedances of partial sums of Markov variables. *Ann. Prob.* **19**:1756–1767.
12. Mott, R. (1992). Maximum-likelihood estimation of the statistical distribution of Smith-Waterman local sequence similarity scores. *Bull. Math. Biol.* **54**:59–75.
13. Altschul, S.F. (1993). A protein alignment scoring system sensitive at all evolutionary distances. *J. Mol. Evol.* **36**:290–300.
14. Arratia, R. & Waterman, M.S. (1994). A phase transition for the score in matching random sequences allowing deletions. *Ann. Appl. Prob.* **4**:200–225.
15. Dembo, A., Karlin, S. & Zeitouni, O. (1994). Limit distribution of maximal non-aligned two-sequence segmental score. *Ann. Prob.* **22**:2022–2039.
16. Waterman, M.S. & Vingron, M. (1994). Sequence comparison significance and Poisson approximation. *Stat. Sci.* **9**:367–381.
17. Altschul, S.F. & Gish, W. (1996). Local alignment statistics. *Meth. Enzymol.* **266**:460–480.
18. Sellers, P.H. (1984). Pattern recognition in genetic sequences by mismatch density. *Bull. Math. Biol.* **46**:501–514
19. Altschul, S.F. & Erickson, B.W. (1986). Locally optimal subalignments using nonlinear similarity functions. *Bull. Math. Biol.* **48**:633–660.
20. Waterman, M.S. & Eggert, M. (1987). A new algorithm for best subsequence alignments with applications to tRNA-rRNA comparisons. *J. Mol. Biol.* **197**:723–728.
21. Huang, X., Hardison, R.C. & Miller, W. (1990). A space-efficient algorithm for local similarities. *CABIOS* **6**:373–381.
22. Karlin, S. & Altschul, S.F. (1993). Applications and statistics for multiple high-scoring segments in molecular sequences. *Proc. Natl. Acad. Sci. U.S.A.* **90**:5873–5877.
23. Gumbel, E.J. (1958). Statistics of extremes. Columbia University Press, New York.
24. Woodsmall, R.M. & Benson, D.A. (1993). Information resources at the National Center for Biotechnology Information. *Bull. Med. Libr. Assoc.* **81**:282–284.
25. Altschul, S.F., Boguski, M.S., Gish, W. & Wootton, J.C. (1994). Issues in searching molecular sequence databases. *Nature Genet.* **6**:119–129.
26. Henikoff, S. & Henikoff, J.G. (1991). Automated assembly of protein blocks for database searching. *Nucleic Acids Res.* **19**:6565–6572.
27. Lawrence, C.E., Altschul, S.F., Boguski, M.S., Liu, J.S., Neuwald, A.F. & Wootton, J.C. (1993). Detecting subtle sequence signals: A Gibbs sampling strategy for multiple alignment. *Science* **262**:208–214.
28. Altschul, S.F. (1991). Amino acid substitution matrices from an information theoretic perspective. *J. Mol. Biol.* **219**:555–565.
29. Gotoh, O. (1982). An improved algorithm for matching biological sequences. *J. Mol. Biol.* **162**:705–708.
30. Fitch, W.M. & Smith, T.F. (1983). Optimal sequence alignments. *Proc. Natl. Acad. Sci. USA* **80**:1382–1386.
31. Altschul, S.F. & Erickson, B.W. (1986). Optimal sequence alignment using affine gap costs. *Bull. Math. Biol.* **48**:603–616.
32. Myers, E.W. & Miller, W. (1988). Optimal alignments in linear space. *CABIOS* **4**:11–17.
33. Henikoff, S. & Henikoff, J.G. (1992). Amino acid substitution matrices from protein blocks. *Proc. Natl. Acad. Sci. USA* **89**:10915–10919.
34. Bairoch, A. & Boeckmann, B. (1994). The SWISS-PROT protein sequence data bank: current status. *Nucleic Acids Res.* **22**:3578–3580.

35. Xu, G.F., O'Connell, P., Viskochil, D., Cawthon, R., Robertson, M., Culver, M., Dunn, D., Stevens, J., Gesteland, R., White, R. & Weiss, R. (1990). The neurofibromatosis type 1 gene encodes a protein related to GAP. *Cell* **62**:599–608.
36. Cvrckova, F. & Nasmyth, K. (1993). Yeast G1 cyclins CLN1 and CLN2 and a GAP-like protein have a role in bud formation. *EMBO J.* **12**:5277–5286.
37. Wootton, J.C. & Federhen, S. (1993). Statistics of local complexity in amino acid sequences and sequence databases. *Comput. Chem.* **17**:149–163.

2

HIDDEN MARKOV MODELS FOR HUMAN GENES

Periodic Patterns in Exon Sequences

Pierre Baldi,[1]* Søren Brunak,[2] Yves Chauvin,[3]† and Anders Krogh[4]

[1]Division of Biology
California Institute of Technology
Pasadena, CA
[2]Center for Biological Sequence Analysis
The Technical University of Denmark
DK-2800 Lyngby, Denmark
[3]Net-ID, Inc.
San Francisco, CA
[4]The Sanger Centre
Hinxton Hall
Hinxton, Cambridgeshire CB10 1RQ, UK

ABSTRACT

We analyse the sequential structure of human genomic DNA by hidden Markov models. We apply models of widely different design: conventional left–right constructs and models with a built-in periodic architecture. The models are trained on segments of DNA sequences extracted such that they cover complete internal exons flanked by introns, or splice sites flanked by coding and non-coding sequence. Together, models of donor site regions, acceptor site regions and flanked internal exons, show that exons — besides the reading frame — hold a specific periodic pattern. The pattern has the consensus: non-T(A/T)G and a minimal periodicity of roughly 10 nucleotides.

1. INTRODUCTION

Due to the superposition of protein encoding signals, regulatory signals and structural features in one and the same DNA sequence periodicities are hard to separate.[1] In partic-

*Jet Propulsion Laboratory, Caltech.
†Department of Psychology, Stanford University.

ular, a specific *oscillatory* pattern easily detectable by one method, may be more or less invisible to others.[2] Two of the most well known periodic codes carried by DNA are the ribosome triplet reading frame (see Ref. 3 and below) and the chromatin code, which provides instructions on the proper placement of nucleosomes along the DNA molecule.[1,2,4,5,6,7,8,9]

When addressing the need for reliably separating coding regions from non-coding regions in unannotated DNA generated by the large genome sequencing projects, intrinsic periodicities are also highly interesting from a computational view point. Gene parsing requires the statistical integration of several weak signals, some of which are poorly known, over length scales of at least several hundred nucleotides. In addition to consensus sequences at the splice sites, there seem to exist a number of other weak signals[10,11,12] embedded in the 100 intron nucleotides upstream and downstream of an exon.

In revealing these patterns a different alternative to conventional algorithms is the use of machine learning approaches. Adaptive algorithms are ideally suited for domains characterized by the presence of large amounts of data and the absence of a comprehensive underlying theory.[13] The fundamental idea behind adaptive algorithms is to learn the theory from the data, through a process of model fitting. Models can be selected from a number of different classes, such as Neural Networks (NNs), Hidden Markov Models (HMMs) or Stochastic Context Free Grammars (SCFGs). Such models are usually characterized by a large number of parameters.

Indeed, in recent years, the parsing problem has also been tackled using Neural Networks[14,15,16,17,18] with encouraging results. Conventional neural networks typically use a fixed window size input, and perhaps are not ideally suited to handle all sorts of elastic deformations introduced by evolutionary tinkering in genetic sequences. Standard neural network architectures are quite successful in handling features which align nicely in a gap free manner. Such features include the discrimination between coding and non-coding DNA involving the recognition of the organism specific (integer) triplet reading frame, but also sequence patterns used when classifying amino acid residues in proteins after their category of secondary structure. Alpha-helices display for example a periodicity of four because they are stabilized by backbone hydrogen bonds between residues with this separation in the sequence window. In contrast, non-integer periodicities are rather hard to detect (and exploit) using standard non-recurrent feed-forward neural network architectures.

Another trend in recent years, has been the casting of DNA and protein sequences problems in terms of formal languages using stochastic context free grammars,[19,20] probabilistic automata and HMMs (see also Ref. 21). HMMs in particular have been used to model protein families and address a number of tasks such as multiple alignments, classification and database searches.[22,23,24,25,26,27,28] It is the success obtained with this method on protein sequences, and the ease with which it can handle insertions and deletions, that naturally suggests its application to human genes.

Thus, the main thrust of this effort is towards the development and application of HMMs and other related adaptive techniques for modeling and parsing human genes and splice sites, and specifically for the detection of new statistical regularities. In Ref. 29, HMMs are applied to the problem of detecting coding/non-coding regions in bacterial DNA (*E. coli*), which is characterized by the absence of introns (like other prokaryotes). Their approach leads to an HMM that integrates both genic and intergenic regions, and can be used to locate genes fairly reliably. A similar approach for human DNA, that is not based on HMMs, but uses dynamic programming and neural networks to combine various gene finding techniques, is described in Ref. 17. Here, we focus on detecting novel features of human exons by HMMs.

In this paper we report on a new periodic pattern found in human exons. The pattern, which is described in statistical terms, has an average period of about 10, and features a fairly strong consensus pattern [$\hat{\text{T}}$] [AT] G. This pattern was found using several different types of HMMs, and it was checked that the pattern is not commensurate with a period of three, i.e., it seems not to stem from the exon reading frame. Because the period is close to the period of the double helix in its B-form we suggest that it may be related to structural properties of the DNA.

2. MATERIALS AND METHODS

2.1. Hidden Markov Models

HMMs are a class of statistical models that have been used in a number of applications, especially speech recognition,[30,31] but also for other problems, such as single ion channel recordings.[32] Here, we first briefly review HMMs and how HMMs are applied to biological sequences.

A first order discrete HMM is completely defined by a set of states **S**, an alphabet of m symbols, a probability transition matrix $\mathbf{T} = (t_{ij})$, and a probability emission matrix $\mathbf{E} = (e_{iX})$. The model is intended to describe a stochastic system that evolves from state to state, while randomly emitting symbols from the alphabet. When the system is in a given state i, it has a probability t_{ij} of moving to state j, and a probability e_{iX} of emitting symbol X. The model is called hidden because what is observed is the output string of symbols from the system and one of the goals is to gather information about the hidden set of transitions that may have led to its production.

In the case of DNA or RNA sequences, with one symbol per nucleotide, we have for the corresponding alphabet, $m = 4$. Common knowledge about evolutionary mechanisms suggests to introduce three classes of states (in addition to the start and end states): the main states, the delete states and the insert states with

$$\mathbf{S} = \{\text{start}, m_1, ..., m_N, i_1, ..., i_{N+1}, d_1, ..., d_N, \text{end}\};$$

N is the length of the model. Usually, it is set equal to the average length of the sequences in the family being modeled. Alternatively, N can be iteratively adjusted during learning, as in Refs. 28 and 29. Prior to any learning, the transition and emission parameters of a model can be initialized uniformly, at random or according to any other desirable distribution. The main and insert states always emit a letter of the alphabet, whereas the delete states are mute.

2.2. HMM Learning Algorithms

The most important aspect of HMMs is that they are adaptive: given a set of training sequences, the parameters of a model can be iteratively modified to optimize the fit of the model to the data according to some measure, usually the product of the likelihoods of the sequences. Different algorithms are available for HMM training, such as the classical Baum–Welch algorithm,[31,33] which is a special case of the more general EM algorithm in statistical estimation,[34] and different forms of gradient descent and their approximations (for instance Ref. 35). Learning algorithms can be distinguished depending on the function being optimized, and how the optimization is being carried out (EM/gradient descent, on-line/off-line, uniform/non-uniform initialization, with or without the Viterbi approximation). Here we give a brief overview of the main algorithms, used in the following experiments, in a Bayesian framework.

Given a set of training sequences $O = \{O_1, ...O_K\}$ and an HMM $M(y)$, where y is a vector of parameters containing all the emission and transition probabilities, a reasonable goal is to find the most likely value of y given O, i.e., to maximize the quantity $p(M(y))|O$. Using Bayes rule,

$$p(M(y))|O) = \frac{p(O|M(y))p(M(y))}{p(O)}$$

where $p(O|M(y))$ is the data likelihood and $p(M(y))$ is the prior on the model. This is the standard framework of maximum a posteriori estimation (MAP). By taking negative logarithms of both sides and noticing that $p(O)$ does not depend on y, our goal becomes the minimization of

$$-\log p(O|M(y)) - \log p(M(y))$$

At this point, it is standard to assume that the sequences are independent and therefore $p(O|M) = p(O_1|M)...p(O_K|M)$. So that finally we want to find

$$\min_y - \sum_k \log p(O_k|M(y)) - \log p(M(y))$$

The effect of the prior is thus equivalent to the introduction of a regularizer in the objective function. If the prior term is neglected, which is course also equivalent to assuming a uniform prior on the models, then the goal reduces to standard maximum likelihood (ML) estimation.

For the time being, let us concentrate on the likelihood term or, equivalently, on ML. In general, there is no way of directly finding a value of y which minimizes the negative log-likelihood of the data. So one has to resort to some iterative procedure. The best known is the Baum–Welch or EM algorithm. The EM update equations are usually computed in batch mode (across all sequences) and given by

$$t_{ij} = \frac{n_{ij}}{n_i} \quad \text{and} \quad e_{iX} = \frac{m_{iX}}{m_i}$$

where $n_i = \sum_j n_{ij}$, $m_i = \sum_X m_{iX}$ and n_{ij} and m_{iX} are the expected counts derived from a double dynamic programming procedure, known as the forward–backward algorithm (see Ref. 31 for additional details). Thus the EM algorithms reset the parameters to their most likely value, given the data and the current value of the parameters. A classical theorem shows that each EM steps increases the likelihood of the sequences, until a (possibly local) maximum is reached.

In the present study, we have also extensively used a form of gradient descent analyzed in Ref. 35. More specifically, we first reparametrize the model using normalized exponentials and a new set of variables w_{ij} and v_{iX}

$$t_{ij} = \frac{e^{w_{ij}}}{\sum_k e^{w_{ik}}} \quad \text{and} \quad e_{iX} = \frac{e^{v_{iX}}}{\sum_Y e^{v_{iY}}}$$

This reparametrization has two advantages: (1) modification of the w's and v's automatically preserves the normalization constraints on the original emission and transition probability distributions; (2) transition and emission probabilities can never reach the absorbing value 0. The on-line gradient descent on the negative log-likelihood are:

$$\Delta w_{ij} = \eta(n_{ij} - n_i t_{ij}) \quad \text{and} \quad \Delta v_{iX} = \eta(m_{iX} - m_i e_{iX})$$

where η is the learning rate, $n_i = \sum_j n_{ij}$ (resp. $m_i = \sum_Y m_{iY}$), n_{ij} (resp. m_{ij}) are again expected counts derived by the forward–backward procedure, this time for each single sequence if the algorithm is to be used on-line (see Ref. 35 for details). On-line algorithms do not require memorizing the contribution of each sequence and have the added advantage of being smoother. They also perform a sort of stochastic gradient descent and, as such, may be able to avoid being trapped in some local minima.

2.3. Viterbi Algorithm

The Viterbi algorithm is a recursive dynamic programming procedure (similar to the forward propagation) that computes the most likely path associated with a given sequence in a given model. This can be done efficiently in $O(N^2)$ steps. In certain applications, especially if the Viterbi paths tend to dominate all the other ones, the likelihood of the Viterbi path is used as an approximation to the full likelihood $p(O|M)$. Alternatively, one may define the likelihood of a sequence as being the likelihood associated with its Viterbi path and proceed accordingly in the previous learning algorithm. This has been particularly effective in the case of protein family modeling, probably because of the particular significance attached with the optimal paths.

The optimal paths of each sequence can be used during training by computing approximate counts of transitions and emission, in batch mode, and then use the EM update equations. Alternatively one can use another algorithm discussed in Ref. 35. At each iteration, transition and emission probabilities are increased along the Viterbi path of the corresponding sequence. Specifically, at each step along a Viterbi path, and for any state i on the path, the parameters of the model are updated according to

$$\Delta w_{ij} = \eta(T_{ij} - t_{ij}) \quad \text{and} \quad \Delta v_{iX} = \eta(E_{iX} - e_{iX})$$

where η is the learning rate. $T_{ij} = 1$ (resp. $E_{iX} = 1$) if the $i \rightarrow j$ transition (resp. emission of X from i) is used, and 0 otherwise. In the case of a loop, as for the insert states, the update must be repeated every time the loop is traversed. The new parameters are therefore updated incrementally, using the discrepancy between the frequencies induced by the training data and the probability parameters of the model. These update rules must be repeated for each training sequence until equilibrium. In Ref. 35, it is shown that this algorithm approximates a gradient descent procedure on the negative log-likelihood of the sequences given the model. As such, it can be expected to also converge to a (possibly local) maximum likelihood estimator.

Even if the Viterbi algorithm is not used for training, it is essential for aligning any sequence to the model, or for aligning two sequences to each other. Both tasks are achieved by calculating the corresponding most likely paths. Regardless of the training algorithm used, once an HMM has been successfully trained on a family of primary sequences, it constitutes a model of the entire family and can be used in a number of different tasks. First, for any given sequence, we can compute its likelihood according to the model and also its most likely path using the Viterbi algorithm. A multiple alignment results immediately from aligning all the optimal paths of the sequences in the family. The model can also be used for classification, i.e., to decide whether a given sequence belongs to the family or not. This can be done by comparing the likelihood of the sequence, according to the model, to the likelihoods of the sequences in the family. Entire databases of sequences can be searched in this fashion.[25,28] The model can also be used to detect motifs, by examining the profile of the emission entropy (Blahut 1987) across the model, and sub-families, by examining how different protein paths cluster throughout the model.

2.4. Regularization

Once an architecture has been chosen, there are still many ways by which prior information can be incorporated into the structure of the weights, their initialization, and/or the learning algorithm. Here, we consider only one of the most simple class of priors one can use for both the emission e_{iX} and transition t_{ij} probabilities of a model, namely Dirichlet priors. Dirichlet priors are also considered in Ref. 28. By definition, a Dirichlet distribution on the probability vector $\vec{p} = (p_1, ..., p_K)$, with parameters α and $\vec{q}$, has the form

$$D_{\alpha\vec{q}}(\vec{p}) = \frac{\Gamma(\alpha)}{\prod_i \Gamma(\alpha q_i)} \prod_i p_i^{\alpha q_i - 1} = \prod_{i=1} p_i^{\alpha q_i - 1} / Z(i)$$

with $\alpha, p_i, q_i \geq 0$ and $\sum p_i = \sum q_i = 1$. For such a Dirichlet distribution, $E(p_i) = q_i$, $\mathrm{Var}(p_i) = q_i(1 - q_i)/(\alpha + 1)$ and $\mathrm{Cov}(p_i p_j) = -q_i q_j/(\alpha + 1)$. The parameters α determines how peaked is the distribution around its mean $\vec{q}$. To allow for maximum flexibility, each state should have its own prior on both emissions and transitions. In the most general case, this prior is not necessarily a Dirichlet distribution but could be more complicated, for instance a mixture of Dirichlets. Assuming as usual that all the individual state priors are independent, the total prior on the model has the form

$$\prod_{i \in \mathcal{S}} D_{\alpha(i)\vec{q}_i} t_{ij} \prod_{i \in \mathcal{E}} D_{\beta(i)\vec{r}_i} e_{iX}$$

where $\mathcal{S}$ (resp. $\mathcal{E}$) is the set of all states (resp. of all emitting states, i.e., main and insert states).

To avoid having too many parameters, it is natural to link the different priors so as to have the same Dirichlet transition distribution out of all main states, and similarly for insert and delete states, as well as for emission Dirichlet for the main and insert states. We then need to choose the 5 corresponding parameters α's (3) and β's (2), and the 5 corresponding parameter vector $\vec{q}$'s (3) and $\vec{r}$'s (2). For the emissions, it is natural to initialize the vectors $\vec{r}$ uniformly, or from the average composition of the family being modeled. Limited experiments seem to indicate that in many cases this choice does not make much of a difference. So that in general we initialize emission parameters uniformly. Notice also that when negative logarithms of the priors are taken, the quantities $\beta r_i - 1$ appear classically as regularization constants. Thus when a Dirichlet prior is centered on a uniform distribution, the strength of the corresponding regularizer can be made arbitrary small by choosing the corresponding β close to $1/r_i = 1/r$ as desired. However, in all the other cases and especially so for a Dirichlet which is centered on a vector $\vec{r}$ far away from uniform, it is impossible to arbitrarily reduce the influence of the corresponding regularizer, on the objective function, by varying β. The same holds true for transition priors.

For the transitions, and again after limited experimentation and for simplicity, we have also often used a uniform Dirichlet prior on the transitions out of the delete and insert states. This in conjunction with a choice of the corresponding α which eliminates altogether the effect of the corresponding prior or regularization term. Alternatively, one can use a Dirichlet prior which favors the transition from a delete, or an insert state, back to the corresponding main state on the backbone. For the main state transitions, however, we have found it essential to introduce a prior which tends to favor the backbone. This is particularly true with the conventional left–right architecture where the main states and the insert states have the same fixed fan-out of 3. Therefore, everything else being equal, there is no preference for emitting symbols from the backbone, rather than from the insert states. This must be corrected at initialization time, and/or by including an appropriate prior. In the following experiments, we have often initialized all the transitions uniformly but used a Dirichlet prior on the main to main transitions, typically with $\alpha = 3.01$ and $\vec{q} = (1.01/3.01, 1/3.01, 1/3.01)$. This the function to be optimized contains a regularizer term associated with the probability of the backbone in the form:

$$\min_{y} - \sum_{k=1}^{K} \log p(O_k|M(y)) - \gamma \sum_{i=0}^{N} \log t_{m_i m_{i+1}}$$

where $t_{m_i m_{i+1}}$ is the transition probability between two consecutive main states (m_0 = start and m_{N+1} = end). γ is the the regularization constant that controls the relative influence of the prior term and the data likelihood term ($\gamma = 0.01$).

2.5. Weight Sharing and Loop Architectures

To build truly periodic models, one must consider architectures containing loops (other than the trivial ones of the insert states) to account optimally for repetitive phenomena. Strictly speaking, architectures with loops are not left–right architectures any more, and therefore need particular treatment.

For simplicity, we will consider here only the case of a single major loop, originated from a particular state along the main backbone of the model that we call the anchor of the loop. The analysis can immediately be extended to the case of several independent loops anchored at different places, along the backbone, as well as to the general case of completely arbitrary HMM architectures. There are no possible deletions associated with the anchor state itself. The anchor state is the starting state of a loop, which itself has the basic structural pattern of the rest of the model, with both main, insert, and delete states. Any path from the start state to the end state which traverses the loop n times, must go through the anchor state $n + 1$ times.

In any case, all the current learning algorithms for HMMs are based on three basic procedures: the computation of most likely paths by dynamic programming (the so-called Viterbi algorithm), the forward propagation and the backward propagation (for instance, Ref. 31). Since the forward and backward propagation are very similar, it is sufficient to study the effect of loops on the Viterbi algorithm and on the forward propagation. The usual learning equations are based on counts or expectations of transitions and emissions associated with the production of the training sequences using the Viterbi algorithm or the forward–backward procedure. It is easy to see that differences in the transition/emission counts or expectations can result in an architecture with loops only because of the existence of paths containing entire loops of delete states.

Notice that if the anchor of the loop is an emitting state, there are no associated path loops consisting only of delete states. In this case, the usual procedures (forward–backward and Viterbi) and associated learning algorithms (EM, gradient descent, Viterbi approximations, ...) can be applied *without any modifications whatsoever.* Therefore, we are now left with the case of a loop anchored on a state which is mute, akin to any other delete state, but which is associated, each time it is traversed, with a decision of whether to enter the loop or not.

Effect on Viterbi algorithm. The key observation, in this case, is that in the computation of a Viterbi path, nothing is to be gained by going through a loop of delete states and therefore such loops of delete states can be disregarded altogether. Specifically, the key step of the implementation of Viterbi algorithms the recursive (with respect to t) computation of two quantities $V_t(i)$ and $S_t(i)$, for each state i in the model. For a given sequence $O = O_1, ...O_T$, $V_t(i)$ is the maximal (over all possible paths) probability of being in state i at time t, having produced $O_1, ..., O_t$,

$$V_t(i) = \max_{\mathcal{P}(\text{start}\rightarrow i)} P(\text{start} \rightarrow i, O_1, ..., O_t)$$

where $\mathcal{P}(\text{start} \rightarrow i)$ represents all possible paths from the start state to state i. $S_t(i) = j$ if and only if j is the *emitting* predecessor of i along the optimal path in the previous equation. (When the optimal path is not unique, we can for instance select one randomly or alphabetically). In other words there exists a "silent path" (no emissions) from $j \rightarrow i$ such that $V_t(i) = V_t(j)P_{dir}(j \rightarrow i)$, $P_{dir}(j \rightarrow i)$ being the probability associated with the most likely direct silent path form $j \rightarrow i$. In the conventional left–right architecture, there is at most a unique silent path form j to i, and so its probability is well defined. This is what we mean by $P_{dir}(j \rightarrow i)$. In the case of an architecture with loops, there may be infinitely many silent paths leading from j to i. In this case, $P_{dir}(j \rightarrow i)$ is the probability of the most probable one, which *necessarily* does not contain any loops. With this simple caveat, the rest of the Viterbi procedure remains the same. Notice that in the implementation, your pointer must recursively keep track, for each i of the entire silent direct path from $j \rightarrow i$. To avoid precision problems, one can work as usual with the logarithms of the probabilities without any additional problem.

Effect on forward procedure. The forward procedure recursively computes, for each state i, the probability $\alpha_t(i)$ of being in state i having produced the partial subsequence $O_1, ..., O_t$. If we evaluate the forward propagation only for the emitting states $\mathcal{E}$, then the recursion has the form

$$\alpha_t(i) = P(O_t|i) \sum_{j\in\mathcal{E}} P_S(j \rightarrow i)\alpha_{t-1}(j) = e_{iO_t} \sum_{j\in\mathcal{E}} P_S(j \rightarrow i)\alpha_{t-1}(j)$$

Here $P_S(j \rightarrow i)$ denotes the cumulative probability of all possible silent paths from the emitting state j to the emitting state i, i.e., all paths from j to i which do not contain any other emitting states. The problem is then to compute $P(j \rightarrow i)$. In the architectures we are considering, there is always a unique direct silent path from $j \rightarrow i$, with associated probability $P_s(j \rightarrow i)$. Either this is the only silent path from $j \rightarrow i$, in which case $P_S(j \rightarrow i) = P_s(j \rightarrow i)$. Or there are infinitely many silent paths from $j \rightarrow i$, each one associated with a number of revolutions around a loop of delete states, in which case

$$P_S(j \rightarrow i) = P_s(j \rightarrow i) + P_s(j \rightarrow i)P(L) + P_s(j \rightarrow i)P(L)^2 + ... = \frac{P_s(j \rightarrow i)}{1 - P(L)}$$

where $P(L)$ is the probability of going through the loop one time silently. In this way, one can easily proceed to compute the $P(j \to i)$ and the recurrence on the $\alpha_t(i)$. For precision problems, the usual scaling procedure can be applied in the same way.

2.6. Data Sets

To train HMMs on human DNA we prepared several data sets of training and testing sequences from GenBank, release 81.0. The aim was to make a large unique set of internal exons. Entries were excluded if: (1) the Feature Table was missing, (2) the ORIGIN Label was missing, (3) the CDS Feature Key was missing, (4) the CDS Feature Key did contain a complement operator, (5) the CDS Feature Key had no operator and no intron Feature Key (assumed to be cDNA), (6) they had alternative splicing, (7) the CDS Feature Keys had overlapping, multiple reading frames. From the remaining set of entries the internal exons only were kept in the set. Exons with no information about acceptor and donor sites were also not included.

The main data set contains 2,019 non-redundant human internal exon sequences and their flanking regions. From this basic set, we extracted different training sets for pure exons, in open or closed reading frame, as well as for flanked exons, or flanked splice sites. On a pure exon experiments, for instance, a typical training set typically contains 500 exon sequences (for the patterns reported we did not notice any important differences with larger training sets). The bulk of the data set contained exons with a length from 100 to 200 nucleotides; most of the experiments were done using exons from this subset only. For full statistical detail on the data set, see Ref. 26.

3. RESULTS

We report first briefly results from experiments where HMM's have been trained on *splice sites*, either as paired sites linked by exons, or separate acceptor and donor sites flanked by intron or exon nucleotides. Secondly, we report results from a large number of experiments on *exons*, where these have been mixed or in one particular reading frame only.

3.1. Acceptor and Donor Sites Linked by Exons

To see whether an HMM would pick up easily known features of human acceptor and donor sites, a model with conventional left–right architecture, was trained on 500 randomly selected flanked internal exons, with the length of the exons restricted to being between 100 and 200 nucleotides only.

The probability of emitting each one of the four nucleotides, across the main states of the model, is plotted in Fig. 1. We see striking periodic patterns, especially present in the exon region, characterized by a minimal period of 10 nucleotides, with A and G in phase, and C and T in anti-phase. In the joint probabilities there is also a clear three base pair periodicity visible especially in C+G, where every third emission corresponds to a local minimum. This is consistent with the reading frame features of human genes,[3] which are strong especially on the third codon position ($\approx 30\%$ C and $\approx 26\%$ G).

By close inspection of the parameters of an HMM trained specifically on flanked acceptor sites we observed that the model learns the acceptor consensus sequence perfectly:

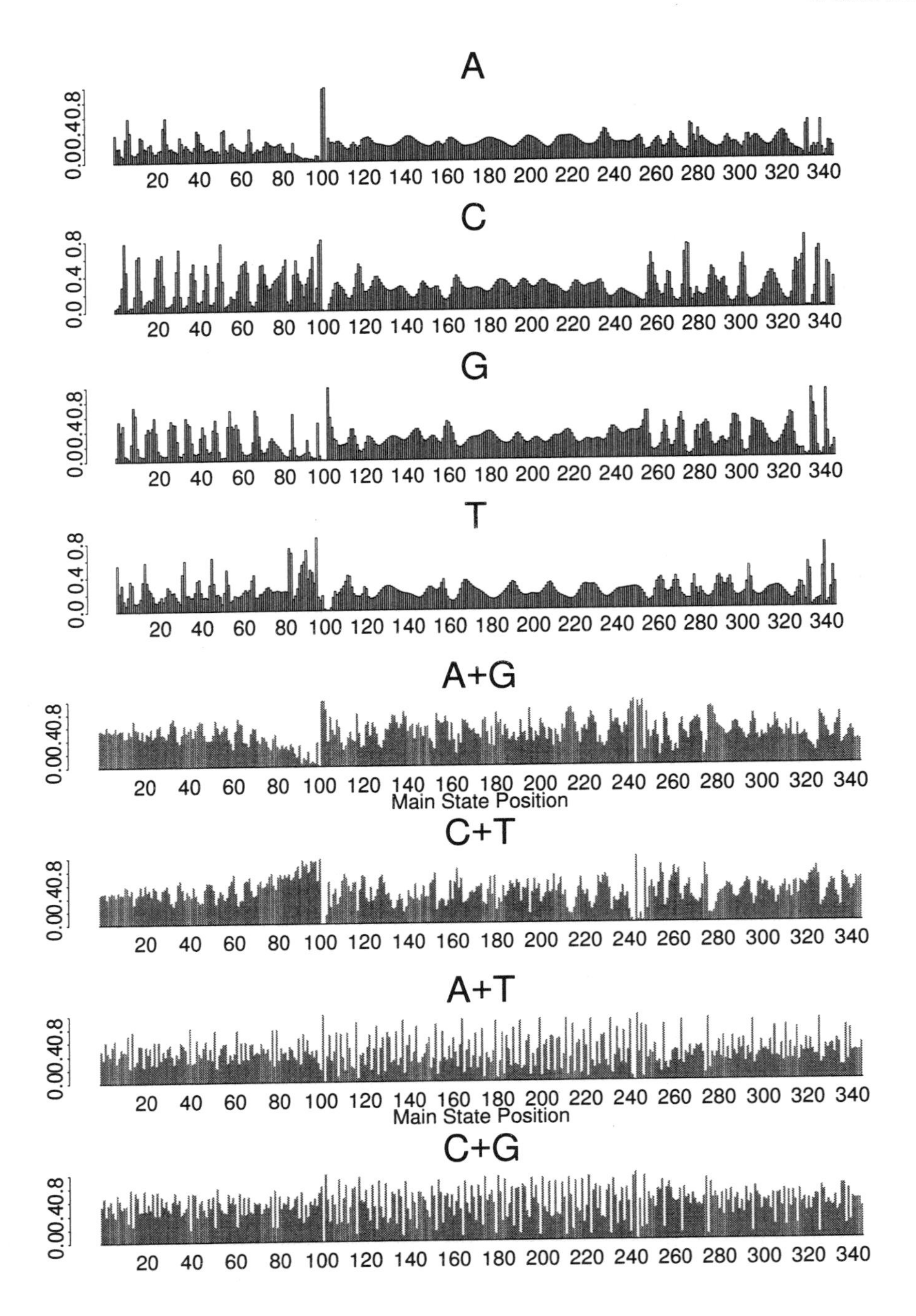

Figure 1. Emission distribution from main states of an HMM model trained on 500 flanked internal exons. The length of the training exons this time is constrained to be between 100 and 200 nucleotides, with average of 142, and fixed intron flanking of 100 on each side. The model was not fully regularized, with no bias favoring the main states backbone path. The donor site is not as clear as the acceptor site. Notice the oscillatory pattern in the exon region, and outside in the single nucleotide emission distribution. The lower set of plots (combined values) clearly shows that the model in the exon part recognizes the reading frame with C+G peaks for every third position. Human genes have a high content of these two nucleotides on the third codon position.

([TC] ... [TC] [N] [CT] [A] [G] [G]). The pyrimidine tract is clearly visible, as were a number of other known weak signals such as a branching (lariat) signal with a high A, in the 3' end of the intron (Fig. 2).

Similarly, the donor sites are also clearly visible in a model trained on flanked donor sites, but much harder to learn than the acceptor sites. The consensus sequence of the donor site is learnt perfectly: ([CA] [A] [G] [G] [T] [AG] [A] [G]), as was the G-rich region,[12] extending roughly 75 bases downstream from the donor site (Fig. 2). The fact that the acceptor site is easier to learn is most likely explained by the more extended nature of acceptor site regions as opposed to donor sites. However, it could also result from the fact that exons in the training sequences are always flanked by *exactly* 100 nucleotides upstream. To test this hypothesis, we trained a similar model using the same sequences, but in *reverse* order. Surprisingly, the model still learns the acceptor site (which is now downstream from the acceptor site) much better than the donor site. The periodic pattern in the reversed exon region is still present. The periods we observe could also be an artifact of the method: for instance, when presented with random training sequences, periodic HMM solutions could appear naturally as local optima of the training procedure. To test this hypothesis, we trained a model using random sequences of similar average composition as the exons and found no distinct oscillatory patterns in the emission distribution. We also tested that our database of exons does not correspond prevalently to the 3.6 amino acid period found in α-helical domains of proteins. This was done simply by computing from the reading frame assignments the amino acid composition and comparing it to the ranking of the helix forming potential of the twenty amino acids.[36]

In summary, after a number of initial experiments, the main results were that: (1) donor sites are harder to learn than acceptor sites; (2) there seem to be some kind of statistical periodicity, at least in the exon regions, with a period of about 10 nucleotides. In the following, we shall try to elucidate (2), by training several architectures, either with off-line Baum–Welch with initialization favoring the backbone, or on-line gradient descent with uniform initialization and backbone regularization. In all cases we have tested, the two training algorithms have given very similar results. To test the periodic patterns, we also use tied and loop architectures, as discussed in the section on methods.

3.2. Exons

The HMMs were trained using a set of non-redundant internal exon sequences, typically 500, without any flanking nucleotides. To avoid any effects due to very short or very long exons, all exons had again length between 100 and 200 nucleotides. The average length (and therefore the length of the models) was typically 142 or 143. The experiments were repeated using several randomly selected sets without any change in the observed patterns in the emission probabilities.

A periodic pattern in the parameters of the models of the form [AT][CG], (or [AT]G) with a periodicity of roughly 10 base pairs, could be seen at positions: 10, 19, 28, 37, 46, 55, 72, 81, 90, 99, 105, 114, 123, 132, 141. Notice that this pattern was detected in the weights of the model, and not directly in the sequences themselves. There is also an apparent TGCA diagonal signal, starting at position 7, which emerges quite consistently across different experiments.

The emission profile of the backbone was compared to the cumulative distributions of two nucleotides jointly (data not shown). The plots of A+G and C+T are considerably smoother than those of A+T and C+G both in the intron and the exon side. The 10 periodicity

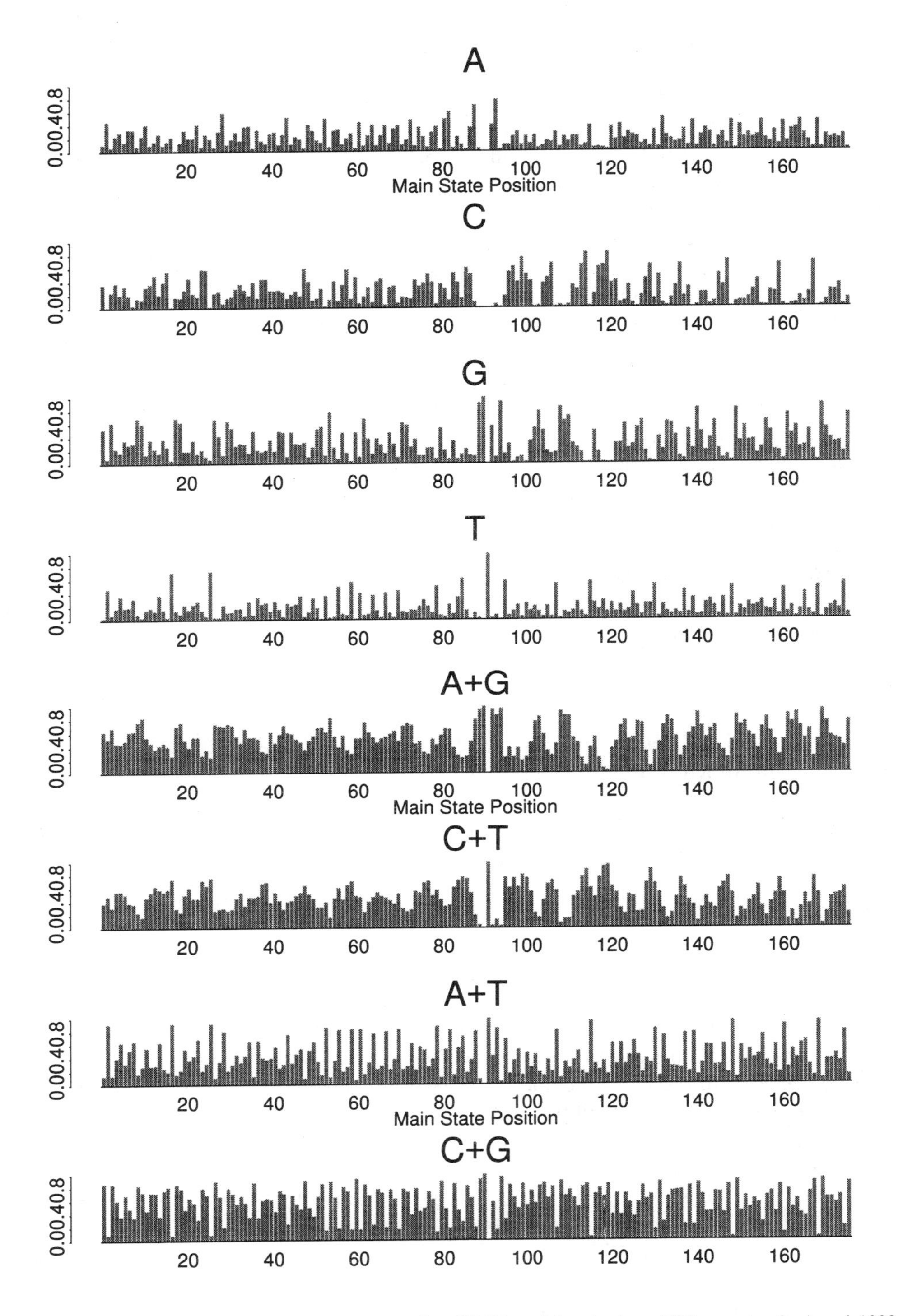

Figure 2. Emission distribution from main states of an HMM model trained on 1000 acceptor (top) and 1000 donor sites (bottom).

is visible both in the smooth phase/antiphase pattern of A+G and C+T, and in the sharp contrast of high A+T followed by high C+G. Again the three base pair periodicity[3] related to the reading frame was clearly visible especially in C+G.

One possibility is to look for possible reading frame effects on the patterns we observe. Therefore we also trained models using 500 exons with identical reading frame. The exon length was again filtered in the $[100, 200]$ interval. The average length was 143. So a model of length 143 was trained as above. Interestingly, we obtain very similar results including the TGCA signal (this time starting at position 8) and 10 periodicity. Therefore the models do not seem to be affected by reading frame effects.

To further test our findings, we trained a "tied" exon model with a hard-wired periodicity of 10, see Ref. 24 and Ref. 37. The tied model consists of 14 identical segments of length 10, and 5 additional positions in the beginning and end of the model, making a total length of 150. During training the segments are kept identical by *tying* of the parameters, i.e., the parameters are constrained to be exactly the same throughout learning, as in the weight sharing procedure for neural networks. The model was trained on 800 internal exon sequences of length between 100 and 200, and it was tested on 262 different sequences. The parameters of the repeated segment after training, are shown in Fig. 3. Emission probabilities are represented by horizontal bars of corresponding proportional length. There is a lot of structure in this segment. The most prominent feature is the regular expression [T̂] [AT] G at position 12–14. The same pattern was often found at positions with very low entropy in the "standard models" described above. In order to test the significance, the tied model was compared to a standard model of the same length. By comparing the average negative log-likelihood they both assign to the exon sequences and to random sequences of similar composition, it was clear that the tied model achieves a level of performance comparable to the standard model, but with significantly less free parameters. Therefore a period of around 10 in the exons seems to be a strong hypothesis.

However, the type of left–right architecture we have used is not the ideal model of an exon, because of the large length variations. It would be desirable to have a model with a loop structure such that the segment can be entered as many times as necessary for any given exon, see Ref. 29 for a loop structure used for *E. coli* DNA.

So we finally trained a different sort of loop model, using a data set of 500 exons. The model was a "wheel" model of length 10, without flanking, without any distinction between main and insert states, and without delete states. Thus there are no problems associated with potential silent loops. Sequences can enter the wheel at any point. The point of entry can of course be determined by dynamic programming. The structure of the model obtained after training with the EM algorithm is shown in Fig. 4. The thickness of the arrows from "outside" represents the probability of starting from the corresponding state. Remarkably, the emission parameters in the wheel have a structure very similar to those found in the repeated segment of the tied model. In particular the pattern [T̂] [AT] G is clearly recognizable.

One obvious question one can ask about the 10 periodicity is how likely is it to arise by pure chance? This question itself is not well defined because the periodicity itself is not well defined. Suppose, for the sake of the argument, that we observe something like [AT][GC] every 10 base pair or so, that is with a positional variability of $+1$ or -1. Suppose also for simplicity that each nucleotide occurs with probability 0.25. If currently we observe [AT] as a starting point of the pattern, there is a 0.5 chance of immediately seeing a [GC] right after. There is a 0.25 chance of seeing the pattern [AT][GC] 9 positions downstream, a 0.25 chance of seeing it 10 positions downstream, and a 0.25 chance of seeing it 11

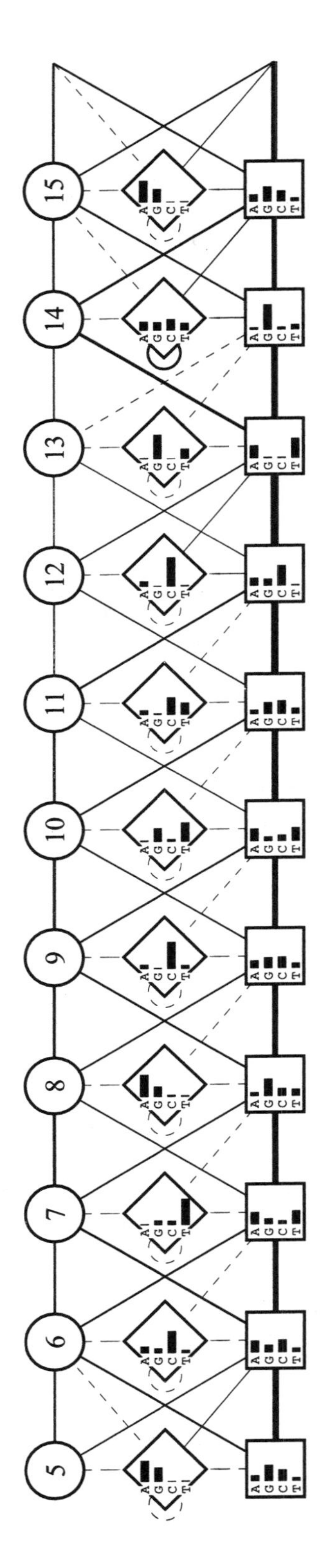

Figure 3. The repeated segment of the tied model. Rectangles represent main state and circles represent delete states. Histograms represent emission distributions from main and insert states. Thickness of connections is proportional to corresponding transition distribution. Note that position 15 is identical to position 5.

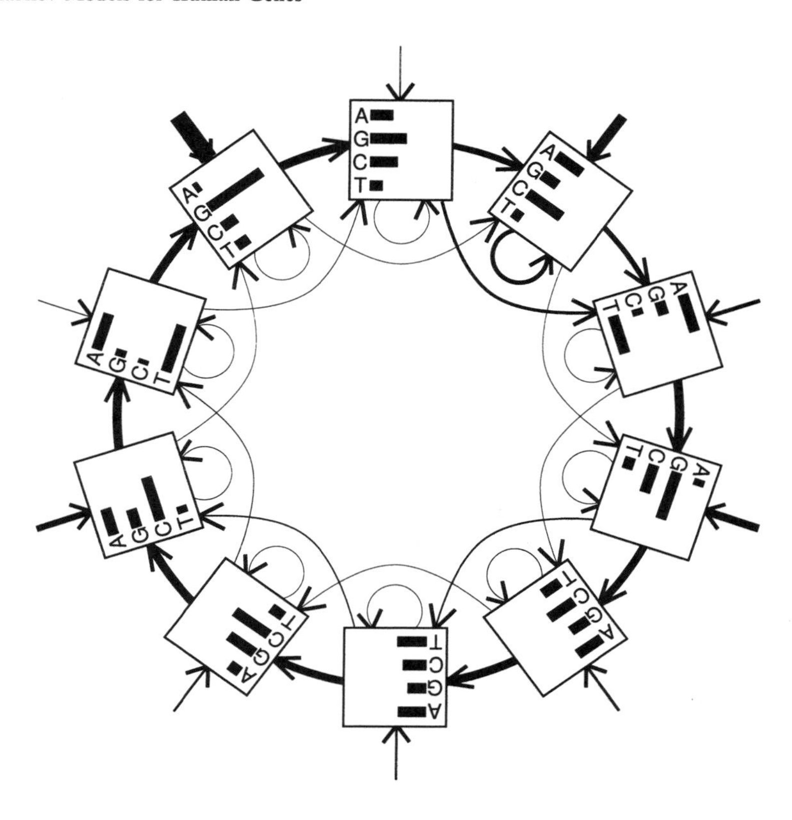

Figure 4. Wheel model trained on 500 exons of length between 100 and 200. Thickness of external arrows show the probability of starting in the corresponding state. Emission probabilities are represented by bars inside boxes.

positions downstream, in a randomly generated sequence. So the total chance of observing a first period, knowing that the starting point is a [AT], is 0.5×0.75. Similarly the chance of seeing n such "periods" is 0.5×0.75^n. In the case of a typical exon $n \approx 13$ or so. This gives a probability of observing the oscillatory pattern in random uniform sequences of approximately $0.5 \times 0.75^{13} \approx 0.01$. Even if we allow for 10 possible different starting positions, we get a probability of 0.1. In other words, the pattern would occur in at most one in ten training sequences, and there is no reason why the HMM should pick it up (given that in a random sequences there are many other "periodic" patterns with the same likelihood, such as the reverse pattern [GC][AT]). These probabilities become even smaller if we use any skewed nucleotide distribution, such as the one found in real exons.

3.3. Introns, Intragenic Regions, and Other Experiments

A number of other experiments have been tried which are not reported here at the present time. For instance, it is known that surviving isolated insertions and deletions in exons are very rare, since they entirely disrupt the local reading frame. Accordingly we have trained architectures where insertions and deletions could only occur, while respecting the triplet reading frame structure. The result are consistent with the ones reported here. Likewise several alignment experiments have been considered.

Table 1. Average negative log-likelihood per nucleotide in the wheel model. Non-coding exon is transcribed, but not translated

Sequence type	Negative log-likelihood
Exons	1.355150
Last coding exon	1.351415
First coding exon	1.357349
First exon coding/non-coding	1.361875
Last exon coding/non-coding	1.374327
Introns	1.397193
Intragenic regions	1.400396
Deep introns	1.402820
Randomized exons	1.402836

As far as the periodic pattern of period 10, it is natural to wonder whether it is confined to exons or exons with their immediate flanking, or also in the middle of large intron regions and in intragenic regions. We are in the process of constructing data sets to check these possibilities. A preliminary experiment was run, starting with a data base of introns with length at least 800. From these sequences, we removed 400 base pair on each side, to remove any proximity effects due to splice sites. We were left with 447 "deep" intron sequences, of length greater than 100. 69 deep intron sequences had length above 200. 400 sequences were selected at random and further cut to match the length distribution of the exon data base to avoid possible length effects. Finally, an left–right HMM as in Fig. 1, was trained by gradient descent with regularization. No oscillations, or other particular patterns, seem to be clearly present. After 6 training cycles, the cumulative probabilities A+G and C+T are smooth, as for exons, but so is also A+T. Overall these curves are less smooth than in the exons and their proximal flanking. After 12 training cycles, all smoothness seem to disappear.

Using the wheel model to estimate the average negative log-likelihood per nucleotide we obtain the values given in Table 1. The figures are computed specifically for various types of sequence, different types of exons, introns and intragenic regions. They strongly indicate that the above described periodic pattern belongs to exons, rather than non-coding deep intron sequence.

4. DISCUSSION

With HMMs we have been able to rapidly recognize the well known pattern and statistics related to exons and splice sites. Examples include the splice site consensus sequences, or the 3 periodicity inside exons. In addition, we are able to detect a new pattern which is a sort of periodicity, with a period of roughly 10. Our experiments indicate that this periodicity exists in the exons, but possibly also in the immediate flanking regions, but not in the deep introns. The period 10 signal is stronger than the 3 periodicity in the sense that models constrained to period 9 are harder to train.

The pattern is best seen in the weights of the model, and is also associated with the smoothness of the cumulative distributions of purines A+G (in phase) and pyrimidines C+T (in antiphase). Plots of A+T and C+G are much more jagged, with a greater tendency towards 3 periodicity. Exon regions seem to be characterized also by larger oscillation amplitudes than the immediately adjacent intron regions. Such patterns would be very difficult to detect with other methods, in part because of exon length variability.

All the tests we have conducted so far, have led to results that are consistent with these patterns. In particular, testing the 10 periodicity has forced us to expand the HMM method, for instance by developing new architectures. These may be useful for other problems, where periodic effects also are important.

ACKNOWLEDGMENTS

We thank C. Kesmir and K. Rapacki for competent programming. This work (SB) was supported by the Danish Natural Science Research Council and the Danish National Research Foundation. The work of PB was supported by grants from the ONR, the AFOSR and a Lew Allen Award at JPL. The work of YC was supported in part by grant number R43 LM05780 from the National Library of Medicine. The contents of this publication are solely the responsibility of the authors and do not necessarily represent the official views of the National Library of Medicine.

REFERENCES

1. Trifonov, E. N. 1989. The Multiple Codes of Nucleotide Sequences, Bull. Math. Biol. 51:417–432.
2. Drew, H. R. and Travers, A. A. 1985. DNA Bending and its Relation to Nucleosome Positioning, J. Mol. Biol. 186:773–790.
3. Trifonov, E. N. 1987. Translation framing code and frame–monitoring mechanism as suggested by the analysis of mRNA and 16S rRNA nucleotide sequences, J. Mol. Biol., 194:643–652.
4. Trifonov, E. N. and Sussman, J. L. 1980. The pitch of chromatin DNA is reflected in its nucleotide sequence, PNAS USA 77:3816–3820.
5. Brendel, V., Beckmann, J. S. and Trifonov, E. N. 1986. Linguistics of Nucleotide Sequences: Morphology and Comparison of Vocabularies, J. Mol. Struct. Dyn. 4:11–21.
6. Goodman, S. D. and Nash, H. A. 1989. Nature, 341:251–254.
7. Crothers, D. M. and Steitz, T. A. in *Transcriptional Regulation eds. McKnight, S. L. and Yamamoto, K. R.*, 501–534 Cold Spring Harbor Laboratory Press, New York, 1992.
8. Haran, T. E., Kahn, J. D. and Crothers, D. M. 1994. Sequence Elements Responsible for DNA Curvature, J. Mol. Biol. 244:135–143.
9. Muyldermans, S. and Travers, A. A. 1994. DNA Sequence Organization in Chromatosomes, J. Mol. Biol., 235:855–870.
10. Senapathy, P., Shapiro, M. B., and Harris, N. L. 1990. Splice Junctions, Branch Point Sites, and Exons: Sequence Statistics, Identification and Applications to Genome Project. Patterns in Nucleic Acid Sequences, Academic Press, 252–278.
11. Nussinov, R. 1989. Strong patterns in homooligomer tracts occurrences in non-coding and in potential regulatory sites in eukaryotic genomes. J. Biomol. Struct. Dyn. 6:985–1000.
12. Engelbrecht, J., Knudsen, S. and Brunak S., 1992. G/C rich tract in 5' end of human introns, J. Mol. Biol., 227:108–113.
13. Rumelhart, D. E., Durbin, R., Golden, R. and Chauvin, Y. 1994. Back-propagation: the Theory. In: Back-propagation: Theory, Architectures and Applications. Y. E. Chauvin and D. E. Rumelhart Editors, Chapter 1, Lawrence Erlbaum Associates, in press.
14. Lapedes, A., Barnes, C., Burks, C., Farber, R. and Sirotkin, K. Application of Neural Networks and Other Machine Learning Algorithms to DNA Sequence Analysis. In G. I. Bell and T. G. Marr, editors. The Proceedings of the Interface Between Computation Science and Nucleic Acid Sequencing Workshop. Proceedings of the Santa Fe Institute, volume VII, pages 157—182. Addison Wesley, Redwood City, CA, 1988.
15. Brunak, S., Engelbrecht, J. and Knudsen, S. 1991. Prediction of Human mRNA Donor and Acceptor Sites from the DNA Sequence. J. Mol. Biol., 220:49–65.

16. Uberbacher, E. C. and Mural, R. J. 1991. Locating Protein-Coding Regions in Human DNA Sequences by a Multiple Sensor-Neural Network Approach. PNAS USA, 88:11261–11265.
17. Snyder, E. E. and Stormo, G. D. 1993. Identification of Coding Regions in Genomic DNA Sequences: an Application of Dynamic Programming and Neural Networks. Nuc. Acids Res., 21:607–613.
18. Xu, Y., Einstein, J. R., Mural, R. J., Shah, M. and Uberbacher, E. C. 1994. An Improved System for Exon Recognition and Gene Modeling in Human DNA Sequences. Proceedings of Second International Conference on Intelligent Systems for Molecular Biology Stanford University. , R. Altman and D. Brutlag and P. Karp and R. Lathrop and D. Searls Editors, AAAI Press, 376–383.
19. Searls, D. B. 1992. The Linguistics of DNA. American Scientist, 80:579–591.
20. Sakakibara, Y., Brown, M., Underwood, R. C., Mian, S. I. and Haussler, D. 1993. Stochastic Context-Free Grammars for Modeling RNA. Technical Report UCSC-CRL-93-16, University of California, Santa Cruz.
21. Churchill, G. A. 1989. Stochastic Models for Heterogeneous DNA Sequences. Bull. Math. Biol., 51:79–94.
22. Baldi, P., Chauvin, Y., Hunkapiller, T. and McClure, M. A. 1993. Hidden Markov Models in Molecular Biology: New Algorithms and Applications. Advances in Neural Information Processing Systems 5:747–754, Morgan Kaufmann Pub.
23. Baldi, P., Chauvin, Y., Hunkapiller, T. and McClure, M. A. 1994a. Hidden Markov Models of Biological Primary Sequence Information. PNAS USA, 91:1059–1063.
24. Baldi, P., Brunak, S., Chauvin, Y., Engelbrecht, J. and Krogh, A. 1994b. Hidden Markov Models of Human Genes. Advances in Neural Information Processing Systems 6:761–768, Morgan Kaufmann Pub.
25. Baldi, P. and Chauvin, Y. 1994b. Hidden Markov Models of the G-Protein Coupled Receptor Family. J. Comp. Biol., 1:311–335.
26. Baldi, P., Brunak, S., Chauvin, Y., Engelbrecht, J. and Krogh, A. 1994c. Hidden Markov Models of Human Genes. CalTech Technical Report. Division of Biology, Caltech.
27. Haussler, D., Krogh, A., Mian, I. S. and Sjölander, K. 1993. Protein Modeling using Hidden Markov Models: Analysis of Globins, Proceedings of the Hawaii International Conference on System Sciences, **1**, IEEE Computer Society Press, Los Alamitos, CA, 792–802.
28. Krogh, A., Brown, M., Mian, I. S., Sjölander, K. and Haussler, D. 1994a. Hidden Markov Models in Computational Biology: Applications to Protein Modeling. J. Mol. Biol. 235:1501—1531.
29. Krogh, A., Mian, I. S. and Haussler, D. 1994b. A Hidden Markov Model that Finds Genes in *E. coli* DNA, Nuc. Acids Res., 22:4768—4778.
30. Levinson, S. E., Rabiner, L. R. and Sondhi, M. M. 1983. An Introduction to the Application of the Theory of Probabilistic Functions of a Markov Process to Automatic Speech Recognition. The Bell Syst. Tech. J., 62:1035–1074.
31. Rabiner, L. R. 1989. A Tutorial on Hidden Markor Models and Selected Applications in Speech Recognition. Proc. IEEE, 77:257–286.
32. Ball, F. G. and Rice, J. A. 1992. Stochastic Models for Ion Channels: Introduction and Bibliography. Mathematical Bioscience.
33. Baum, L. E. 1972. An Inequality and Associated Maximization Technique in Statistical Estimation for Probabilistic Functions of Markov Processes. Inequalities, 3:1–8.
34. Dempster, A. P., Laird, N. M. and Rubin, D. B. 1977. Maximum Likelihood from Incomplete Data via the EM Algorithm. J. Roy. Stat. Soc., B39:1–22.
35. Baldi, P. and Chauvin, Y. 1994a. Smooth On-Line Learning Algorithms for Hidden Markov Models. Neural Comp., 6:305–316.
36. Creighton, T. E. 1993. Proteins, Structures and Molecular Properties, W. H. Freeman, New York.
37. Baldi, P., Brunak, S., Chauvin, Y., Engelbrecht, J. & Krogh, A. 1995. Periodic sequence patterns in human exons. In *Proc. of the Third Int. Conf. on Intelligent Systems for Mol. Biol.*, (Rawlings, C., Clark, D., Altman, R., Hunter, L., Lengauer, T. & Wodak, S. eds.), pp. 30–38. AAAI Press, Menlo Park.
38. Zhurkin, V. B. 1983. Specific alignment of nucleosomes on DNA correlates with periodic distribution of purine-pyrimidine and pyrimidine–purine dimers, FEBS Lett. 158:293–297.

3

IDENTIFICATION OF MUSCLE-SPECIFIC TRANSCRIPTIONAL REGULATORY REGIONS

James W. Fickett

Theoretical Biology and Biophysics Group, MS K710
Los Alamos National Laboratory, Los Alamos, New Mexico 87545
E-mail jwf@t10.lanl.gov
Telephone: +1 505 667–7510. Fax: +1 505 665–3493

ABSTRACT

We are working to develop an algorithm to analyze windows of a few hundred base pairs of DNA sequence, and discriminate those that contain a transcriptional regulatory region (TRR) able to drive skeletal-muscle specific expression. Here we report partial progress towards this goal.

INTRODUCTION

An attractive open problem in computational molecular biology is the algorithmic characterization of transcriptional regulatory regions active in a particular context. Transcriptional regulation is, of course, one of the most active current areas of basic experimental research, with close connections to developmental biology and cancer. Computationally, the problem is attractive for the possibilities of (1) advancing gene identification by making promoter recognition more reliable and (2) helping to establish a tentative function for genes by identifying their transcriptional context.

The choice of a particular transcriptional context to study is to some extent arbitrary. We chose to study the point at which myoblasts terminally differentiate to form multinucleated myotubes in skeletal muscle (for a recent and lucid review see [Ludolph and Konieczny 1996]). This context is attractive first because two families of "master switch" transcriptional regulators are known, namely the MyoD family and MEF2 family, that can, when transfected into a variety of cell types, cause them to differentiate into myotubes (for the first discovery in this area see [Davis, Weintraub & Lassar 1987]). Second, it is an attractive context because there are myoblast cell lines that can be caused to terminally differentiate in culture (and, of course, this change is easy to assess on the basis of morphological changes). Thus there is an abundance of experimental data.

Theoretical and Computational Methods in Genome Research, edited by Suhai
Plenum Press, New York, 1997

We hope to provide a formal description of what is unique about skeletal specific transcriptional regulatory regions (TRR; including both promoters and enhancers). It is not trivial to see where to begin on this problem. The state of the chromatin (location of insulators, degree of methylation, and location of histones, for example) obviously plays an important role. Nevertheless, it seems that the single most critical determinant of transcriptional activity is the pattern of transcription factor binding sites (see for example [Koleske & Young 1995]). We are attempting to characterize such patterns by working on three fronts: characterization of transcription factor DNA binding specificity; compilation of a muscle transcription information resource; and discovery of muscle-specific patterns of coordinate transcription factor binding.

TRANSCRIPTION FACTOR DNA BINDING SPECIFICITY

The DNA-binding specificity of most transcription factors is poorly characterized. Usually a "consensus sequence" has been determined on the basis of a few known sites. For the factors we are interested in, we have begun with the sequences of a large number of oligonucleotides selected on the basis of their ability to bind the factor (e.g. [Wright, Binder & Funk 1991]), and used the Gibbs sampling algorithm [Lawrence et al. 1993] to align putative binding sites and extract a position weight matrix (PWM). For MEF2, one of the key transcription factors involved in muscle differentiation, a fairly extensive benchmarking effort showed that the PWM was considerably more reliable than the consensus sequence approach in assessing putative sites for the ability to bind MEF2 [Fickett 1996a].

MUSCLE TRANSCRIPTION INFORMATION RESOURCE (MTIR)

There is now an extensive and complex literature on the muscle-specific transcriptional regulation of a large number of genes. However, although there is a great deal of information, it is not easy to find and retrieve the data most relevant to analyzing transcriptional regulation in a particular gene. To provide a single entry point to the literature for this purpose, we are surveying the literature and writing summaries on a number of regulatory regions [Lspez & Fickett, submitted].

Purpose

The purpose of MTIR is to summarize what is known about specific mechanisms of muscle-specific transcriptional regulation for each gene where such regulation has been individually studied. We could not, of course, include all the information in all the papers covered. Rather, the aim is to provide a reference work that allows one to quickly find answers to questions like, "Has anyone shown that myogenin plays a role in the expression of MRF4?", or "What is the minimal promoter region for the myogenin gene that still gives muscle-specific expression?". In general, enough experimental evidence is presented to (1) give the reader an overview of what kind of work has been done, and (2) direct the reader to those papers most relevant for answering more detailed questions.

Completeness

There are 30–40 genes for which muscle-specific transcriptional regulation has been studied at least to the level of mapping the boundaries of promoters or enhancers (of course the list keeps growing). MTIR, currently 152 pages long, covers information from 147 articles on 22 genes. New genes are being added at the rate of about one every two weeks. In addition, as new information appears on genes already covered, it is added to the resource.

Several groups have begun to build databases of transcriptional elements in general. These databases include TFD [Ghosh 1990], Transfac [Knueppel et al. 1994], and TRRD [Kel et al. 1995]. However since these collections are attacking the entire transcriptional regulation literature, their broad coverage is often balanced by incomplete information on particular genes. Thus we chose to gain complete coverage in at least one small area, by narrowing the focus.

Form

The information is textual. Although of course a true database will be valuable in the long run (this will probably be achieved by collaborating with the above-mentioned database groups), it was decided to delay the formalization of data structure until the uses and applications of these data were more clearly defined. The primary organization is by gene. Within each gene summary there are a few main categories of information (see the sample gene entry in the Appendix):

- The protein: a very brief description of the function of the protein. Very closely related proteins with essentially the same function but different expression patterns may be mentioned.
- Nucleotide sequences: Database accession numbers for genomic or cDNA sequence data.
- When and where the gene is expressed: A brief summary of the conditions under which the gene is known to be expressed, or conditions known to affect transcription.
- Transcriptional regulatory regions: Here all regions of the genome known to be involved in the regulation of the gene are described.
- Transcription factor binding sites. Grouped by regulatory regions, all known transcription factor binding sites are described, and evidence is summarize concerning the cognate transcription factors and their in vivo functional role.
- The transcription event. Finally, the synergy of the various factors and DNA binding sites in the overall event of transcription initiation is discussed.

In each of the last three of these categories, the current state of knowledge is summarized in a few paragraphs, and following each paragraph a number of bullets summarize the experimental evidence for the assertions made.

Availability

The resource is maintained as an on-line document in FrameMaker (TM). Whenever the document is updated, the new version is converted to a World Wide Web page using WebMaker (TM). Thus the latest version is always available both on-line and as a normal document. The Web version may be reached (using, for example, Netscape (TM)) at the address http://synapse.lanl.gov/muscle/HomePage.html. A hardcopy version is available upon request.

MUSCLE-SPECIFIC PATTERNS OF COORDINATE TRANSCRIPTION FACTOR BINDING

It is thought that transcriptional specificity is most often mediated by a combination of co-occurring factors, where the combination is much more limited in distribution than any one of the factors alone. In the case of skeletal muscle-specific regulation it has been thought for some time that the MyoD and MEF2 families might act synergistically. As a first step towards characterizing muscle-specific TRR, we have shown [Fickett 1996b] that for pairs of sites from these two families,

- Spacing in naturally occurring pairs is conserved between organisms
- The spacing follows a natural geometric rule (which also holds for some of the synthetic oligos selected for myogenin binding; cf. [Funk & Wright 1992])
- The hypothesis that coordinate binding according to this geometric rule is biologically significant can explain some difficult data
- For the four genes where the two families of factors are both known to play a role, pairs of sites spaced according to the geometric rule are found only in the TRR
- If one searches the database for very good matching sites spaced according to the rule, such are only found in muscle-specific TRR

Thus properly spaced pairs of MEF2 and MyoD-family binding sites begin to define a pattern characteristic of skeletal muscle-specific transcriptional regulation.

ACKNOWLEDGMENTS

This work was supported by Public Health Service grant #HG00981–01A1 from the National Center for Human Genome Research.

REFERENCES

1. Andres, V., Fisher, S., Wearsch, P., and Walsh, K. (1995). Regulation of GAX homeobox gene transcription by a combination of positive factors including myocyte- specific enhancer factor 2. *Mol. Cell. Biol.* **15**, 4272–81.
2. Davis, R., Weintraub, H., and Lassar, A.B. (1987) Expression of a single transfected cDNA converts fibroblasts to myoblasts. *Cell* 51, 987–1000.
3. Fickett, J.W. (1996a) Quantitative discrimination of MEF2 sites (1996) *Mol. Cell. Biol.* **16,** 437–441.
4. Fickett, J.W. (1996b) Coordinate positioning of MEF2 and myogenin sites. *GeneCOMBIS http://www.elsevier.nl/locate/genecombis; and Gene* **172,** GC19–32.
5. Funk, W.D. & Wright, W.E (1992) Cyclic amplification and selection of targets for multicomponent complexes: myogenin interacts with factors recognizing binding sites for basic helix-loop-helix, nuclear factor 1, myocyte-specific enhancer-binding factor 2, and COMP1 factor. *Proc. Natl. Acad. Sci. USA* **89,** 9484–9488.
6. Ghosh, D. (1990) A relational database of transcription factors. *Nucl. Acids. Res.* **18**, 1749–1756.

7. Gorski, D.H., LePage, D.F., Patel, C.V., Copeland, N.G., Jenkins, N.A., and Walsh, K. (1993). Molecular cloning of a diverged homeobox gene that is rapidly down- regulated during the G0/G1 transition in vascular smooth muscle cells. *Mol. Cell. Biol.* **13**, 3722–33.
8. Kel, O.V., Romachenko, A.G., Kel, A.E., Naumochkin, A. & Kolchanov, N.A. (1995) Structure of data representation in TRRD - database of transcription regulatory regions on eukaryotic genomes. In Proceedings of the 28th Annual Hawaii International Conference on System Sciences (HICSS) Wailea, Hawaii, January 4–7, 1995.
9. Knueppel, R., Dietze, P., Lehnberg, W., Frech, K. & Wingender, E. (1994) TRANSFAC retrieval program: A network model database of eukaryotic transcription regulating sequences and proteins. *J. Comp. Biol.* **1**, 191–198.
10. Koleske, A.J. & Young, R.A. (1995) The RNA polymerase II holoenzyme and its implications for gene regulation. *Trends Biochem. Sci.* **20**, 113–116.
11. Lawrence, C.E., Altschul, S.F., Boguski, M.S., Liu, J.S., Neuwald, A. & Wootton, J.C. (1993) Detecting subtle sequence signals: a Gibbs sampling strategy for multiple alignment. *Science* **262**, 208–214.
12. LePage, D.F., Altomare, D.A., Testa, J.F., and Walsh, K. (1994). Molecular cloning and localization of the human GAX gene to 7p21. *Genomics* **24**, 535–40.
13. Ludolph, D.C. and Konieczny, S.F. Transcription factor families: muscling in on the myogenic program. FASEB J. 9, 1595–1604 (1996)
14. Wright, W.E., Binder, M. & Funk, W. (1991) Cyclic amplification and selection of targets (CASTing) for the myogenin binding site. Mol. Cell. Biol. 11, 4104–4110.

APPENDIX: SAMPLE GENE SUMMARY: GAX

Some genes have been studied over a number of years by several groups, while for others there is only one research article extant. Below a summary of intermediate complexity, for the GAX gene, is given as an example.

The Protein

GAX is a member of the homeobox family of genes (encoding transcription factors that contain a characteristic 60-amino-acid DNA-binding domain). GAX may play a role in maintaining the differentiated state of cardiovascular tissue.

Nucleotide Sequences

Human [LePage et al. 1994] accession number L36328; and rat [Gorski et al. 1993] Z17223.

When and Where the Gene Is Expressed

The GAX gene is widely expressed in the developing embryo; in the adult, it is mainly expressed in cardiovascular tissue including heart, lung, kidney, and blood vessels. GAX is down-regulated with proliferation stimulated by mitogens.

- The induction of serum growth factors causes down-regulation of GAX by 15-fold within 2 hours, but normal levels return by 24 to 48 hours. Removal of growth serum from growing cells induced GAX expression 5-fold within 24 hours. "Gax is likely to have a regulatory function in the G0-G1 transition of the cell cycle in vascular smooth muscle cells" [Gorski et al. 1993]*.

Transcriptional Regulatory Regions

There is 83% homology at the nucleotide level, and 98% at the amino acid level, between human and rat GAX [LePage et al. 1994]*.

GAX promoter: The GAX 138-bp promoter is located between -125 and +13 and contains multiple elements necessary for promoter activity.

- Deletion analysis located a 138-bp minimal GAX promoter fragment between - 125 and +13. Transient transfection assays with A10 rat smooth muscle cells showed that the GAX promoter fragment from -3829 to +13 activated expression of the luciferase reporter gene. Deletion of sequences up to -125 had little effect on promoter activity, and deletion to -110 reduced activity by ~75%. Further deletion to -74 almost eliminated luciferase activity. The longer GAX promoter (-3829 to +13) shows highest level of expression in 10T1/2 fibroblasts, the lowest in CV1 cells, and intermediate expression in A10 rat smooth muscle cells, aorta smooth muscle and C2C12 myoblast cultures [Andres et al. 1995]*.

Transcription Factor Binding Sites

Binding sites in the GAX promoter: There are three elements located within the core GAX promoter, HRF-1 Left, Sp1—overlapping with HRF-1 Right—and MEF2. (See Figure 1.)

The GAX promoter contains an inverted palindrome motif, designated HRF-1 left and right, at approximately -110 which binds homeobox-regulating factor 1 (HRF-1).

- Electrophoretic mobility shift assays with a radiolabeled probe (extending from approximately -124 to -97) identified a HRF-1 -binding site at the palindrome

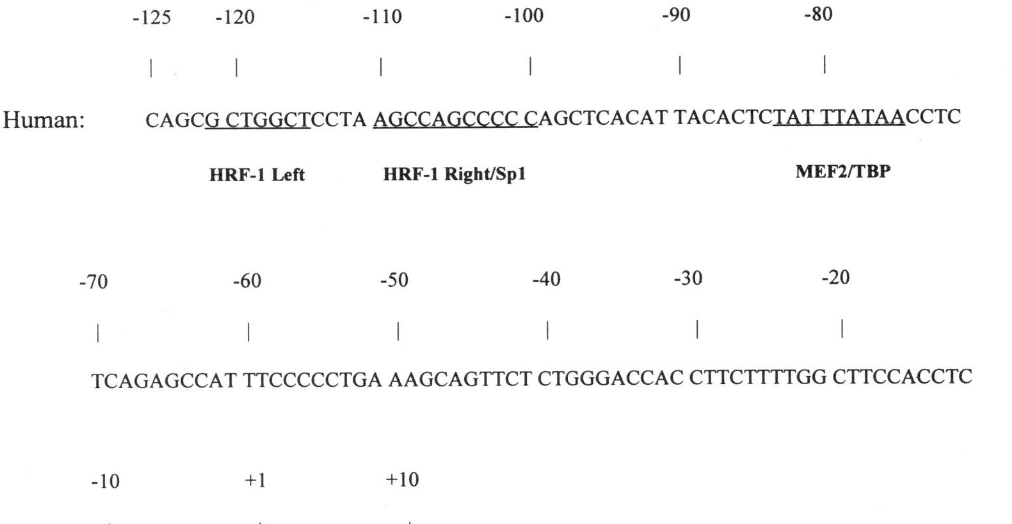

Figure 1.

motif in A10, 10T1/2 and primary rat aorta cells. Deletion of the HRF-1 left motif abolished HRF-1 binding in A10 cells. Mutation of this site (from gctggct to TTtAAAt) resulted in a reduction of activity by ~20% from wild type in A10 cells [Andres et al. 1995]*.

The Sp1 site (-109 to -100), which distinctly binds Sp1, overlaps with the HRF-1 right motif (located between -110 to -103) which binds the HRF-1 protein.

- Deletion of sequences between -125 and -100 (containing a HRF-1 right motif at -110 and an Sp1-binding site at -109) from the 138-bp GAX promoter/luciferase construct (-125 to +13 of the promoter linked to a luciferase reporter gene) caused a decrease in luciferase expression to ~13% of that expressed by the 138-bp promoter in A10 cells. Similar results occurred in primary rat aorta cells and embryonic chicken hearts. Sp1 binds to the Sp1- binding site and is distinguished from the HRF-1 nucleoprotein complex. Competition experiments showed that the Sp1 oligonucleotide could not efficiently compete for binding of HRF-1 to the GAX probe (consisting of sequences between -132 and -94) [Andres et al. 1995]*.
- Mutation of the Sp1-binding site (from gccagcccc to gccagcAAA) and of the HRF-1 Right site (from agccagcccc to aTTTaAAccc) resulted in a loss of ~75-80% activity from wild type in A10 cells [Andres et al. 1995]*.

The MADS-box transcription factors MEF2 and TBP bind to overlapping sites, between -84 and -75, within the GAX promoter. "MEF2 contributes to the regulation of GAX promoter activity in a cell type-specific manner but does not confer strict tissue-specific expression" [Andres et al. 1995]*.

- Competition experiments show that the proteins MEF2 and TBP bind to overlapping sequences between -84 to -75 in the GAX promoter. Cotransfection experiments showed that MEF2 is capable of transactivating GAX expression in A10 cells in a sequence-specific manner; exogenous MEF2 transactivated expression 3.8-fold in transfected C2C12 myoblasts [Andres et al. 1995]*.
- The intact MEF2/TBP site (ctatttataa)M+T+, as well as the site with mutated TBP-binding sequences (ctaAAAataa)M+T- effectively competed for MEF2-^ binding in extracts from A10 cells, and 4-fold higher in C2C12 cells. Mutation of only the MEF2-binding sequences (from ctatttataa to AATtCtataa)M-T+ reduced activity by ~45% but TBP-binding was still possible. Mutation of both the MEF2 and TBP binding sequences (from ctatttataa to cGaAttCtaa)M-T- resulted in a ~50% decrease in activity from wild type levels in A10 cells, and prevented binding of TBP and transactivation by MEF2 in C2C12 myocytes (but not myoblasts or 10T1/2 fibroblasts) [Andres et al. 1995]*.

Transcription Event

The elements essential for activity of the 138-bp GAX promoter—HRF-1, Sp1, and MEF2 -binding sites—occur within a 51-bp fragment (between -125 and -75). The synergy between these three positive transcription elements is largely undefined. Andres et al. [1995]* showed that the two halves of the palindrome (HRF-1 left and right) are necessary for efficient binding of HRF-1; mutation of either half reduced DNA-binding affinity of HRF-1 significantly.

4

A SYSTEMATIC ANALYSIS OF GENE FUNCTIONS BY THE METABOLIC PATHWAY DATABASE

Minoru Kanehisa and Susumu Goto

Institute for Chemical Research
Kyoto University
Uji, Kyoto 611, Japan

1. INTRODUCTION

The genome sequencing projects of different organisms are fast producing catalogs of genes and gene products. The next obvious step is to understand functional implications, namely, to decipher both experimentally and computationally when, where, and how genes and molecules function in living organisms. In fact, our knowledge on the functioning of genes and molecules is also rapidly expanding owing to the advancement of experimental technologies in wide areas of molecular and cellular biology. In order to make full use of the information obtained by genome projects, it is essential that such functional data are properly computerized.

The functional data that relate to sequence information are currently stored, for example, in features tables of the sequence databases and in motif libraries such as Prosite. However, these basically represent sequence-function relationships of single molecules, i.e., individual components of a biological system, and they do not contain higher level information, i.e., wiring diagrams, of genetic interactions and molecular interactions. We have thus initiated a project to computerize molecular/genetic information pathways for all known aspects of living organisms, especially in the form of a deductive database where a pair of interacting molecules or genes is represented as a binary relation. As the first step of this project, we are currently working on the metabolic pathways.

One may have an impression that the metabolic pathways are already well characterized and there is not much left to uncover. However, this is not really the case. Table 1 summarizes the numbers of enzymes with assigned EC numbers (LIGAND), enzymes identified in the known metabolic pathways (KEGG), enzymes with known amino acid sequences (SWISS-PROT and PIR), and enzymes with known 3D structures (PDB). Assuming that an EC number is assigned to a chemical reaction that presumes the existence of a specific enzyme (it is possible that one enzyme can catalyze multiple reactions), then only 40% of the enzymes are currently identified in the SWISS-PROT and PIR databases. Fur-

Theoretical and Computational Methods in Genome Research, edited by Suhai
Plenum Press, New York, 1997

Table 1. The number of enzymes appearing in each database

	LIGAND	KEGG	SwissProt	PIR	PDB
Oxidoreductase (EC1)	899	360	303	327	59
Transferase (EC2)	1002	306	332	319	47
Hydrolase (EC3)	992	172	428	370	107
Lyase (EC4)	332	160	129	126	21
Isomerase (EC5)	145	66	54	54	13
Ligase (EC6)	119	59	68	61	13
Total	3489	1123	1314	1257	260

LIGAND: enzymatic reactions
KEGG: metabolic pathways
SwissProt and PIR: amino acid sequences
PDB: 3D structures

thermore, as shown in Table 2 one half of those enzymes that are known on the metabolic pathways are not identified in the sequence databases.

The *Saccharomyces cerevisiae* sequencing project is nearing to a completion, but the biological function of 3,000 genes out of 7,000 are still unknown. It is possible that many of these genes code for proteins in secondary metabolism [1]. By computerizing current knowledge of metabolic pathways in different organisms, and by mapping each gene of a specific organism onto the known pathways, it will become feasible to identify missing elements and missing paths which can in turn be utilized for functional identification of unknown genes.

2. MATERIALS AND METHODS

2.1. DBGET Integrated Database Retrieval System

DBGET is a simple database retrieval system for finding and obtaining specific entries of diverse databases. Here a database is simply considered as a sequential collection of entries, which may be stored in a single text file or multiple text files. Most of the existing molecular biology databases including those for bibliographic data, genetic/physical maps, nucleotide sequences, amino acid sequences, three-dimensional structures, sequence motifs, enzyme reactions, protein mutations, amino acid indices, and genetic diseases, can be treated in this simplified manner, or as so-called flat-file databases. Because each entry of such a database is given an unique identifier, i.e., an entry name or an accession

Table 2. The number of enzymes identified in the KEGG metabolic pathway database

		SwissProt		PIR		PDB	
	KEGG	Yes	No	Yes	No	Yes	No
Oxidoreductase (EC1)	360	186	117	200	127	43	16
Transferase (EC2)	306	179	153	179	140	27	20
Hydrolase (EC3)	172	89	339	82	288	17	90
Lyase (EC4)	160	90	39	88	38	13	8
Isomerase (EC5)	66	35	19	35	19	9	4
Ligase (EC6)	59	40	28	38	23	3	10
Total	1123	619	695	622	635	112	148

number, a number of databases in the world can be retrieved uniformly by the combination of a database name and an entry name (or an accession number):

dbname:identifier

DBGET is an extension of the now defunct IDEAS system [2] and forms the basis of the GenomeNet database service (http://www.genome.ad.jp). At present, DBGET supports seventeen databases, and six of them are the products of the Japanese GenomeNet community.

2.2. LinkDB Database

In recent years it has become a common practice to cross-reference related data among a number of molecular biology databases. We organize and maintain a database of database links called LinkDB [3], which is a collection of binary relations in the form of:

dbname1:identifier1 → dbname2:identifier2

There are three types of links: original links given in each database, reverse links obtained by reversing the original links, and indirect links computed from multiple links. LinkDB is daily updated and is an integral part of the DBGET system.

2.3. Kyoto Encyclopedia of Genes and Genomes (KEGG)

KEGG (Kyoto Encyclopedia of Genes and Genomes) is our attempt to computerize known pathways and to correlate them with gene catalogs. KEGG is constructed for use in the World Wide Web (WWW) and a preliminary version is made publicly available through the GenomeNet WWW server. KEGG consists of three types of data:

Genes	—	hierarchical text data
Molecules	—	hierarchical text data
Pathways	—	graphics data

that are linked with each other and with the existing databases as shown in Figure 1.

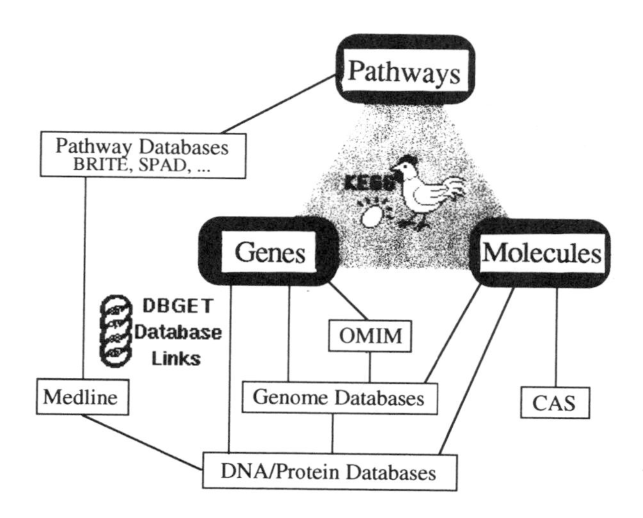

Figure 1. A conceptual illustration of KEGG, Kyoto Encyclopedia of Genes and Genomes.

The main feature of KEGG is the pathways section, which currently contains about 100 diagrams (maps) of metabolic pathways. The data were entered from the compilation of the Japanese Biochemical Society [4] together with the Boehringer Biochemical Pathways [5]. A map contains linked graphics objects of enzymes represented as rectangles with EC numbers inside. This object is clickable to retrieve the corresponding entry of the LIGAND database and related entries from other databases through DBGET.

The genes section is a collection of gene catalogs for selected organisms containing gene symbols, gene product names, EC numbers for enzymes, and links to other databases. The gene catalog is represented in a hierarchical tree, for example, according to Riley's classification scheme [6], and the user may expand or collapse selected branches of the tree by simply clicking on the headings and subheadings. We are working on a hierarchical classification of gene functions based on the pathway data being entered. Thus, the gene classification is linked to the pathway maps and genes identified in each organism can be seen easily by coloring corresponding enzymes on the pathway maps.

The tables of the molecules section are also represented by hierarchical text data similar to the gene catalogs. Currently, three types of enzyme classifications are available; they are based on the function (reaction) of EC numbers, the sequence similarity taken from the PIR superfamily classification, and the 3D structural similarity taken from the SCOP database. More hierarchical tables will be added and they will be used in the process of query relaxation [7] in pathway computation.

2.4. Deductive Database for Metabolic Pathways

In order to compute possible pathways from a given list of enzymes, we are organizing a deductive database using CORAL [8]. Suppose enzyme E catalyzes a chemical reaction with substrate X and product Y.

$$reaction(E, X, Y).$$

When the conversion of compound X to compound Y is a multistep process consisting of a number of enzymes, the enzymatic pathway is represented by:

$$path(X, Y, [E]) \leftarrow reaction\ (E, X, Y).$$

$$path(X, Y, [E \mid EL]) \leftarrow reaction\ (E, X, Z), path(Z, Y, EL).$$

Given a catalog of enzyme genes, we can calculate possible pathways catalyzed by the gene products. Thus, for example, the correctness of gene identification in the genome sequencing project can be checked against the degree of completeness of the derived pathways. It is also possible to compare pathway diagrams and search for local similarities, in addition to analyzing sequence and structural similarities of enzymes on the pathway diagrams.

3. RESULTS AND DISCUSSION

3.1. Linking Existing Databases to KEGG

As of March 21, 1996, the public release of KEGG contains 82 metabolic pathway diagrams. A pathway may be chosen from the text menu or from the clickable diagram shown in Figure 2. This collection of pathways is linked to the DBGET system through the LIGAND database [9] for enzyme reactions; LIGAND now has a link to the pathway diagrams and, conversely, an enzyme (boxed object) on a pathway diagram is linked to a LIGAND entry. Thus, KEGG may be considered another flat file database containing graphical objects rather than text data, which can easily be integrated in the WWW version of DBGET.

Figure 3 shows a query example of KEGG, where the user asked all the enzymes known to have β/α barrel (TIM barrel) to be identified on all the metabolic pathways. The user chose the 3D fold classification table according to SCOP database [10] (Fig. 3(a)), expanded the hierarchy for α/β proteins, copied the EC numbers for β/α barrel proteins, chose the pathway search option, pasted the EC numbers in the pathway search window (Fig. 3(b)), executed the query, and obtained the list (Fig. 3(c)). The user may then examine the actual locations of the enzymes on each pathway. Figure 4 shows the results for the pathways of glycolysis (Fig. 4(a)) and tryptophan biosynthesis (Fig. 4(c)), where β/α barrel proteins appear clustered in localized positions [11]. From these figures, however, it is not possible to tell what fraction of enzymes in a particular pathway have β/α barrels because 3D structures of all proteins are not known. By clicking the title box in each pathway diagram, one can invoke an option to check which enzymes are already represented in the existing databases. As shown in Figures 4(b) and (d), many enzymes are already in the PDB, i.e., their 3D structures are known, for the glycolytic pathway, while this is not the case for the pathway of phenylalanine, tyrosine, and tryptophan biosynthesis.

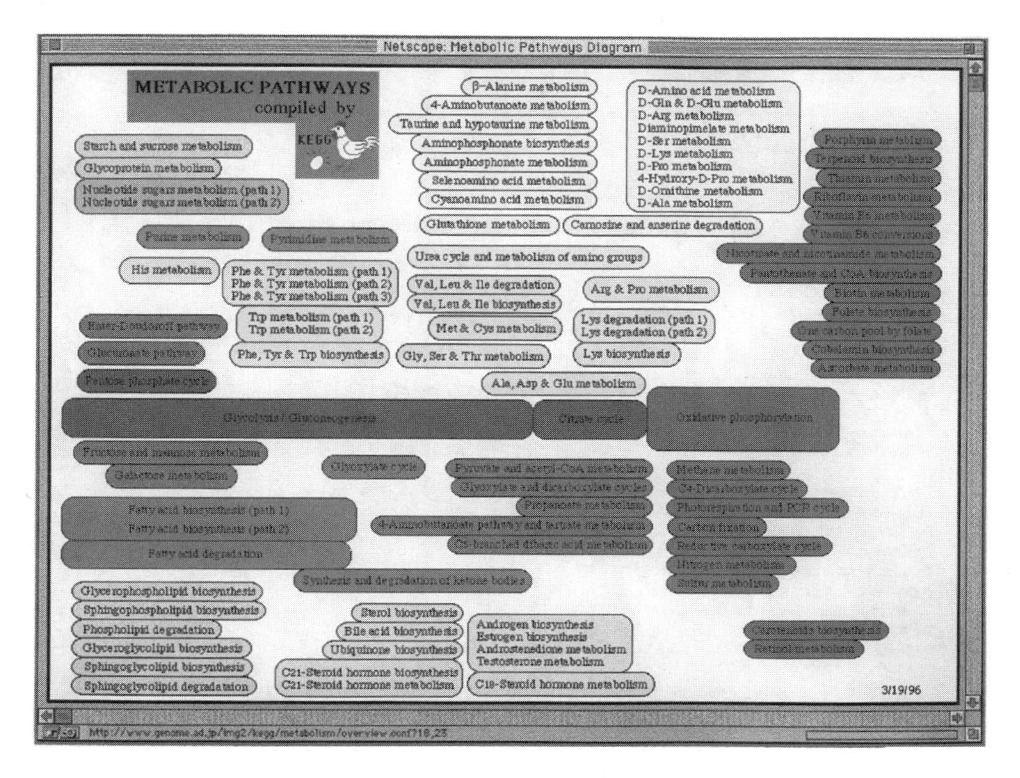

Figure 2. An entry for selection of metabolic pathway diagrams in KEGG.

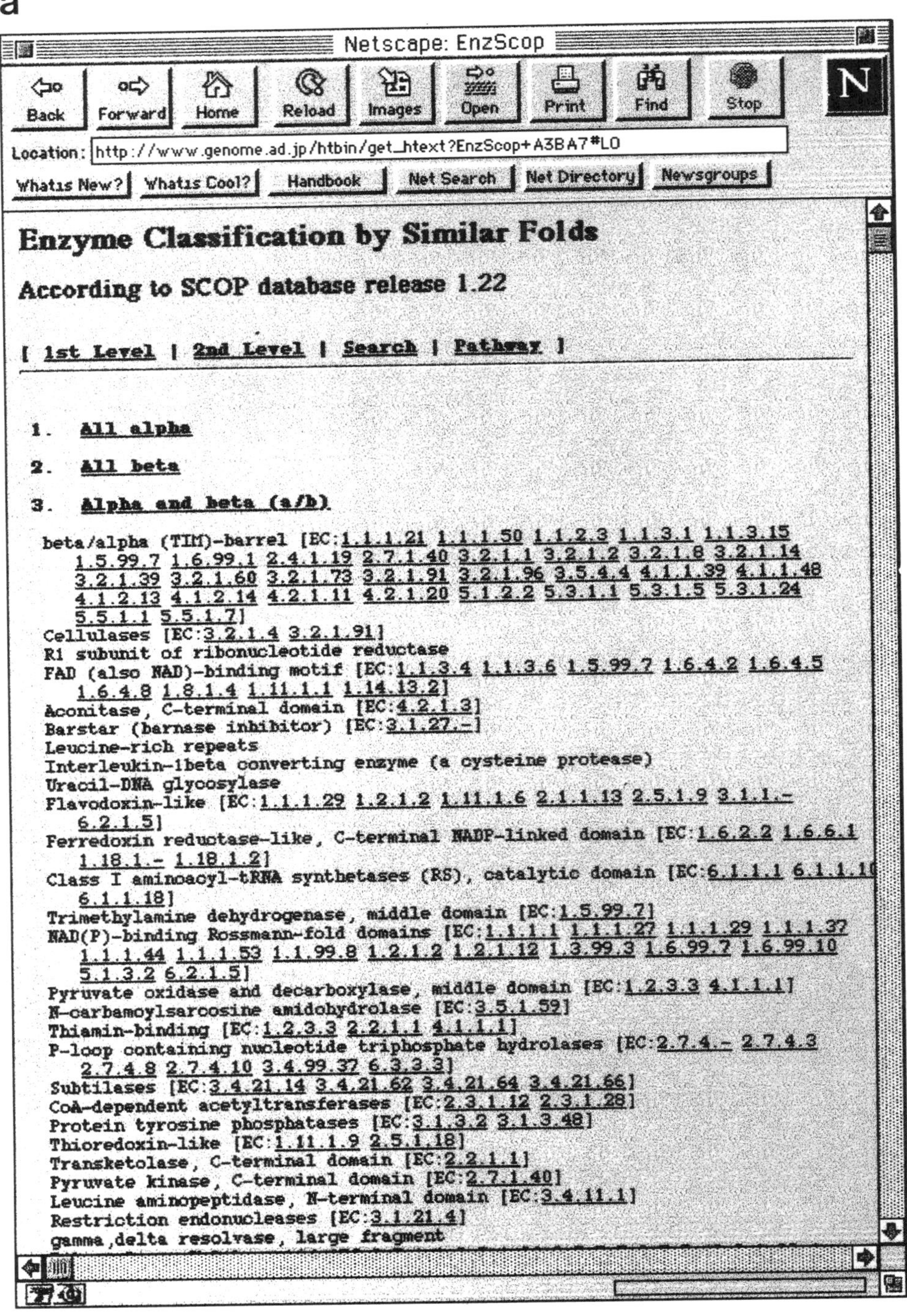

Figure 3. (a) Enzyme classification table by similar folds; (b) pathway search window; and (c) search result window.

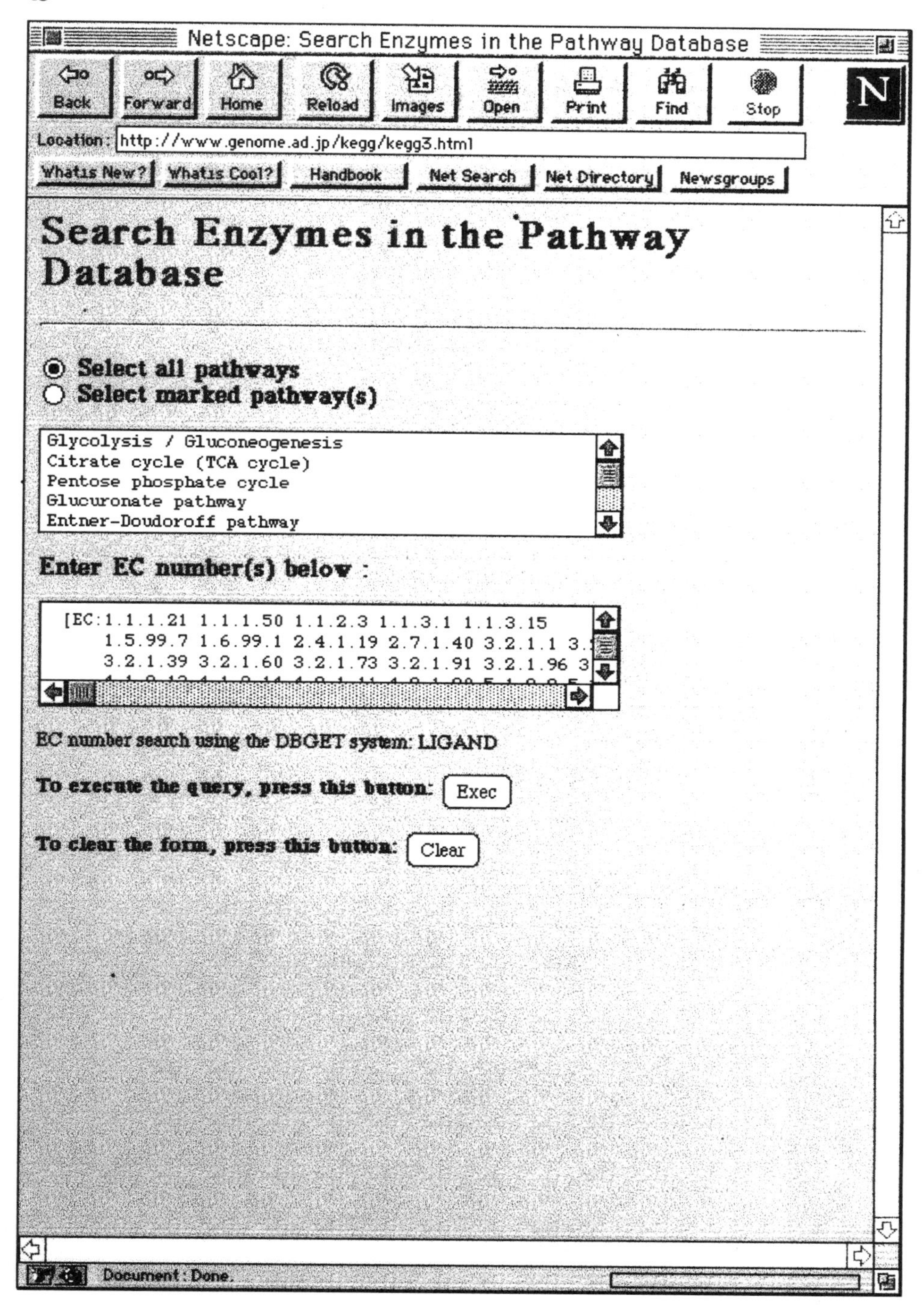

Figure 3. *(Continued)*

C

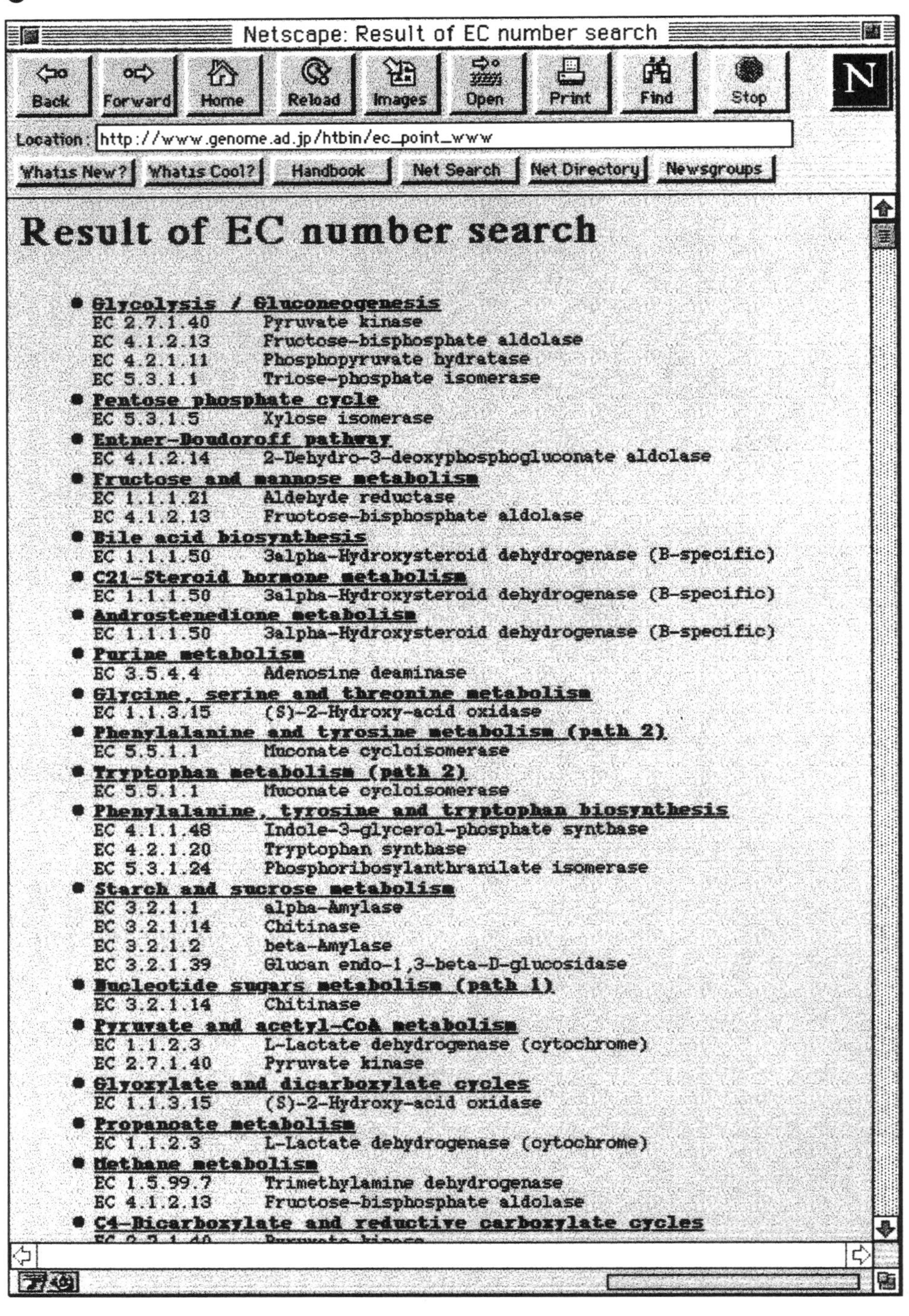

Figure 3. (*Continued*)

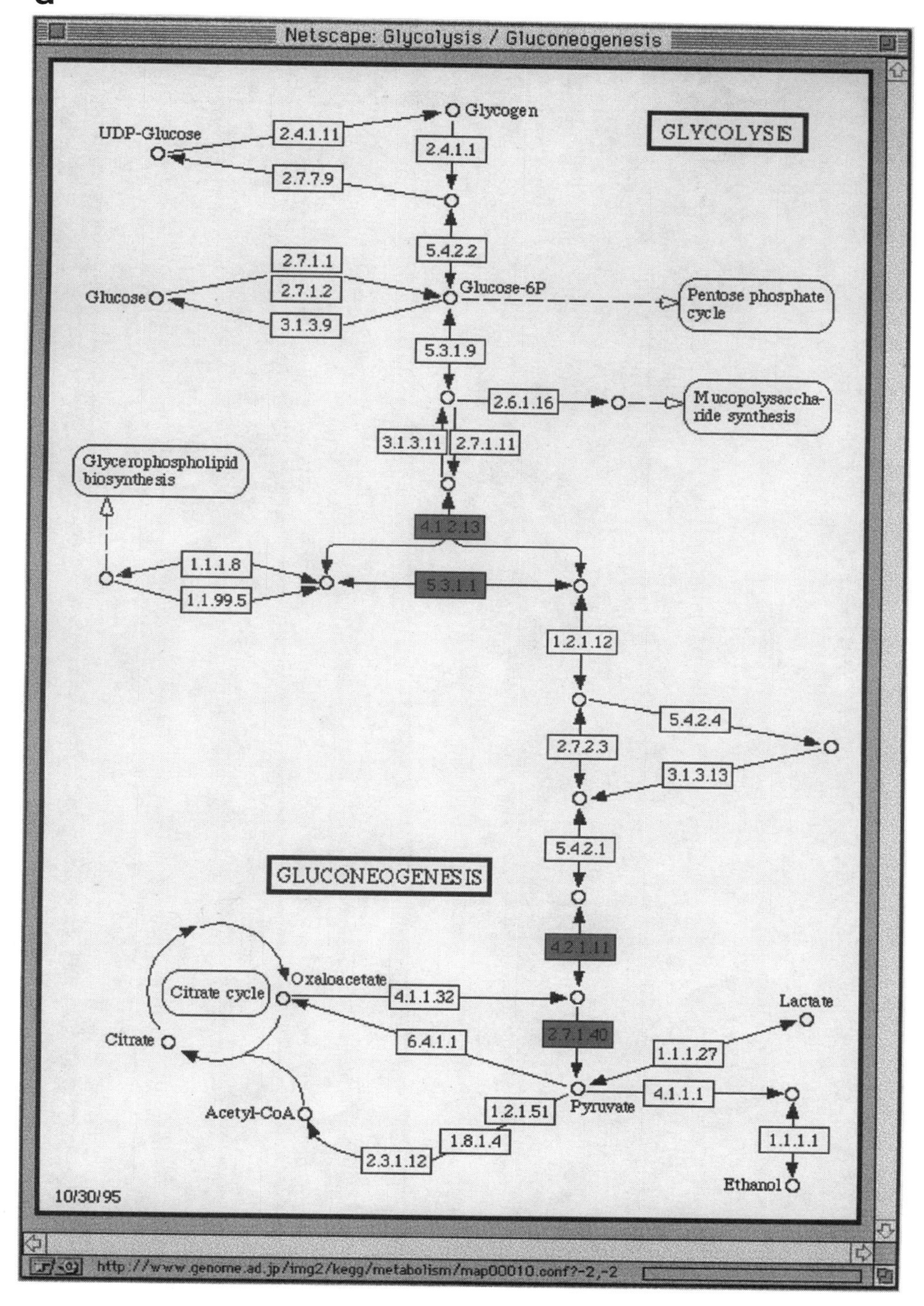

Figure 4. Display of proteins having β/α barrels in comparison with structurally known proteins. (a) β/α barrels and (b) all known structures for glycolysis, and (c) β/α barrels and (d) all known structures for phenylalanine, tyrosine, and tryptophan synthesis.

b

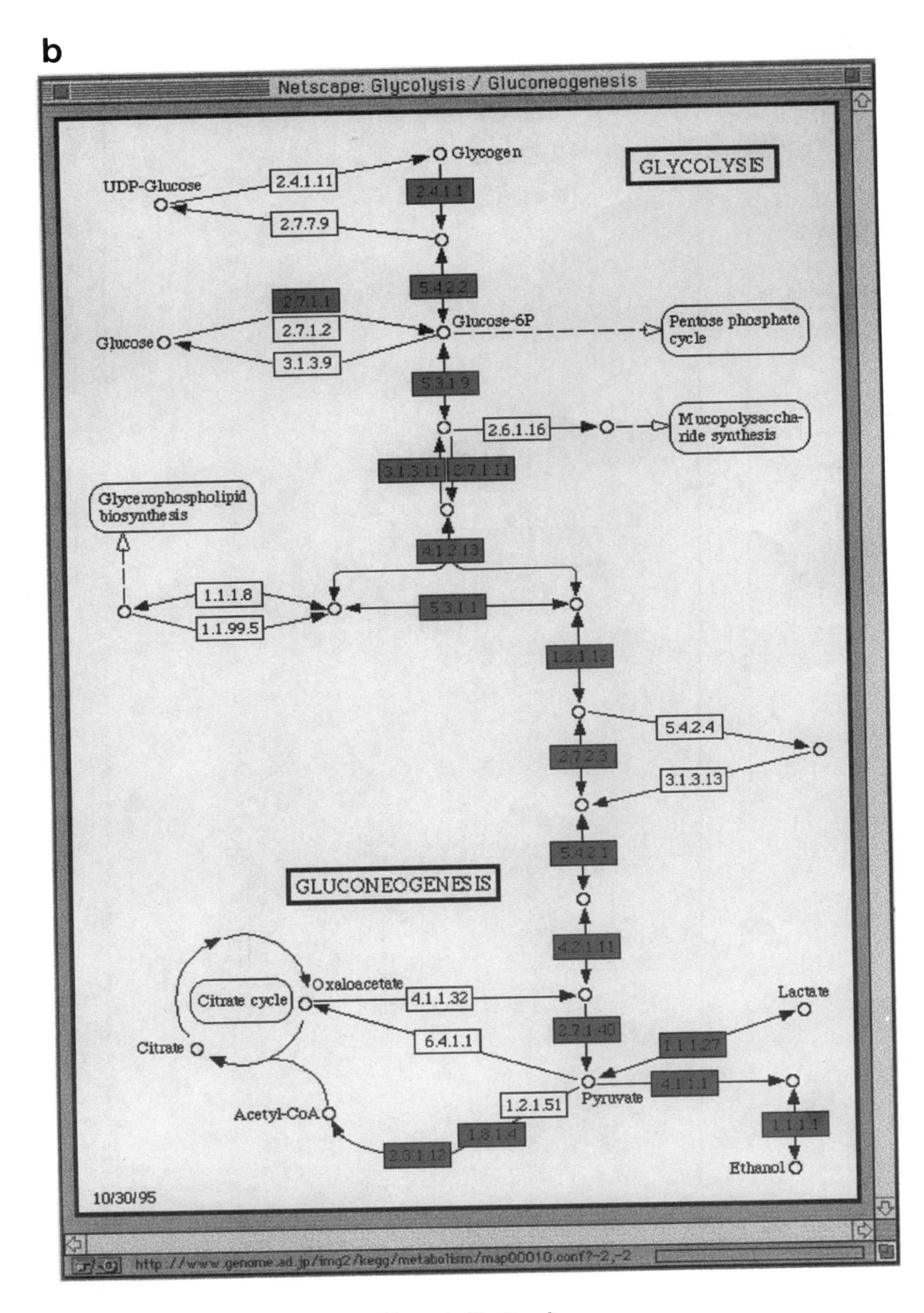

Figure 4. (*Continued*)

c

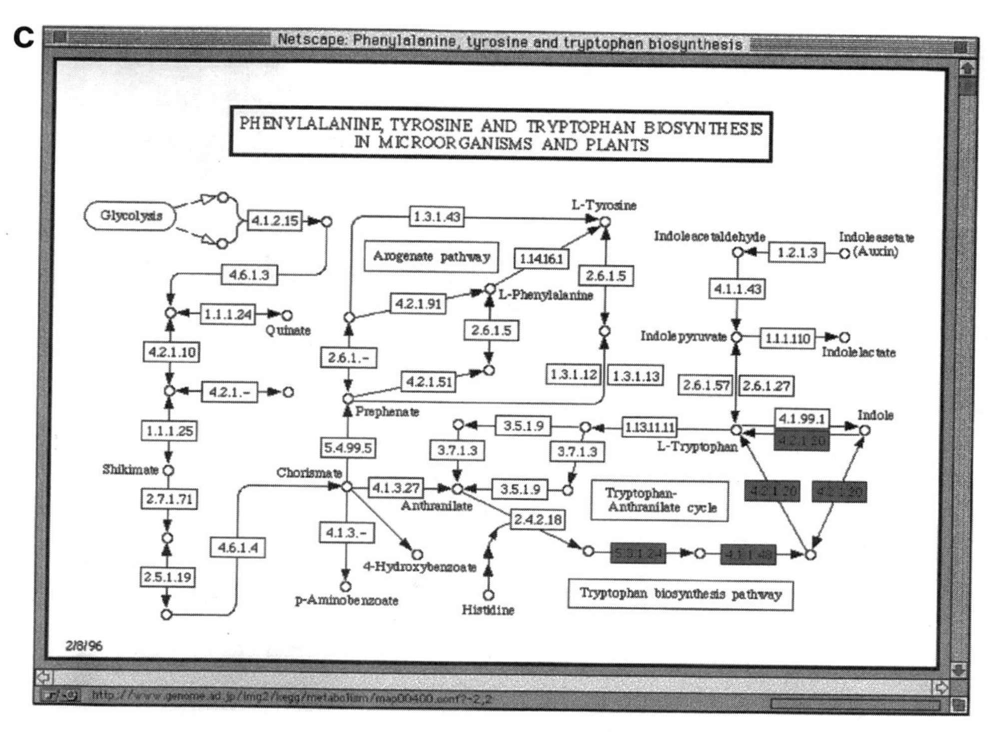

d

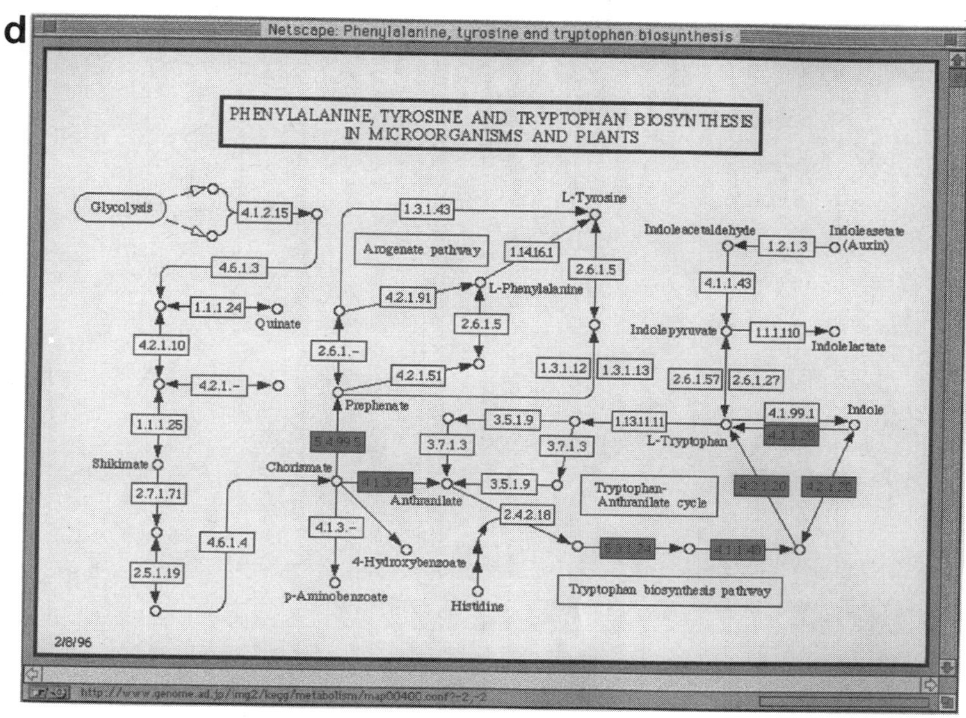

Figure 4. *(Continued)*

a

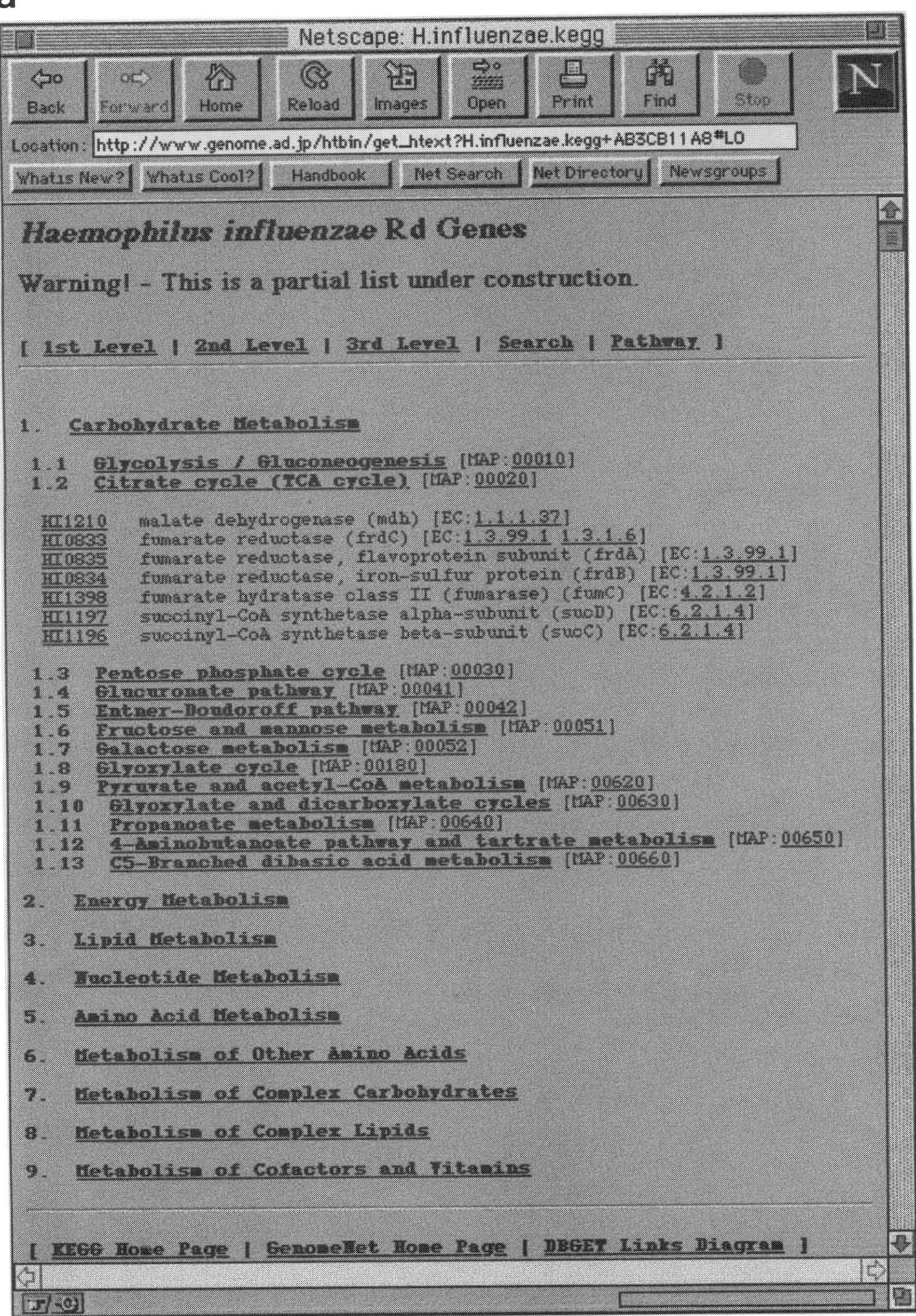

Figure 5. (a) Gene classification for *Haemophilus influenzae*; (b) genes identified for the TCA cycle; and (c) the result of path computation.

3.2. Linking Gene Catalogs to Metabolic Pathways

As of March 21, 1996, KEGG contains gene tables for four organisms hierarchically classified according to the metabolic pathway data: *Escherichia coli*, *Bacillus subtilis*, *Haemophilus influenzae*, and *Mycoplasma genitalium*. In addition, other types of classifications, mostly based on physical maps, are available for 13 organisms and human diseases.

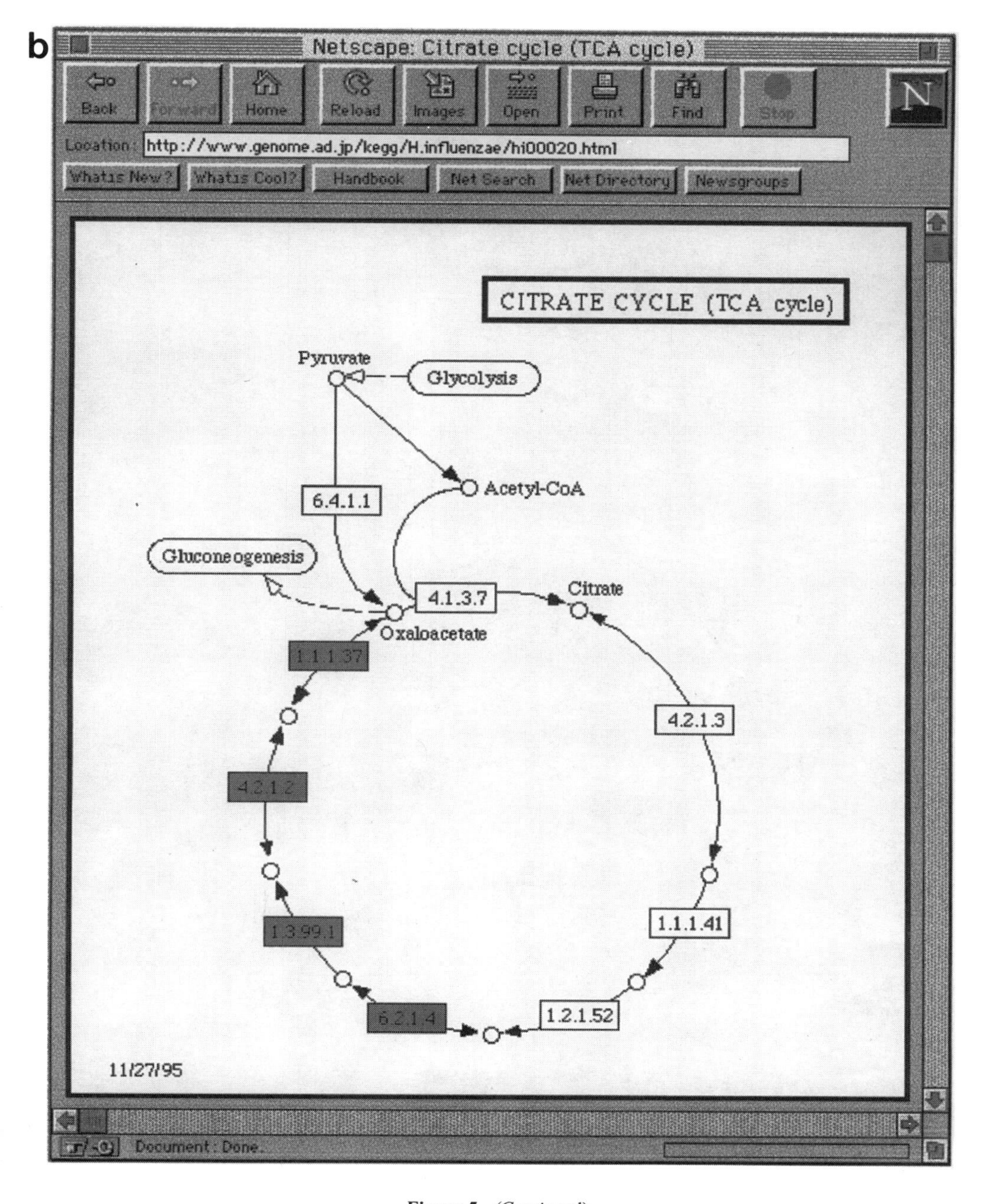

Figure 5. *(Continued)*

Note that KEGG does not contain most of the information stored in the existing genome databases and DNA/protein databases. The KEGG tables are used to link the existing databases to the pathway data being entered and to provide functional views of gene and gene products.

Figure 5 shows an example of viewing a gene catalog of *Haemophilus influenzae* in comparison with the consensus pathway. First, the hierarchical gene classification table

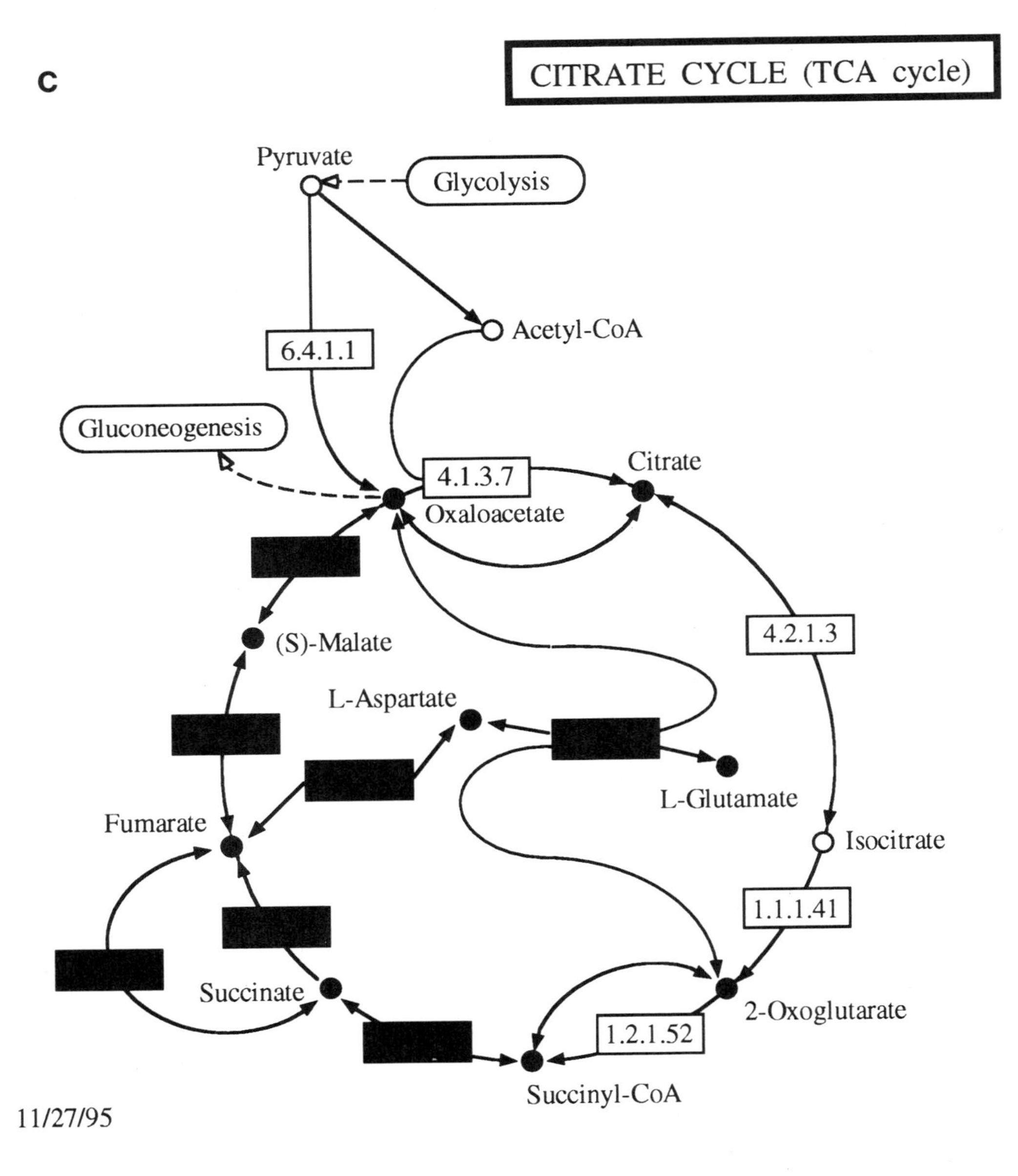

Figure 5. *(Continued)*

(Fig. 5(a)) for *Haemophilus* was chosen and expanded to see the section for the TCA cycle. Then by clicking on the MAP link marked 00020, the pathway diagram was obtained (Fig. 5(b)) showing the gene products identified in *Haemophilus* by color. It is apparent from this figure that this bacteria does not have four enzymes: EC 4.1.3.7 (citrate synthase), EC 4.2.1.3 (aconitate hydratase), EC 1.1.1.41 (isocitrate dehydrogenase), and EC 1.2.1.52 (oxoglutarate dehydrogenase) in the consensus TCA cycle. Since the last one can be replaced by EC 1.2.4.2 (oxoglutalate dehydrogenase) and EC 2.3.1.6 (dihydrolipoamide succinyltransferase) as an alternative pathway, the number of missing enzymes is three [12].

3.3. Path Computation

We then examined if a cycle could be made by considering additional enzymes from the gene catalog of *Haemophilus*. We utilized the deductive database for path computation and found that by adding EC 2.6.1.1 (aspartate transaminase) a cycle could actually be made as shown in Figure 5(c).

The path computation capability has not been implemented in the publicly available version of KEGG. There are still many problems to overcome, including representation and calculation of multi-substrate, multi-product reactions, and hierarchical classification of metabolic compounds for use in query relaxation. We believe this type of analysis will become increasingly important in functional identification of genes being determined by the genome projects.

ACKNOWLEDGMENTS

This work was supported by Kazusa DNA Research Institute and by a grant from the Ministry of Education, Science, Sports and Culture. The computation time was provided by the Supercomputer Laboratory, Institute for Chemical Research, Kyoto University.

REFERENCES

1. Oliver, S.G. (1996) Nature 379, 597–600.
2. Kanehisa, M. (1982) Nucleic Acids Res. 10, 183–196.
3. Goto, S., Akiyama, A., and Kanehisa, M. (1995) MIMBD-95: Second Meeting on the Interconnection of Molecular Biology Databases. http://www.ai.sri.com/people/pkarp/mimbd/95/abstracts/goto.html.
4. Nishizuka, Y. (1980) "Metabolic Maps", Biochemical Society of Japan (in Japanese).
5. Michal, G. (1992) "Biochemical Pathways", Third Edition, Boehringer Mannheim.
6. Riley, M. (1993) Microbiol. Rev. 57, 862–952.
7. Gaasterland, T., Godfrey, P., and Minker, J. (1991) Relaxation as a platform of cooperative answering. Proc. Workshop on Non-Standard Queries and Non-Standard Answers, pp. 101–120.
8. Ramakrishnan, R., Srivastava, D., and Sudarshan, S. (1992) Proc. Int. Conf. on Very Large Databases.
9. Suyama, M., Ogiwara, A., Nishioka, T., and Oda, J. (1993) Comput. Appl. Biosci. 9, 9–15.
10. Murzin, A.G., Brenner, S.E., Hubbard, T., and Chothia, C. (1995) J. Mol. Biol. 247, 536–540.
11. Wilmanns, M. and Eisenberg, D. (1993) Proc. Natl. Acad. Sci. USA 90, 1379–1383.
12. Fleischmann, R.D., et al. (1995) Science 269, 496–512.

5

POLYMER DYNAMICS OF DNA, CHROMATIN, AND CHROMOSOMES

Jörg Langowski, Lutz Ehrlich, Markus Hammermann, Christian Münkel, and Gero Wedemann

Division Biophysics of Macromolecules (0830)
Deutsches Krebsforschungszentrum
Im Neuenheimer Feld 280, D-69120 Heidelberg, Germany
Interdisciplinary Center for Scientific Computing (IWR)
Im Neuenheimer Feld 368, D-69120 Heidelberg, Germany

ABSTRACT

DNA and the chromatin fiber are long flexible polymers whose structure and dynamics are determined by elastic, electrostatic and hydrodynamic interactions. We are using Monte-Carlo and Brownian dynamics (BD) models to describe the structure and dynamics of DNA and chromatin. Using known values for structural parameters of DNA and the nucleosome (i.e. bending and torsional elasticity, hydrodynamic radii and electrostatic interactions), we compute experimentally accessible parameters such as diffusion coefficient and radius of gyration. The simulations agree very well with measurements on kb-length superhelical DNAs.

The model shows that structural fluctuations in superhelical DNA are slow: after inducing a 90° bend in the center of an interwound 2000 bp DNA, the structure takes up to several milliseconds to reach its new equilibrium with the bend in one of the end loops. Bend- induced structural rearrangements may play a role in determining long-range interactions in gene regulation.

Chromatin fibers were modeled as a DNA chain with nucleosome-size beads attached and the angle between the linker DNA arms fixed at 40° from comparison with experimental data on di- and tetranucleosomes. The model predicts folding of a stretched 25-nucleosome fiber over several 100 μs into a zigzag structure, with a considerable amount of disorder due to thermal fluctuations. Computed structural and hydrodynamic properties of the model agree very well with experimental data on oligonucleosomes and SV40 minichromosomes.

The organization of human chromosomes in interphase was simulated by a new chromosome model using a parallel Monte-Carlo algorithm. The model comprises excluded volume interactions of the chromatin fiber and distinct folding according to the

Theoretical and Computational Methods in Genome Research, edited by Suhai
Plenum Press, New York, 1997

band structure of chromosomes. Random behavior on scales up to some Mbp and a functional organization of chromosomes at the level of chromosome bands are shown. Computed average interphase distances between markers agree with experimental data. Finally we show simulations of chromosome painting or labeling and subsequent image acquisition by light microscopy.

1. INTRODUCTION

The spatial organization of DNA in the eukaryotic nucleus plays a prominent role in determining gene activity. On the level of local DNA structure (some 100–1000 bp), the action of transcription factors is often mediated by DNA looping between the regulatory site and the promoter, and the structure of the intervening DNA has a decisive influence on the strength of the interaction between the transcription factor and the RNA polymerase initiation complex (as reviewed in[1]). On a higher level of organization, a great number of cases have been described where the expression of a gene is determined by the packaging of its DNA into a more or less compact chromatin structure (for a review see[2]).

Finally, interphase chromosomes are not a random criss-cross arrangement of chromatin fibers inside the nuclear envelope, but occupy distinct territories[3]. Recent fluorescence in situ hybridization (FISH) experiments have given evidence that the majority of actively transcribed genes are localized at the periphery of chromosome territories, and it has been suggested that intranuclear transport of macromolecules takes place in the space between these territories[3]. Also, evidence exists that the relative positioning of genes and enhancer sequences inside territories can influence gene activity over quite long distances: for instance, in Burkitt's lymphoma the myc oncogene is put under the control of an immunoglobulin enhancer by a long-range interaction involving a chromatin loop of 250 kB (P. Lichter, personal communication).

We see therefore that there is a lot of evidence from molecular and cellular biology techniques that supports the view of a dynamic arrangement of the interphase chromatin fiber into chromosome territories. It is worthwhile to develop a physical model for this arrangement at its various levels of organization (Fig. 1).

The models that we review in this paper describe the DNA or the chromatin fiber as a simple elastic filament. The molecular details of the DNA or of the proteins bound to it are disregarded in such an approach. We shall show that many of the known structural and dynamic aspects of interphase chromosomes, chromatin fibers and DNA can be explained by these models, and that predictions can be made about the mechanisms and time scales of intramolecular rearrangements.

2. THEORY

2.1. The Genome as an Elastic Chain

The coarse-grained view of the genome as an elastic filament is *sufficient* as long as one is interested in questions like the kinetics of approach of distant pieces of DNA, the structural fluctuations of a supercoil or a chromatin fiber, or the three-dimensional organization of a whole chromosome. Such processes occur on length scales much greater than interatomic distances, typically > 10 nm, and on time scales much slower than molecular vibrations, typically > 1 ns; thus, atomic detail may be safely neglected. For describing the

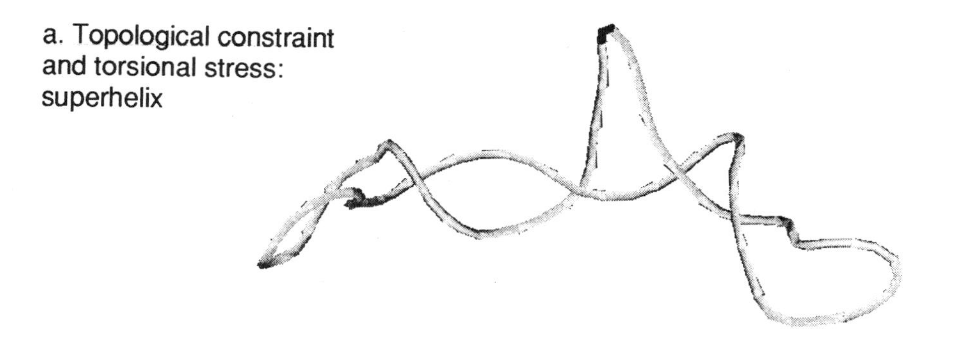

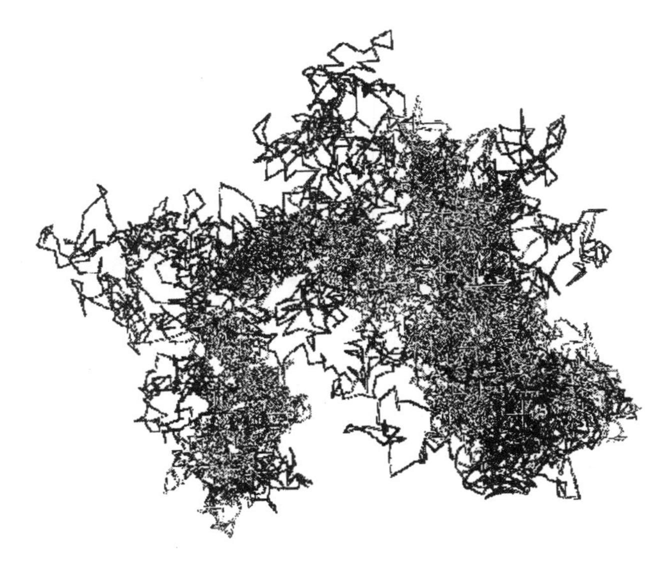

Figure 1. Organization of the genome as a polymer chain on the level of DNA (a), chromatin fiber (b) and chromosomes (c).

large-scale motions of a long DNA molecule (several hundreds or thousands of base pairs), the elastic-filament approximation is also *necessary*: atomic level calculations on such a large system, including solvent water molecules, are unthinkable with present-day computing equipment.

2.1.1. Elastic Parameters. On a local scale the filament may be characterized by three simple mechanical quantities, the bending, twisting and stretching elasticity.

The bending elasticity of a stiff polymer can be expressed as the *bending persistence length* p_B. It is defined as the distance over which the average cosine of the angle between the two chain ends has decayed to 1/e. Qualitatively, molecules shorter than p_B behave approximately like a rigid rod, while longer chains show significant internal flexibility. If one defines the bending elasticity A as the energy required to bend a polymer segment of unit length over an angle of 1 radian, $p_B = A/k_BT$ with k_B being Boltzmann's constant and T the absolute temperature. For B-DNA, p_B = 50 nm over a broad range of ionic conditions[4]. A value for p_B of chromatin has recently been estimated to 200 nm [5].

Analogous to the bending persistence length, one may define a *torsional persistence length* p_T which is the correlation length of the torsional orientation of a vector normal to the chain axis. Again, the torsional elasticity C, defined as the energy required to twist a polymer segment of unit length through an angle of 1 radian, is related to p_T by $p_T = C/k_BT$. C has been measured by various techniques, including fluorescence polarization anisotropy decay[6–8] and cyclization of DNA fragments[9–11], and the published values converge on a torsional persistence length of p_T = 65 nm.

The stretching elasticity of DNA has long remained inaccessible to experimental techniques, and usually some 'rather stiff' chain was used in models[12, 13]. Recently, however, several papers have been published that measured the behavior of single linear DNAs under external stretching forces[14,46]. From this work, one may estimate the stretching elasticity of DNA, which is given by a Young's modulus of $(3.46\pm0.3)\cdot10^8$ Pa. This value would correspond to a mean squared stretching fluctuation of a 37 bp DNA segment of $\delta = 0.016$, which is coincidentally not too far from the value $\delta = 0.008$ that was assumed in earlier simulation work[13].

The torsional and stretching rigidities of chromatin fibers are unknown; the chromosome model developed here therefore does not take into account these parameters.

2.1.2. Topological Constraints. When the two ends of a polymer chain are not free to rotate with respect to each other, e.g. in a circular DNA, internal torsional strain can cause the helix axis to bend into a 'superhelical' equilibrium configuration (Fig. 1a).

Such internal torsional strain is caused by a deviation of the *linking number* Lk from its equilibrium value. Lk is equal to the number of times the two single strands of a circular DNA are wound around each other if the circular DNA were spread out into a plane. The equilibrium value for a B-DNA of N base pairs is $Lk° = N/10.5$. The *linking number difference* is simply $\Delta Lk = Lk - Lk°$. One defines the *superhelical density* σ as the specific linking number difference: $\sigma = \Delta Lk/Lk°$. Typical values for superhelical DNA isolated from native sources are σ = -0.05…-0.07. This torsional strain is divided into two contributions: One due to bending of the DNA, the *writhe* Wr, which is a measure of the number of windings of the helix axis in space, and one due to twist around the local helix axis, the *twist difference* ΔTw, which is the difference in *local* helix turns relative to the equilibrium B-DNA conformation. White[15] has shown that

$$\Delta Lk = Wr + \Delta Tw \tag{1}$$

Thus, any local structural change that causes a change in ΔTw such as helix unwinding, B to Z transitions, and local bending of the DNA is coupled to the writhe of the superhelix through eq. (1) and therefore to its global shape. Furthermore, local structural changes that change Wr, such as sequence- and protein-induced bending, will in turn influence the global twist and therefore the overall shape of the superhelix. Theoretical[16] and experimental[17, 18] evidence suggests that a permanent bend can position the end loop of a superhelix, and influence the global dynamics of superhelical DNA[19].

2.1.3. Electrostatic Interactions. A major factor determining the configuration of a large DNA is the electrostatic repulsion between the negatively charged phosphate groups of the DNA, which is given by a screened Coulomb potential of the form $U(r) \propto \exp(-\kappa r)/r$. Here, κ is the Debye screening length which is 1 nm at 0.1M Na^+, and inversely proportional to the square root of the ionic strength of the medium. In addition, the effective charge of a phosphate group in DNA is only approx. -0.25 because of ion condensation close to the surface of the DNA molecule[20].

2.1.4. Hydrodynamic Interactions. Finally, the dynamics of a macromolecule in water are dependent on the viscous drag that it experiences when moving through the medium. The motion of the solvent induced by displacing part of the molecule against this viscous friction will induce forces on the rest of the molecule which are inversely proportional to the distance between the interacting parts. Such hydrodynamic interactions determine the Brownian motion of the entire molecule. They can be easily calculated for the case of two interacting spheres (see below).

2.2. Brownian Dynamics and Monte-Carlo Simulations of Superhelical DNA

Based on the three types of intramolecular forces described above we can now develop 'coarse-grained' models of DNA and chromatin chains. When atomic detail is neglected, the water can be taken as a homogeneous viscous fluid, and the macromolecule as an arrangement of cylindrical segments that are connected through the elastic potentials. Each model segment may correspond to several tens of base pairs, as long as it is small enough compared to the torsion and bending persistence length. The task is then to compute the equilibrium conformation of such a model and its dynamics.

A large amount of work has gone into computing the elastic equilibrium properties of DNA chains (for a review, see[21]). Although knowledge of the minimum energy structure of short DNA pieces is important, e.g., for understanding how DNA wraps around a nucleosome or how a 200 bp DNA circle can be formed, for longer DNA the fluctuations around the minimum energy state are so dominant that it becomes necessary to compute the *minimum free energy* rather than the *minimum energy* state[22]. Such a computation can be done by a Monte Carlo (MC) simulation using the Metropolis algorithm, an approach first taken by Frank-Kamenetskii, Vologodskii and co-workers[23, 24] and later applied in numerous studies; for instance, the influence of a bend on the structure of a superhelical DNA has been computed using such a model[16]. It was shown there that a sequence-dependent or protein-induced bend can inhibit structural fluctuations in the superhelix and define the interactions between segments of DNA which are separated by several hundreds of base pairs.

For calculating the dynamics of DNA, equations of motion for the segmented DNA chain have to be set up using the intramolecular interaction potentials described above and including the thermal motion through a random force. This is the Brownian Dynamics (BD) method[25] which several groups applied in interpreting experimental data from fluorescence depolarization[26], dynamic light scattering[13, 27], or triplet anisotropy decay[28]. Circular DNA has recently also been modeled by a BD approach[29]. Starting from our earlier model[13], we added a twisting potential for DNA, defined boundary conditions for the covalent closure of the DNA circle, and derived the expressions for the coupling of torsion along the local DNA axis with translational displacement of the DNA ('torsion-bending coupling'). The model allows to predict the kinetics of supercoiling and the internal motions of superhelical DNA over a time range of several milliseconds.

The details of the DNA models used in the work described here have been published, for the Monte-Carlo (MC) model in[16] and for the Brownian dynamics model in[13, 29]. The most important parameters and potentials used shall be repeated here. The DNA segment length is 12.7 nm, corresponding to 37 base pairs. The bending and torsional potentials are:

$$U_b = \frac{K_B T}{2\Psi^2} \sum_{i=0}^{N} (\beta_i - \beta_{0,i})^2 \tag{2}$$

$$U_t = \frac{K_B T}{2\xi^2} \sum_{i=0}^{N} (\alpha_i + \gamma_i)^2 \tag{3}$$

where α_i, β_i and γ_i are the Euler angles for rotating segment i into segment i+1 and $\beta_{0,i}$ the equilibrium bending angle between the segments. Ψ and ξ are the elastic constants between segments which can be calculated from the known elastic parameters of the DNA as described in[13]. The stretching constant was set to a mean square fluctuation of the segment length of $\delta = 0.008$ (see above). Control simulations with the recently published measured value of the stretching constant, $\delta = 0.016$, showed no significant difference to the data presented here.

A permanent bend between segments i and i+1 is modelled by setting $\beta_{0,i}$ to a nonzero value. The bend direction is restricted into a plane through a 'kink potential' of the form

$$U_k = \frac{K_B T}{2\zeta^2} \sum_{i=0}^{N} (\alpha_i - \alpha_{0,i})^2 \sin^2(\beta_{0,i}) \tag{4}$$

where the plane of the bend is defined by the value of $\alpha_{0,i}$ and ζ is a parameter that determines the bending anisotropy[29].

Using the potentials from eqs. (2–4) one can then derive the equations of motions under the influence of a stochastic force (Langevin equations) for the Brownian dynamics of DNA in a viscous fluid[24,29]. The simulation of DNA motion in solution is performed by numerically integrating the Langevin equations and using their discretized form. In the model[29] used here, the chain is described by a string of N segments with center of mass positions $\{\mathbf{r}_i\}_{i=1,N}$, and the torsion ϕ_i of each segment relative to its equilibrium structure is described by a unit vector $\mathbf{f}$ fixed on the segment and perpendicular to the segment direction[28]. The discrete equations of motion in the solvent for the segment positions and torsions are then[30]:

$$\delta r_i(t) = \delta t \textstyle\sum_j \underline{\mathbf{D}}_{ij}(t)\, \mathbf{F}_j(t) / (K_B T) + R_i(t)$$

$$\delta\phi_i(t) = \delta t\, D_r\, T_i(t) / (K_B T) + S_i(t) \tag{1}$$

where δt is the iteration time step, $\underline{\mathbf{D}}_{ij}$ the hydrodynamic interaction matrix, $\mathbf{F}_j(t)$ and $T_i(t)$ are the forces and torques acting on the beads, and the random displacements $R_i(t)$ and $S_i(t)$ are sampled from Gaussian distributions with the following momenta:

$$<R_i(t)> = 0 \; ; <R_i(t) : R_j(t)> = 2\ \delta t\ \underline{\mathbf{D}}_{ij}(t)$$

$$<S_i(t)> = 0 \; ; <S_i(t)\ S_j(t)> = 2\ \delta t\ D_r \tag{2}$$

For superhelical DNAs we assume periodic boundary conditions: $\mathbf{r}_{i+N} = \mathbf{r}_i$ and $\mathbf{f}_{i+N} = \mathbf{f}_i$. The (3N x 3N) $\underline{\mathbf{D}}_{ij}$ matrix in the first line of Eqs.1,2 is the Rotne-Prager generalization of the Oseen tensor[31]. which characterizes the hydrodynamic interaction between two spherical beads (eq. 5 and 6 in[13]) For calculation of this matrix, the cylindrical DNA segments were approximated by beads with radius $r_b = 2.53$ nm. The rotational diffusion coefficient around the long axis of the DNA segment is $D_r = k_B T / (4\pi\eta\ a^2 b_0)$, where b_0 is the bond length and a the segment radius.

2.3. The Chromatin Model

Recent revisions of the existing structural data on chromatin fibers[32, 33] suggest a local 'zig-zag' arrangement of the nucleosomes which is folded up into a random-walk higher order structure. Randomness can occur either through non-uniform linker DNA length, causing a twist of one nucleosome with respect to the other, or by thermal fluctuations. Based on these ideas, one can develop a BD model for the chromatin structure through some simple extensions of the elastic chain DNA model.

Our model (Figs. 1b,2) consists of a linker DNA backbone with a constant intranucleosomal length of 72 base pairs in two segments. A nucleosome is attached as a sphere of 5.95 nm radius at every other segment joint. The equilibrium bending angle at the joint is equal to the angle between the linker DNA arms entering and exiting the nucleosome; this angle and the torsion angle between nucleosomes are adjustable parameters. Since the precise nature of the electrostatic interactions in chromatin is not known, we assume that the potentials that determine the total energy of the linker DNA chain—bending and twisting elasticity, electrostatic interactions—are the same as for naked DNA; the interactions between nucleosomes and between nucleosomes and DNA are modelled by hard-core repulsive (Lennard-Jones) potentials.

For the chromatin model, as for the free DNA, one can then set up the BD equations of motion, using a modified hydrodynamic interaction matrix that takes into account the different sizes of the nucleosomes and the DNA segments.

2.4. R-G Loop Model of Interphase Chromosomes

The basic structural element of our chromosome model is the chromatin fiber. Although the explicit structure of chromatin fibers in human interphase nuclei is still under discussion[32, 33], their average geometrical properties on length scales above some thousand base pairs can be described by a polymer consisting of rigid segments of a certain length (Kuhn length) and an excluded volume interaction representing the diameter of the chromatin fiber. Motivated by the different properties of bands, e.g. density of genes or time of replication, we assumed a different organization of chromosome bands in interphase as well. Each 'activated' band (e.g. a gene rich R-band) forms a single giant loop in the 3–10 Mbp range, while chromatin within 'inactivated' bands (e.g. G- or C-bands) is folded into several small loops about 100 kbp each.

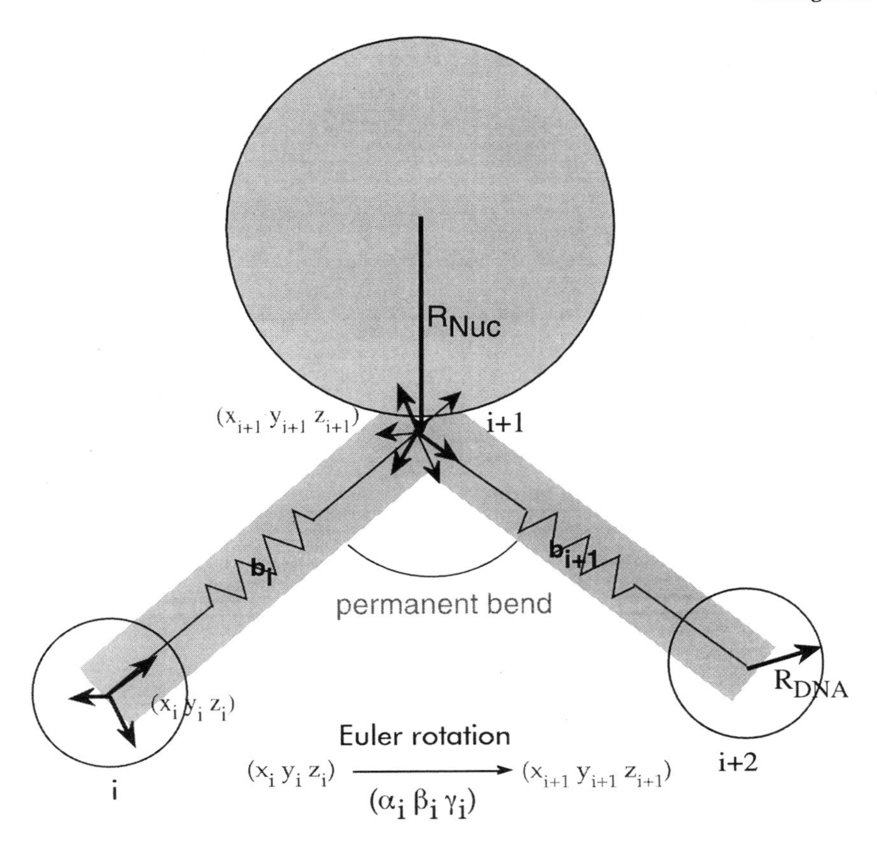

Figure 2. Basic unit of the model chromatin fiber. The nucleosome is a 5.95 nm sphere with two linker DNA segments attached to its periphery. The forces between the segment are the same elastic and electrostatic potentials as for free DNA (see text). The charge of the nucleosome is assumed to be zero.

The amount of chromatin in each band was assumed to be proportional to the size of the band in high resolution metaphase studies[34]. Fig. 1c shows the polymer structure of a simulated human chromosome 3. The chromatin fiber of R-bands is shown in a darker grey than that of G- and C-bands. R-band domains occupy a larger volume and can be found preferentially at the periphery of chromosome territories.

3. RESULTS AND DISCUSSION

3.1. Structure and Dynamics of Superhelical DNA

Using the BD model we computed trajectories of superhelical DNA of 1870 base pairs for simulation times of several milliseconds at its native superhelical density, σ = -0.056[19].. The chains were constructed from 50 segments of equal length. In some of the trajectories a permanent bend was introduced by setting the β_0 angles between four consecutive segments to 25°. The simulation time step was 0.19 ns and the CPU time necessary for these simulations was about 1 hr per µs of simulation time on a 150 MHz SGI Indy R4400. The results are representative for the dynamic behavior of a superhelix, but better statistics can only be expected by running many more trajectories on parallel computers; this work is currently underway in our group.

3.1.1. Phenomenology of Supercoiling and Writhe Relaxation. When one starts the dynamics simulation of a 1870 bp superhelical DNA from a flat circle with $\Delta Lk = Tw = -10$ and $Wr = 0$, the writhing number changes in about 10μs towards a value close to equilibrium ($Wr \approx -5$). During the first microsecond, the circle assumes a locally toroidal form; thermal fluctuations then induce the nucleation of end loops of interwound regions, and in general a branched interwound conformation is formed in this first phase (Fig. 3).

The dominant equilibrium form for an 1870 bp DNA is an unbranched interwound structure[35]. The relaxation towards this form increases the writhe only by a small amount and is much slower than the initial relaxation, as shown in Fig. 4c.

When the simulation is initiated from a straight interwound structure, the writhing number relaxes during the first 10 μs from its initial value -8 towards its equilibrium (Fig. 4b). When the starting configuration is sampled from a Monte Carlo simulation, i.e., already from an equilibrium distribution, the writhing number does not show any indication of an initial fast relaxation (Fig. 4a). In all three cases, the equilibrium value of Wr is $\cong -6.5\pm0.5$, without any effect of the permanent bend. The bend influences the *dynamics* of the chain: the writhe fluctuations are faster for the DNA without a permanent bend than with the bend[19]. This finding is in agreement with our earlier observation that a permanent bend decreases the structural fluctuations of the superhelix as observed in dynamic light scattering[18].

3.1.2. Diffusion Coefficients. The translational diffusion coefficient of 1870 bp superhelical DNA at its native superhelical density has been measured to $D_t = (7.0\pm0.3)\cdot10^{-12}$ $m^2\,s^{-1}$ [36]. This same quantity can be computed from the simulations, either on each single configuration in the trajectory by a rigid-body treatment[37] or from the center of mass motion of the molecule. The diffusion coefficients evaluated this way are reported in Table I. The value from the rigid body treatment is $(6.85\pm 0.2)\cdot 10^{-12}\ m^2\,s^{-1}$. The center of mass motion leads to D_t values with larger uncertainty and with an average $D_t = (7.2\pm0.7)\cdot 10^{-12}$ $m^2\,s^{-1}$, in reasonable agreement with the rigid-body result.

Simulations on other systems (superhelical DNAs of different lengths, relaxed circular DNA) lead to similar good agreement between observed diffusion coefficients from dynamic light scattering and the modelling results[38].

3.1.3. Kinetics of Loop Migration. The first part of the trajectory in Fig. 3 was started from a flat circle in the absence of permanent bends (at t= 450 μs a bend was inserted; see below). During the first 10 μs the torsional stress and thermal fluctuations drive parts of the chain out of the plane. These positions then act as nucleation point for loops. After approx. 50μs, some loops are absorbed into others and 3 to 4 loops remain. During the first 450 μs we never observed less than three loops, whose positions shifted by ≈ 10

Table 1. Translational diffusion coefficients in $10^{-12}\ m^2/s$ units (from [19]). Errors reported are the uncertainties of the parameters in the least-squares fit

Trajectory type (no bend)	D_t center-of-mass	D_t rigid body	Trajectory type (+ bend)	D_t center-of-mass	D_t rigid body
Circle	8.2 ± 0.4	6.9 ± 0.4	circle	7.2 ± 0.3	6.85 ± 0.4
Interwound	6.4 ± 0.6	6.7 ± 0.35	interwound	6.4 ± 0.6	6.6 ± 0.4
Monte Carlo	7.2 ± 0.2	7.0 ± 0.3	Monte Carlo	7.9 ± 0.6	6.7 ± 0.3

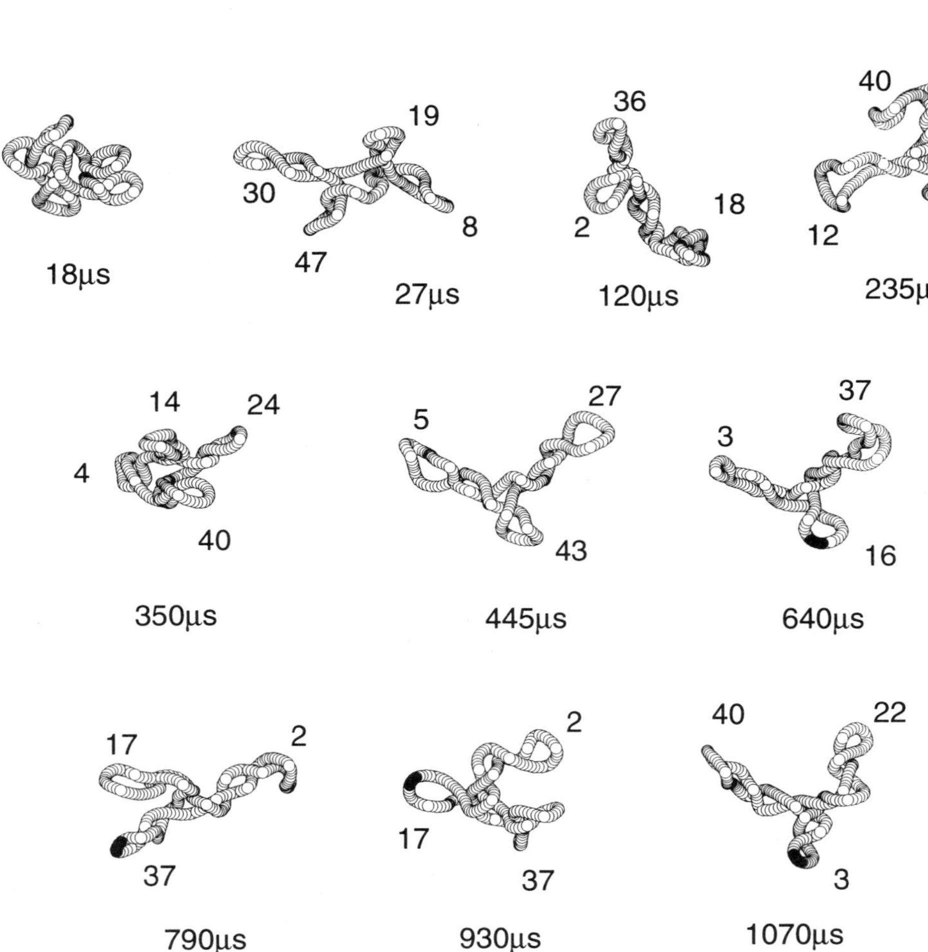

Figure 3. Configurations from the supercoiling trajectory of an1870 bp DNA with ΔLk = -10 starting from a flat circle. The simulation times are printed next to each configuration. The numbers designate the positions of the end loops. At t = 450 μs a permanent bend of 100° was inserted at position 15 (segments marked in black). Data taken from[19].

segments (i.e. 130 nm) over 200–300 μs. When starting from a straight interwound configuration, the molecule's initial regular structure is distorted by thermal motion, but only after approx. 300 μs we see a substantial rearrangement in the loop positions. Thus, the structural fluctuations are characterized by the formation and disappearance of loops on a time scale of several 100 μs. Similar behavior is found if the output of a Monte Carlo simulation is used as the starting configuration.

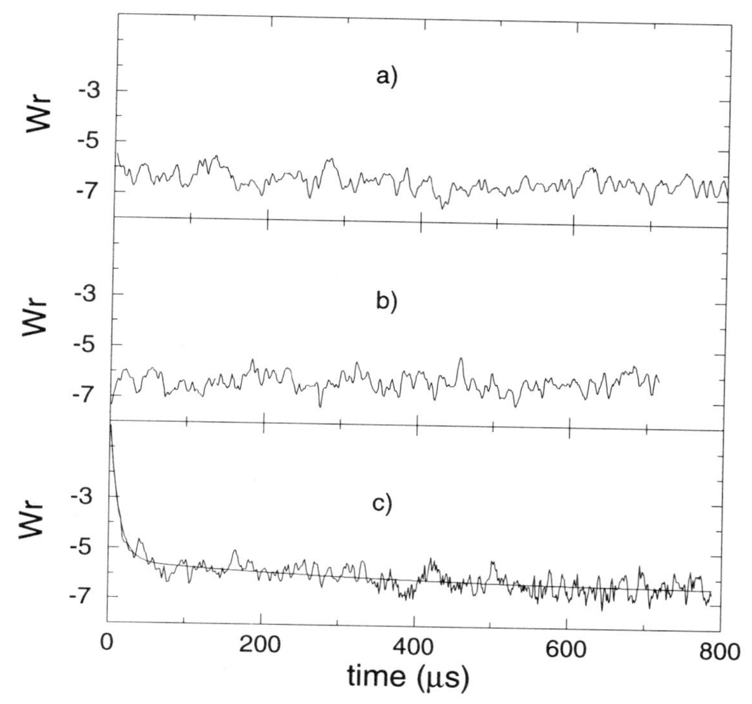

Figure 4. Writhing kinetics for trajectories starting from a Monte Carlo-generated conformation flat circle (a), a straight interwound structure (b) and a flat circle (c) (from [19].). The straight line in c) is a double-exponential fit to the kinetic curve.

As an example for the effect of DNA curvature, we introduced a 100° planar bend during the course of the trajectory by setting the equilibrium bending angles $\beta_{o,i} = 25°$ and $\alpha_{o,i} = 0°$ for four consecutive segments i = m-1...m+2, where m is the center of the bend. The simulation started from a 50 segment straight interwound chain containing a 100° bend which is initially at position 25 and shifted to position 15 at t=200 μs. Fig. 5 gives the time dependence of the loop positions for this simulation: In the first 200 μs the loops remain at positions 0 and 25, and a branch starts to extrude after ≈ 100 μs at position 40. After the shift of the bend to position 15, a new loop appears there, and the loop in position 25 is slowly (over 400 μs) coalescing with the new one. At the same time the loop at position 2 drifts to position 40.

In general, structural changes induced by a bend seem to occur on a time scale of 300–800 μs when the bend is far away from the initial loop position, and in about 50 μs when the distance over which the loop has to move is only of the order of 4 segments (140 bp)[19].

3.2. Chromatin Modelling

The simulations are compared to the measured structural properties of oligonucleosomes. Yao et al.[39] measured an increase of the diffusion coefficients of isolated dinucleosomes with ionic strength, which they interpreted as an increased bending of the linker DNA. We used their data to compare with the computed diffusion coefficients from a model which consisted of two nucleosome spheres connected by two 36 bp DNA seg-

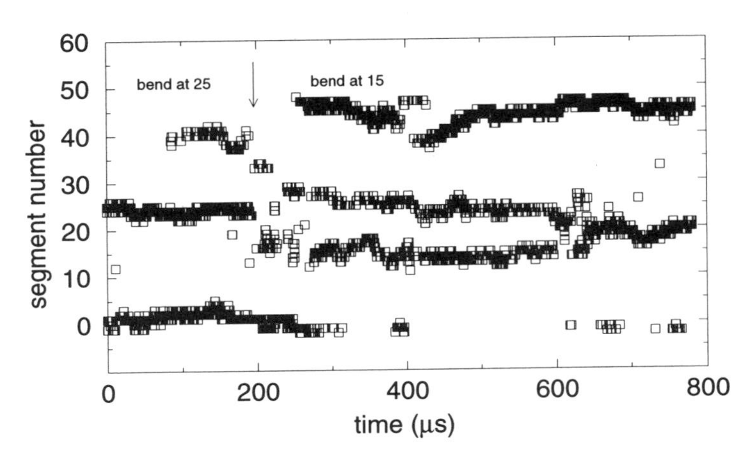

Figure 5. End loop positions vs. time for a trajectory starting from an interwound superhelix of an 1870 bp DNA with $\Delta Lk = -10$ and a permanent bend inserted at position 25. At t=200 μs, the bend was displaced to position 15 (from[19]).

ments and two more such segments protruding to the exterior of each nucleosome. The computed diffusion coefficients depend significantly on the bending angle of the linker DNA Θ at the nucleosome, but only slightly on the torsion angle ϕ between nucleosomes. Values of $\Theta = 40°$ and $\phi = 0°$ gave best agreement with the experimental data: Both the absolute values of D and its ionic strength dependence are predicted within experimental error (Fig. 6). From a more detailed analysis of the generated trajectories, we conclude that the reason for the observed increase in D with ionic strength is a change in the average distance between the protruding linker DNA arms, rather than a bending of the linker DNA.

Another data set that we used for comparison with the simulation results was the ionic-strength dependent sedimentation data on tetranucleosomes given by Thomas et al.[40] Again, best agreement was obtained using the same values for the bending and twisting angles of the DNA at the nucleosome as in the dinucleosome case (data not shown). We conclude therefore that the chromatin model can be used to predict the dynamics of a chromatin fiber in a realistic way. First simulations on larger nucleosome chains and SV 40 minichromosomes[41] show that a 25 nucleosome fiber relaxes from a stretched configuration into a compact structure over a time of approx. 250 μs.

3.3. Chromosome Organization

Yokota et al.[42] and Sachs et al[43] measured the interphase distances of genetic markers as a function of their separation on the genetic map. These data were interpreted to suggest a random walk configuration of the interphase chromosome[5]. From our simulations, we computed the same data as the average of 400 simulated chromosome 4 configurations according to the R-G band model described above. As Fig. 7 shows, the band model gives rather good agreement, indicating that the earlier interpretation of interphase distances is not unique. While the interphase chromosome displays random organization above the 10Mbp scale, distinct subdomains of bands are still present. We also computed the average distance of markers in a 4 Mbp telomer near R-band of chromosome 4, which were used in the measurements by van den Engh et al.[44] The observed increase of dis-

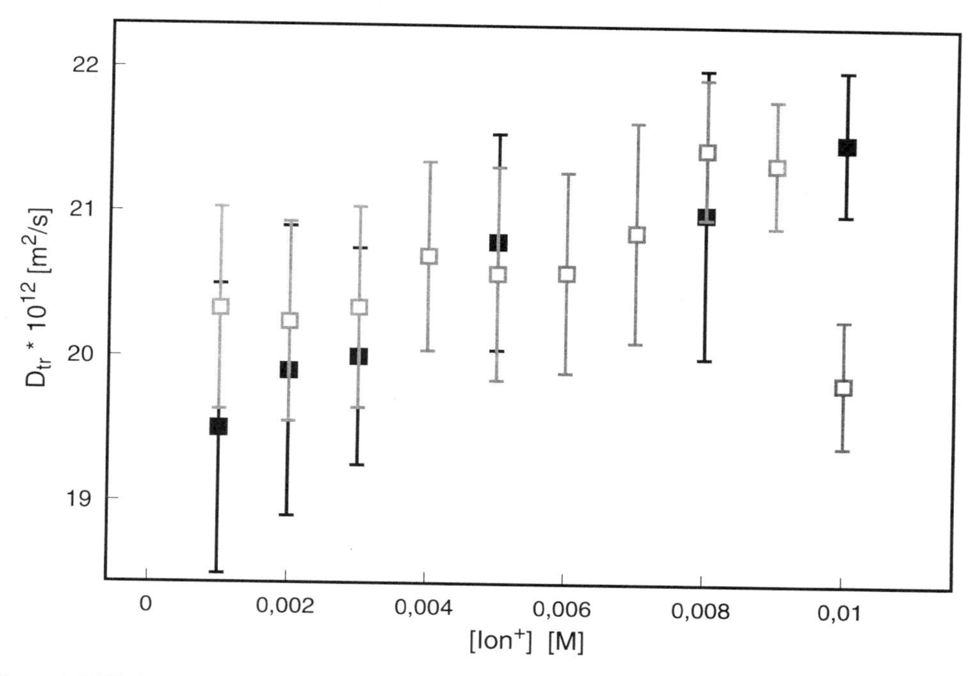

Figure 6. Diffusion coefficients of dinucleosomes as measured by Yao et al.[39] as a function of ionic strength (■). (□) D calculated from a simulated trajectory.

tances as a function of genomic separation, as well as the subsequent decrease, are reproduced very well (data not shown).

While mean interphase distances are explained by the random-walk/giant-loop model of Sachs et al.[43] as well as our polymer model presented here, the organization of subdomains is quite different in both models. Therefore, the model should also be compared with experiments where chromosomal subdomains are analyzed. For this purpose we generated 'virtual' multi-color images of FISH experiments by simulating the scanning of fluorescent markers attached to the chromatin in a confocal microscope, which allows us to quantify the structure of chromosome territory boundaries, subdomains etc. For instance, the painting of chromosome 3 with specific microdissection probes from p and q arms was simulated and the computed images compared with series of confocal sections from experiments (Dietzel et al., in preparation). Fig. 8 shows black and white reproductions of the same central section with p-arm and q-arm specific channels separated. In accordance to experimental findings, 'virtual' confocal sections of our simulated structures show very little overlap of chromosome arms. In contrast, most simulated random-walk/giant-loop chromosomes overlap extensively. In other simulations, R- and G-bands of chromosome 15 were labeled in red and green. Computed confocal sections agree well with recent experimental results, where early and late replicating bands of chromosomes were marked with modified bases during replication (Zink et al., in preparation). Another advantage of our method is the prediction of results for confocal microscopy with different resolutions and the appearance of painted chromosomes using planned next generations of light microscopes with a resolution in the 100nm range.

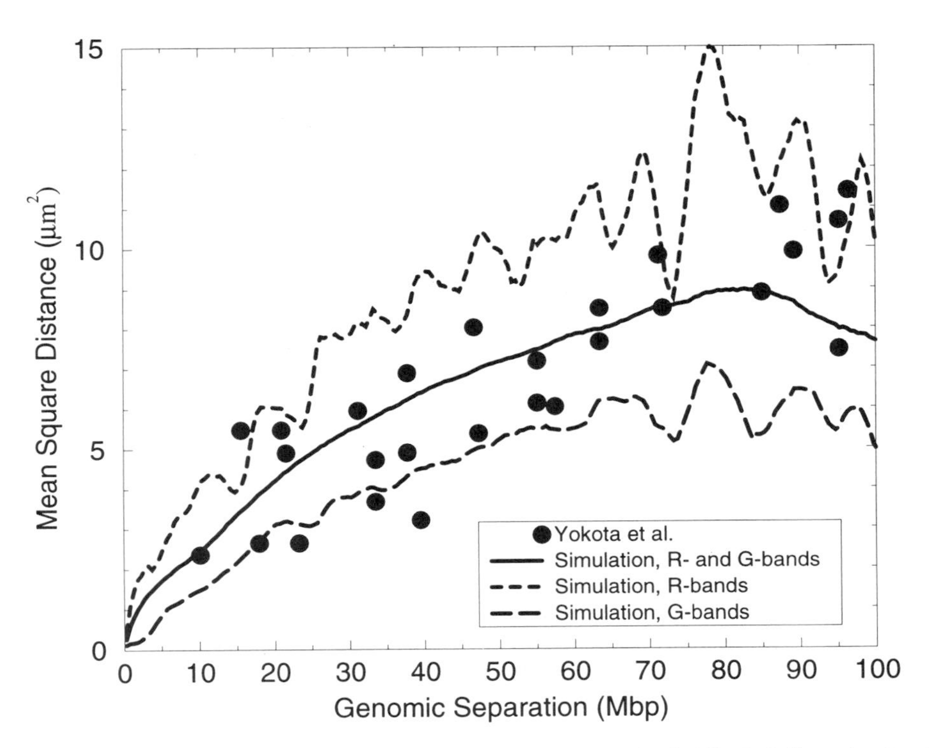

Figure 7. Spatial distances of genetic markers as a function of their separation in Mbp in 2-dimensional projections of interphase nuclei. (●) experimental data by Yokota et al., solid line: model data, averaged over R- and G-bands, dashed lines: R-bands only (upper), G-bands only (lower).

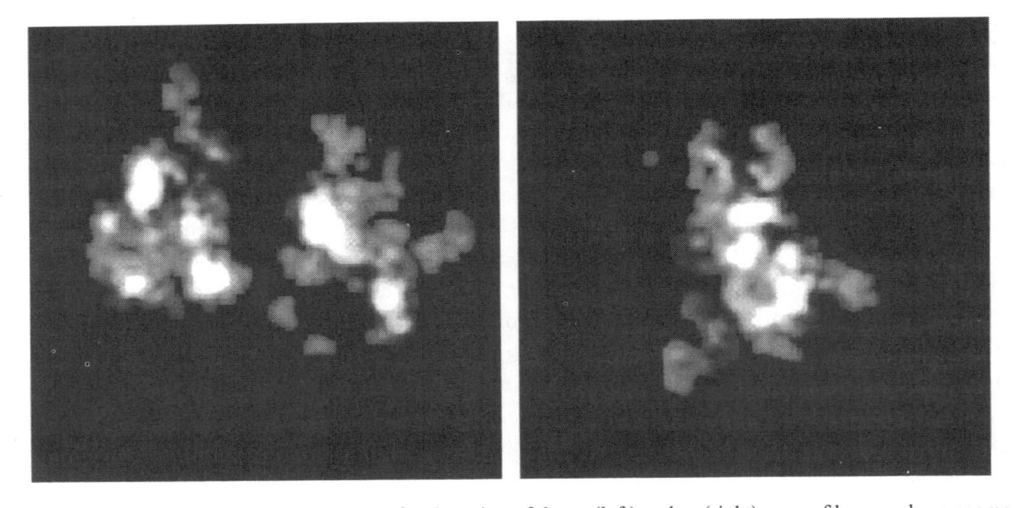

Figure 8. Simulated three-dimensional confocal section of the p- (left) and q- (right) arms of human chromosome 3. The p-arm shows up in two parts, since the central part is behind the image plane. The separation of the two arm regions predicted from the model is clearly visible.

The chromosome simulations by the polymer model will be expanded to whole nuclei, although this still requires enormous computing times. However, on the scale of whole nuclei much of the chromosome subdomain structure is less important than other geometrical constraints regarding the shape of chromosome territories. We therefore modeled chromosome territories as flexible compartments of a nucleus. For various quantities like surface area, relative chromosome positions, distance to the nuclear center or translocation probability we could compute the dependence on chromosome size in agreement with our previous study[45].

4. ACKNOWLEDGMENTS

We thank Katalin Tóth for critical reading of the manuscript. This work was supported by DFG grant no. La 500/4–2 to J.L.

5. REFERENCES

1. Rippe, K., von Hippel, P.H. & Langowski, J. *Trends in Biochemical Sciences* **20**, 500–506 (1995).
2. Elgin, S.C.R. *Current Opinion in Genetics and Development* **6**, 193–202 (1996).
3. Cremer, T., *et al. Cold Spring Harbor Symposium on Quantitative Biology* **58**, 777–792 (1993).
4. Hagerman, P.J. *Annual Review of Biophysics and Biophysical Chemistry* **17**, 265–286 (1988).
5. Ostashevsky, J.Y. & Lange, C.S. *Journal of Biomolecular Structure & Dynamics* **11**, 813–820 (1994).
6. Schurr, J.M., Fujimoto, B.S., Wu, P. & Song, L. in *Topics in Fluorescence Spectroscopy* (ed. Lakowicz, J.R.) 137–229 (Plenum Press, New York, 1992).
7. Barkley, M.D. & Zimm, B.H. *Journal of Chemical Physics* **70**, 2991–3007 (1979).
8. Fujimoto, B.S. & Schurr, J.M. *Nature* **344**, 175–178 (1990).
9. Taylor, W.H. & Hagerman, P.J. *J. Mol. Biol.* **212**, 363–376 (1990).
10. Shore, D. & Baldwin, R.L. *Journal of Molecular Biology* **179**, 957–981 (1983).
11. Horowitz, D.S. & Wang, J.C. *J. Mol. Biol.* **173**, 75–91 (1984).
12. Allison, S.A. & McCammon, J.A. *Biopolymers* **23**, 167–187 (1984).
13. Chirico, G. & Langowski, J. *Macromolecules* **25**, 769–775 (1992).
14. Smith, S., Cui, Y. & Bustamante, C. *Science* **271**, 795–799 (1996).
15. White, J.H. in *Mathematical methods for DNA sequences* (ed. Waterman, M.S.) (CRC Press, Boca Raton, 1989).
16. Klenin, K., Frank-Kamenetskii, M.D. & Langowski, J. *Biophysical Journal* **68**, 81–88 (1995).
17. Laundon, C.H. & Griffith, J.D. *Cell* **52**, 545–549 (1988).
18. Kremer, W., Klenin, K., Diekmann, S. & Langowski, J. *The EMBO Journal* **12**, 4407–4412 (1993).
19. Chirico, G. & Langowski, J. *Biophysical Journal* **71**, *in press* (1996).
20. Manning, G.S. *Quart. Rev. Biophys.* **11**, 179–246 (1970).
21. Olson, W.K. *Curr Opin Struct Biol* **6**, 242–256 (1996).
22. Langowski, J., *et al. Trends Biochem Sci* **21**, 50 (1996).
23. Vologodskii, A.V., Anshelevich, V.V., Lukashin, A.V. & Frank-Kamenetskii, M.D. *Nature* **280**, 294–298 (1979).
24. Frank-Kamenetskii, M.D., Lukashin, A.V. & Vologodskii, A.V. *Nature* **258**, 398–402 (1975).
25. Ermak, D.L. & McCammon, J.A. *Journal of Chemical Physics* **69**, 1352–1359 (1978).
26. Allison, S.A. *Macromolecules* **19**, 118–124 (1986).
27. Allison, S.A., Sorlie, S.S. & Pecora, R. *Macromolecules* **23**, 1110–1118 (1990).
28. Allison, S.A., Austin, R. & Hogan, M. *Journal of Chemical Physics* **90**, 3843–3854 (1989).
29. Chirico, G. & Langowski, J. *Biopolymers* **34**, 415–433 (1994).
30. Dickinson, E., Allison, S.A. & McCammon, J.A. *J. Chem. Soc. Faraday Trans. 2* **81**, 591–601 (1985).
31. Rotne, J. & Prager, S. *The Journal of Chemical Physics* **50**, 4831–4837 (1969).
32. van Holde, K. & Zlatanova, J. *The Journal of Biological Chemistry* **270**, 8373–8376 (1995).
33. Woodcock, C.L. & Horowitz, R.A. *Trends in Cell Biology* **5**, 272–277 (1995).
34. Francke, U. *Cytogenet. Cell. Genet.* **65**, 206–219 (1994).

35. Boles, T.C., White, J.H. & Cozzarelli, N.R. *J. Mol. Biol.* **213**, 931–951 (1990).
36. Steinmaier, C., Diplomarbeit, Fakultät für Physik und Astronomie (Ruprecht-Karls-Universität, Heidelberg, 1995).
37. Garcia de la Torre, J. & Bloomfield, V.A. *Biopolymers* **16**, 1747–1763 (1977).
38. Hammermann, M., Diplomarbeit, Fakultät für Physik und Astronomie (Ruprecht-Karls-Universität, Heidelberg, 1996).
39. Yao, J., Lowary, P.T. & Widom, J. *Proceedings of the National Academy of Sciences of the USA* **87**, 7603–7607 (1990).
40. Thomas, J.O., Rees, C. & Butler, P.J. *European Journal of Biochemistry* **154**, 343–348 (1986).
41. Ehrlich, L., Diplomarbeit, Fakultät für Physik und Astronomie (Ruprecht-Karls-Universität, Heidelberg, 1996).
42. Yokota, H., van den Engh, G., Hearst, J., Sachs, R.K. & Trask, B.J. *The Journal of Cell Biology* **130**, 1239–1249 (1995).
43. Sachs, R.K., van den Engh, G., Trask, B., Yokota, H. & Hearst, J.E. *Proceedings of the National Academy of Sciences of the USA* **92**, 2710–2714 (1995).
44. van den Engh, G., Sachs, R. & Trask, B.J. *Science* **257**, 1410–1412 (1992).
45. Münkel, C., *et al. Bioimaging* **3**, 108–120 (1996).
46. Cluzel, P., *et al. Science* **271**, 792–794 (1996).

6

IS WHOLE HUMAN GENOME SEQUENCING FEASIBLE?*

Eugene W. Myers[1†] and James L. Weber[2]

[1]Department of Computer Science
University of Arizona
Tucson, Arizona 85721-0077
[2]Center for Medical Genetics
Marshfield Medical Research Foundation
Marshfield, Wisconsin 54449

1. SHOTGUN SEQUENCING AND ITS VARIATIONS

1.1 Basic Shotgun Sequencing

As the time to sequence the entire human genome approaches, it appears that the predominant method of collecting this data will entail some variation of the shotgun sequencing strategy [1]. The startpoint for the basic shotgun experiment is a pure sample of a large number of copies of a particular stretch of DNA of some length G, say 100,000 base pairs ($100Kbp$) for the purposes of illustration. The sample is either sonicated or nebulated, randomly partitioning each copy into pieces called *inserts*. The resulting pool of inserts is then size-selected so that fragments that are too large or too small are removed from further consideration. The inserts that remain thus represent a random sampling of segments of the source sequence of a given approximate size. A sub-sample of the inserts are then cloned via insertion into a viral phage, called a *vector*, and subsequent infection of a bacterial host. The cloning process results in the production of a pure sample of a given insert so that it may then be sequenced. Typically this is done via the method of Sanger et al. [2] which produces a ladder-like pattern on an electropheretic gel. Generally, only the first 300 to 800 base pairs of the insert can be interpreted from this experiment. This data is called a *read* and is a contiguous subsequence of the source sequence. For a source sequence of length $G = 100Kbp$, an investigator might collect, say $R = 1250$ reads, where the average length of the reads is typically $\overline{L}_R = 400$. In summary, one make think of

* This work was supported in part by NLM Grant R01 LM04960, and DOE Grant DE-FG03-94ER61911.

† Corresponding author: gene@cs.arizona.edu, +1 (520) 621-6612, fax: +1 (520) 621-4246.

Theoretical and Computational Methods in Genome Research, edited by Suhai
Plenum Press, New York, 1997

the shotgun approach as delivering a collection of R reads that constitute a random sample of contiguous subsequences of the source sequence of length approximately$\overline{L}_R$.

1.2 The Fragment Assembly Problem

Given the reads obtained from a shotgun protocol, the computational problem, called *fragment assembly*, is to infer the source sequence given the collection of reads. For our running example, note that altogether $N = R\overline{L}_R = 500Kbp$ basepairs of data have been collected. Thus one has on average sequenced every basepair in the source $\overline{c}$ = N/G = 5 times. The quantity is called the average sequencing redundancy. In practice, an investigator will decide on a given level of redundancy and then clone and sequence inserts until a total of N = G base pairs of data have been collected. In designing algorithms and software for fragment assembly one must account for the following characteristics of the data:

- Incomplete Coverage: Not every source basepair is sequenced exactly $\overline{c}$ times due to the stochastic nature of the sampling, thus some portions of the source may be covered by more than $\overline{c}$ reads and others may not be covered at all. In general, there can be several such gaps or maximal contiguous regions where the source sequence has not been sampled. Gaps necessarily dictate a fragmented, incomplete solution to the problem.
- Sequencing Errors: The gel-electropheretic experiment yielding a read, like most physical experiments, is prone to error especially near the end of a read where the blots of the ladder-like gel pattern are compressed together. Typically the maximum error rate ε is in the range of 5–10%.
- Unknown Orientation: The source sample is double-stranded DNA. Which of the two strands is actually read depends on the arbitrary way the given insert oriented itself in the vector. Thus one does not know whether to use a read or its Watson-Crick complement[‡] in the reconstruction.

In most common systems for fragment assembly the problem is solved in three phases [3,4]. In the first *overlap phase*, every read is compared against every other read and its complement to determine if they overlap. Due to the presence of sequencing errors, an overlap is necessarily approximate in that not all characters in the overlapping region coincide. This problem is a variation on traditional sequence comparison where the degree of difference permitted is bounded by ε. In the second *layout phase*, the relative position of each read is determined by selecting an approximate overlap that links it to the rest That is, a spanning forest of the graph of all overlaps is selected, in effect producing an arrangement or layout of the fragments. The selection of the spanning forest is driven by some optimization criterion for the layout, most often to minimize length [4], but maximizing likelihood has also been advocated [5]. Given a layout of the fragments, the final *consensus phase*, produces a multi-alignment of the reads in regions where the coverage is 3 or greater. The multi-alignment results in a *consensus sequence* which is reported as the reconstructed source sequence.

[‡] The Watson-Crick complement $(a_1 a_2 \ldots a_n)^c$ of a sequence $a_1 a_2 \ldots a_n$ is $wc(a_n) \ldots wc(a_2)\, wc(a_1)$ where $wc(A) = T$, $wc(T) = A$, $wc(C) = G$, and $wc(G) = C$.

1.3 Current Computational Developments

Ultra-Large Problems. As sequencing efforts have scaled up for human genome sequencing, some new facets of the problem have become evident. The first fact is the rapid increase in the size G of the source sequences that are being shotgunned. In the early 1980's, when shotgun sequencing was first proposed, a typical source sequence size was $3Kbp$. By 1990, experimentalists were shotgun sequencing an entire cosmid at a time for which $G \approx 40Kbp$. In 1995, Ventner reported shotgun sequencing the entirety of *H. Influenza* for which $G \approx 1{,}800Kbp$. It is not clear how far this trend will continue, but the size of the source sequences being directly shotgunned has far exceeded any reasonable projection that would have been made five years ago. It is thus clear that one must design algorithms that work on ultra-large problems: ones so large that memory and time requirements become significant concerns.

Repeats. Another significant development is the extent to which repetitive elements occur within a target sequence. Such occurrences were relatively rare previously because (a) G was small, and (b) the inherently simpler genomes of low-order organisms were the source. It has now become apparent that high-order genomes, such as that of humans, have a rich repeat structure. Combined with the very large stretches being tackled in a single shotgun effort, the chance of encountering not one but several repetitive substructures in the source is almost certain. Moreover repeats occur at several scales. For example, in the T-cell receptor locus of humans there is a 5-fold repeat of a trypsinogen gene that is $4Kbp$ long and varies 5–10% between copies. Three of these were close enough together that they appeared in a single shotgun-sequenced cosmid source [6]. Such large scale repeats are problematic for shotgun approaches as reads with unique portions outside the repeat cannot span it. Smaller elements such as palindromic Alu's of length approximately $300bp$, do not share this feature but are still problematic as they can constitute up to 50–60% of the target sequence [7,8]. Finally, in telomeric and centromeric regions, microsatellite repeats of the form x^n, where the repeated string x is three to six bases long and n is very large, are common [7] Thus one must design software solutions that can resolve assemblies involving repetitive DNA.

Constraints. While experimentalists have ambitiously increased the size of the source sequences, the technology for obtaining a read has not improved the average length of a read $\overline{L}_R$ at a corresponding rate, leading to smaller and smaller ratios of $(= \overline{L}_R / G$. A seminal paper by Waterman and Lander [9] showed that the expected number of gaps (unsampled regions) grows as $Re^{-(R}$. Thus the fragmentation of the solution into a collection of gap-separated "contigs", giving the covered intervals of the source, is increasing significantly. This combined with the increasing difficulty of correctly resolving repetitive elements in the source, has induced investigators to develop hybrid or alternative strategies to reduce these problems Some examples are as follows:

- *PCR Gap Closure*: While gaps can be frequent it is also the case that they are generally quite small, $150bp$ or less. Thus a common technique is to shotgun sequence up to a given redundancy and then "close" gaps by designing PCR probes to amplify the sequence in the gap [10]. The reads obtained from the PCR experiments are known to overlap with reads adjacent to the gaps.
- *Double-Barreled Shotgun Sequencing*: Inserts are size selected so that the average length $\overline{L}_I$ is roughly $2\overline{L}_R$ or longer and *both* ends of the insert are sequenced [11].

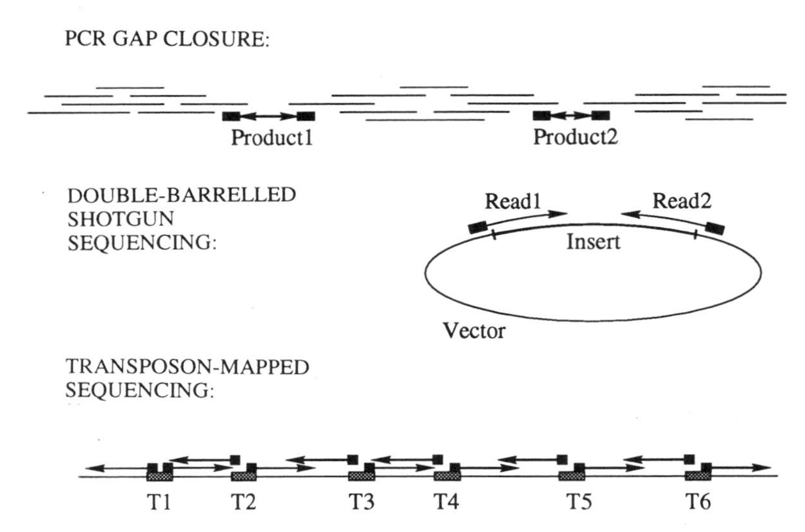

Figure 1. Sequencing protocols involving constraints on the relative positions of reads in terms of overlap, orientation, and distance.

This gives rise to a pair of reads that are known (1) to be in opposite orientations and (2) to be at a distance from each other approximately equal to the insert length.

- *Transposon-Mapped Sequencing*: A carefully engineered transposon is repeatedly inserted into the source and a map of the insertion location is created via restriction enzyme digestion of the unique cut site built into the transposon. Then ladder sequencing is performed in both directions from an inserted transposon with the aid of primers built into both ends of the transposon [12]. This gives rise to a set of overlapping reads as shown in Figure 1. In this scenario it is known that the pair of reads emanating from a given transposon (1) overlap by 4 or 5 base pairs (due to a replication that occurs during transposon insertion), (2) are in the opposite orientation, and (3) that the forward read from transposon i overlaps that of the reverse read from transposon $i+1$.

In all three cases, there is additional information about how the reads should be put together in a layout. We have proposed that this information should be specified to an assembler as a collection of orientation, overlap, and/or distance constraints that we have demonstrated are sufficient to describe all protocols in practice today [13,14]. Regardless of the method of specification, it is clear that modern assembly software will have to handle shotgun data that is augmented by additional constraint information.

1.4 Variations on Shotgun Sequencing

In the previous subsection, we began to explore variations on the basic shotgun strategy, the objectives of which were to alleviate the problems of achieving closure and of resolving sequence repeats for large scale projects on higher-order organisms. Before exploring these further some definitions, illustrated in Figure 2, are required. A *contig* is a collection of reads that cover a contiguous region of the source. Note that contigs are mutually non-overlapping being separated from each other by *gaps*. Furthermore, without ad-

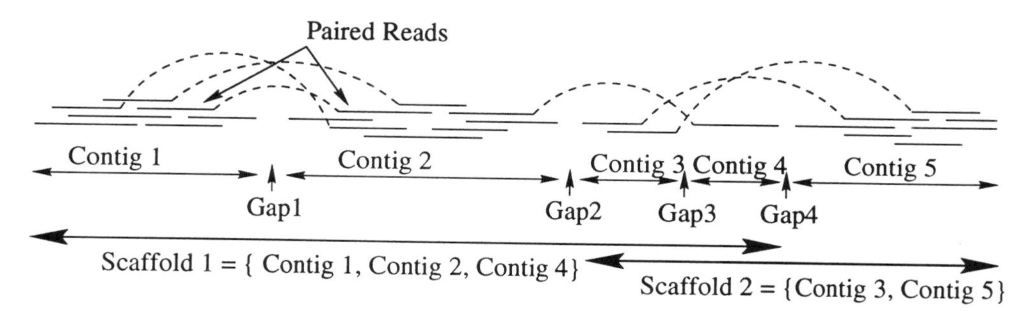

Figure 2. Illustration of contigs, gaps, and scaffolds.

ditional information it is impossible to tell how to order the contigs along the source sequence. Consider next, projects in which at least some percentage of the reads are from opposing ends of suitably long inserts. Often these pairs are in different contigs, which consequently, orders them and gives the approximate distance between them. A maximal collection of contigs so linked is termed a *scaffold* [15]. Note, as illustrated in Figure 2, that scaffolds can and do overlap with the contigs of one scaffold interspersed between those of another. Thus one cannot sensicaly speak of "gaps" between scaffolds. Instead the relevant property of scaffolds is the frequency with which regions of the genome are linked by one.

Double-Barreled Shotgun Sequencing. In a "pure" version of this approach each of I inserts of average length $\overline{L}_I$ is end-sequenced. One can then define the *map redundancy* [15] of the project as $\overline{m} = I\overline{L}_I/G$. Now in general $\overline{m}$ is larger than the sequence redundancy $\overline{c} = \overline{m}\,(2\,\overline{L}_R / \overline{L}_I)$ when $\overline{L}_I \geq 2\overline{L}_R$, i.e., when the inserts are size selected so the reads don't overlap. Thus the expected number of gaps between *insert* contigs, $^1/_2Re^{-\sigma 1/2R}$ where $\sigma = \overline{L}_I/G$, is progressively smaller the larger the insert size $\overline{L}_I$. In other words, double-barreled shotgun sequencing tends to produce very large scaffolds This positions sequence contigs with respect to each other, and so facilitates the determination of PCR primer pairs needed for gap closure. Furthermore, for sufficiently long inserts, it is unlikely that both reads of an insert will lie in a large scale repeat of the source. Thus the read in the relatively unique portion of the source effectively determines which copy of a repeat its mate is in. Because of this, double-barreled shotgun sequencing has been found to be a very powerful way of resolving repeats.

Recent simulation studies [15] have indicated that from a purely informatic perspective there is advantage in using longer inserts and no advantage in having some percentage of the reads be unpaired. However, this must be tempered against the experimental fact, that because of the different cloning vehicles involved as the insert becomes larger (e.g. plasmid $\rightarrow \lambda \rightarrow$ cosmid $\rightarrow$ PAC,BAC,P1 $\rightarrow$ YAC), it is more difficult to sequence the ends of long inserts and there is greater chance of chimerism (two or more inserts ending up in the same cloning vehicle). Thus there is counter-balancing economic pressure to use single reads and shorter inserts. Fortunately one loses little of the benefits of having long end-sequenced inserts in hybrid schemas where a sizable fraction of a project is single reads, and where the paired reads are from inserts over a distribution of insert lengths skewed to the shorter lengths.

Ordered Shotgun Sequencing (OSS). While one may have several repeats in a long source sequence and a shotgun assembly may have many gaps, it is rare that either of these problems occur for source sequences on the order of the length of a λ-clone, e.g *G (* 10*Kbp*. To take advantage of this observation, Chen et al. [16] devised the following two-tiered approach to sequence a large source sequence of length G >> 10*Kbp*. For $\overline{m} \approx 10$, begin by sequencing both ends of $I = G\,\overline{m}/\overline{L}_I$ λ-sized inserts whose average length is $\overline{L}_I \approx$ 10*Kbp*. Initially, let the current fragment set be the set of reads sequenced above, and set the sequenced intervals of the source to the empty set. Repeat the series of steps below until nothing new can be added to the sequenced intervals.

a. Assemble the current set of fragments and sequenced intervals, treating each sequenced interval as a single read.
b. Select from the assembly of (a) the scaffold that spans the largest unsequenced interval of the source. For the first iteration, this criterion reduces to selecting the longest scaffold.
c. Select the smallest subset of inserts of the scaffold of (b) that cover the unsequenced interval, i.e., select a *minimum tiling set*.
d. Shotgun sequence each insert in the minimum tiling set of (c).
e. Use the sequence determined in (d) to update the sequenced intervals. Remove all reads involved in the scaffold of (b) from the current set of reads as well as all those now discovered to be completely within a sequenced interval of the source.

Figure 3 illustrates this process. Note that with each round of the steps above, new overlaps with end reads occur because the interior of several λ-inserts is sequenced. Because the map redundancy is high, it is thus very likely that the process will result in the sequence of all of the source. As a final point, not that while towards the later rounds, it is typical to discover maximal scaffolds extending from the ends of distinct sequenced intervals. One may pursue all such scaffolds in a single round as their tiling sets cannot overlap (e.g. see "Round 3" of Figure 3).

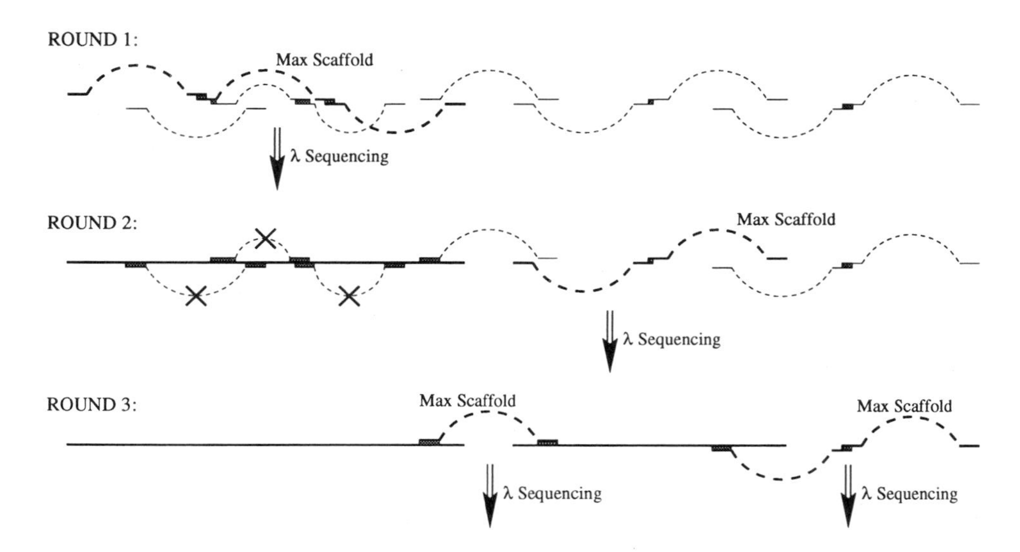

Figure 3. Illustration of Ordered Shotgun Sequencing.

Whole Genome Ordered Shotgun Sequencing. Recently some investigators have effectively proposed using the OSS strategy on the scale of the entire human genome. In this case the size of the source is G = 3,000*Mbp*, has been increased to 20, and instead of λ-clones, BACs of size I = 400 *Kbp* will be end-sequenced. One should see that the problem and potential solutions are identical to the above, save that the proposers are confident that they can shotgun sequence BACs using the double-barreled approach. As proposed investigators would select some seed BACs at random to sequence. Then minimally overlapping BACs at each end of a seed will be found and shotgun sequenced to extend the seed sequence until eventually the entire genome is covered. This is tantamount to OSS where one dispenses with finding maximal scaffolds, and simply proceeds with a pair of extending inserts in each iteration.

Map-Based Whole Genome Sequencing. A final approach to whole genome sequencing is based on the observation that already available today are 100–200*Kbp* resolution maps of sequence tagged sites (STSs) of the human genome [17]. An STS is a contiguous sequence of the genome of typically 300*bp* that is unique, sequence-wise, with respect to rest of the genome and which can be consistently be PCR'd with a pair of primers from each end, typically each about 18bp long. The available STS maps order some 15,000 of these sites along each chromosome of the genome along with a rough estimate of the distance between them. Given the linking and repeat resolving power of double-barreled sequencing the following possibility arises. Directly shotgun the entire human genome at a given coverage $\bar{c}$, say 5–10, sequencing both ends of every insert sampled. Applying a fragment assembly algorithm to the resulting data set of roughly 30–60 million reads would likely by a futile exercise. However, given the map of STS markers, the problem of determining the sequence of the genome reduces to the tractable one of finding a scaffold that spans between a pair of markers roughly 100–200Kbp apart. We call this the inter-marker assembly problem. We will show that, for a wide range of insert size mixes, such a scaffold exists over 99% of the time, and when it does, one may then assemble over 99% of the sequence between the markers yielding exactly the same result that one would obtain if one had shotgun sequenced the stretch between the markers at a redundancy of $\bar{c}$. The remainder of this paper will be devoted to supporting the claims just made via simulation and to the presentation of an algorithm for solving the inter-marker assembly problem.

2. A SIMULATION STUDY OF MAP-BASED WHOLE GENOME SEQUENCING

2.1 Nature of the Simulation

To test the feasibility of the whole genome sequencing approach, we developed a simulator with the following features. For our model of the genome we decided to assume that it contains two sizes of repetitives elements, short repeats and long repeats, whose lengths and number (expressed as a percent of the genome) are inputs to each simulation trial. The idea is simply that the short repeats model the relatively abundant small interspersed nucleic elements (SINEs) like Alu's and the long repeats model long interspersed nucleic element (LINEs) and the exons of repeated genes. The length of these elements are controlled by the parameters SINE_LEN and LINE_LEN while the length of the genome for a given trial is specified by the parameter GENOME_LEN. The relative abundance of these elements is determined by the parameters SINE_PERCENT and LINE_PERCENT which specify the percentage of the genome covered by the given element. That is, for a given trial, SINE_PERCENT * GENOME_LEN / SINE_LEN short repeats and LINE_PERCENT * GENOME_LEN / LINE_LEN long repeats are generated and placed with uniform probability across the genome. The generated elements are guaranteed not to overlap but they can and do end up being adjacent to each other. We term the sum, SINE_PERCENT + LINE_PERCENT, of the two percentages, the repetitiveness of the genome.

With regard to the sequencing protocol, two types of inserts are assumed to be sequenced: long inserts for which both ends are sequenced, and short inserts whose length is selected so that end reads overlap by 100 bases, effectively giving a single long read of twice the average read length less 100. The length of a long insert, LONG_LEN, the length of a read, READ_LEN, and the ratio of long to short reads sampled, LS_RATIO, are parameters that can be set for each simulation trial. From these parameters the length of a short insert for a given trial is inferred to be 2*READ_LEN -100. Finally, the sequencing redundancy for a sequencing trial is set by the parameter REDUNDANCY Given settings for these parameters,

.5*REDUNDANCY*GENOME_LEN / ((1+LS_RATIO)*READ_LEN - 50)

short inserts are sampled uniformly from the genome, as are

.5*REDUNDANCY*GENOME_LEN*LS_RATIO / ((1+LS_RATIO)*READ_LEN - 50)

long inserts in a given trial.

In, for example, the analysis of Waterman and Lander it was assumed that every entity was of exactly the length specified by its controlling parameter. However, in reality, insert lengths, read lengths, and repetitive element lengths (of a given type) range over some distribution of lengths. As in Waterman and Lander's paper, we also argue that such variation, if limited to say 15–25% of the average, does not significantly change the outcomes of the simulations. Indeed, some initial experimentation on small models confirmed this. Given the size of the genomes to be simulated, the number of repetitive elements and especially the number of inserts involved is huge and stresses the memory capacity of most machines. Modeling variability in insert length doubles the memory required, and further allowing variability in the read length at each end doubles the memory required yet again. Thus in the production simulator, we adopt the compromise of modeling variability in insert and repetitive element lengths, but assume all reads are of exactly length READ_LEN. Length variation is controlled by the parameter LEN_VARIATION: the length of inserts and repeats is chosen uniformly from the interval 1 ± LEN_VARIATION times the target length.

Given the genome and sequenced inserts for a specific trial, the simulator examines the nature of the contigs, scaffolds, gaps, map connectivity, and other features of the assembly that could result from such data. A false overlap between two reads is one where the overlapping portion of their sequences are identical to within error rate e but the reads were sampled from non-overlapping regions of the underlying genome, i.e., melding the fragments would lead to an incorrect assembly. Note that the perspective in the simulation is distinctly different from that faced by an assembly program operating on real data. For a given trial we know exactly which segment of the genome a read comes from and exactly where every repetitive element is in this genome. Thus the simulator knows exactly which overlaps are true and false, whereas a hypothetical assembler would have to decide this. False overlaps are of two types. In coincidental overlaps the overlapping portions of the two reads are from different portions of the genome that are similar just by chance. In repeat overlaps the two segments of the genome are similar because they involve copies of the same repetitive element. Whole human genome sequencing would involve at most 75 million fragments, for a total of 10^{16} comparisons. If one insists on at least MIN_OVERLAP = 35 basepairs of overlap, then one should expect fewer than one overlap in roughly 10^{15} to 10^{18} to occur coincidentally. That is, apart from computing at most a few coincidental overlaps all other false overlaps computed by a hypothetical assembler would be due to repetitive elements. In subsection 2.3 we will discuss the issue of how an assembler might distinguish repeat overlaps from true overlaps in the context of solving an inter-marker assembly problem. For now we focus on the issue of whether or not there is enough overlaps or connectivity between the rea ollected. To that end, we finesse the issue of repeat overlaps by throwing them out along with any true overlaps involving a repeat. That is the simulator will consider read A to overlap read B if there are not less than MIN_OVERLAP bases of the overlap that are *not within a repetitive element* of the genome. This definition is conservative with respect to connectivity in that the contigs and scaffolds obtained are the sparsest one could imagine from a whole genome data set. This is so as it essentially says that *R*% of the genome is unusable for connecting fragments in an R% repetitive genome.

An insert is *useless* if it can't possibly be involved in an overlap as defined above. That is, both its reads do not have a segment of MIN_OVERLAP or more base pairs lying outside of a repetitive element. Such inserts are removed from further consideration in our simulation analysis. We define the *effective coverage* of the genome as the percentage of the genome covered by the reads of useful inserts. Note that while one read of a long insert may be completely in a repeat, if the other is not, then the insert is useful and both reads are deemed to cover their respective parts of the genome. This is sensible as the exact utility of dual end sequencing is that the end read in the "unique" portion of the genome resolves the repeat that the other end read may find itself in.

Many simulations results are about to be presented. In most plots we show the effect of varying one or two parameters with the remainder being fixed. In such cases it is assumed that the fixed parameters are at the following default values unless otherwise stated:

GENOME_LEN	500Mbp	LONG_LEN	10Kbp
SINE_LEN	300bp	READ_LEN	400bp
LINE_LEN	1500bp	LS_RATIO	1.0
SINE_PERCENT	20%	REDUNDANCY	10.0
LINE_PERCENT	5%	LEN_VARIATION	15%
MIN_OVERLAP	35bp		

Furthermore, in the default case SINE_PERCENT will be exactly four times LINE_PERCENT, so that when we say a genome is *R%* repetitive, LINE_PERCENT = .2**R* and SINE_PERCENT = .8**R*.

2.2 Simulation Results

For each simulation trial, we measured effective coverage, $\overline{e}$, the average length or span of a contig, $\overline{ctg}$, the average length of a gap, , the number of contigs formed, *C*, and the maximum observed gap length, gap_{max}. The first observation, is the confirmation of the following theoretical predictions. Consider holding all parameters constant except for genome length G. The observables above behave as follows as a function of G:

1. $\overline{e}(G)$ = a constant = $O(1)$
2. $\overline{ctg}(G)$ = a constant = $O(1)$
3. $\overline{gap}(G) = \overline{ctg}(1 - \overline{e}) / = O(1)$
4. $C(G) = G / (\overline{ctg} + \overline{gap}) = O(G)$
5. $E[gap]_{max} = O(ln\ G)$

For example, equation (1) says that if at the default parameter settings effective coverage is 99.3% for a genome of say 100*Mbp*, then it is 99.3% for one of 500*Mbp* or 3000*Mbp*. Often this statement is met with surprise as a common misconception of the Lander and Waterman analysis is that one cannot effectively shotgun genomes that are too large. What is actually true is that the number of contigs grows directly proportional to G (i.e., equation (4)), thus there is considerable fragmentation in terms of gaps. But the number of gaps that will need to be PCR'd or resolved over a series of 30*Kbp* cosmids spanning a genome will be the same as the number of gaps that will need to be resolved if one shotgun sequenced the entire genome. The only drawback is that while the average gap length is the same in both scenarios (i.e., equation (3)), the expected value of the maximum observed gap does grow logarithmically (i.e., equation 5 [18]) and so will be an order of magnitude larger in the whole-genome case. Thus one may encounter a few large gaps in the whole genome data set that one does not encounter in a cosmid-by-cosmid or BAC-by-BAC data set. Fortunately, average gap length is generally quite small, e.g., on the order of less than 100*bp*, when the redundancy is sufficient to guarantee 99% or greater effective coverage In a whole genome data set we observe a maximum gap of length 2400*bp*, and a handful over 1000*bp*. Thus one would expect PCR closure to fail in just a very limited number of cases.

Figure 4a shows a number of plots of expected coverage as function of sequencing redundancy. One plot is for a genome with no repetitive elements, effectively summarizing the Lander-Waterman analysis. The other three assume a genome that is 50% repetitive implying by our convention for defaults that 40% of the genome is covered by SINEs and 10% by LINEs. For these cases the mix of long-to-short inserts affects the results as short inserts are useless with much greater probability than long inserts. To explain the results, first observe that SINEs of length 300 do not create any obstacle to coverage as they are not long enough to make a read useless (i.e., READ_LEN ≥ SINE_LEN + 2*MIN_OVERLAP). However, LINEs of length 1500 cannot be effectively covered by short inserts which at the default settings leaves the inner 10% of each LINE uncovered on average. Thus the ''All short'' curve tops out at 99% effective coverage. Note, however, that with even a 1–1 ratio of long to short inserts, there is only a slight degradation of coverage compared to the ''All long'' curve. We may thus conclude that at a sequencing redundancy of 5x or greater and a 1–1 or greater mix of long to short inserts, a whole genome data set will effectively cover more than 99% of the genome unless it contains many more long repetitive elements than are currently being observed in the human genome.

Figure 4b further shows that the assertion above is true for a wide range of read length assumptions and overlap assumptions. Assuming quite short reads of 300*bp* and requiring 55*bp* of overlap to connect reads, does not significantly degrade effective coverage as shown in the lower of the two curves in the figure.

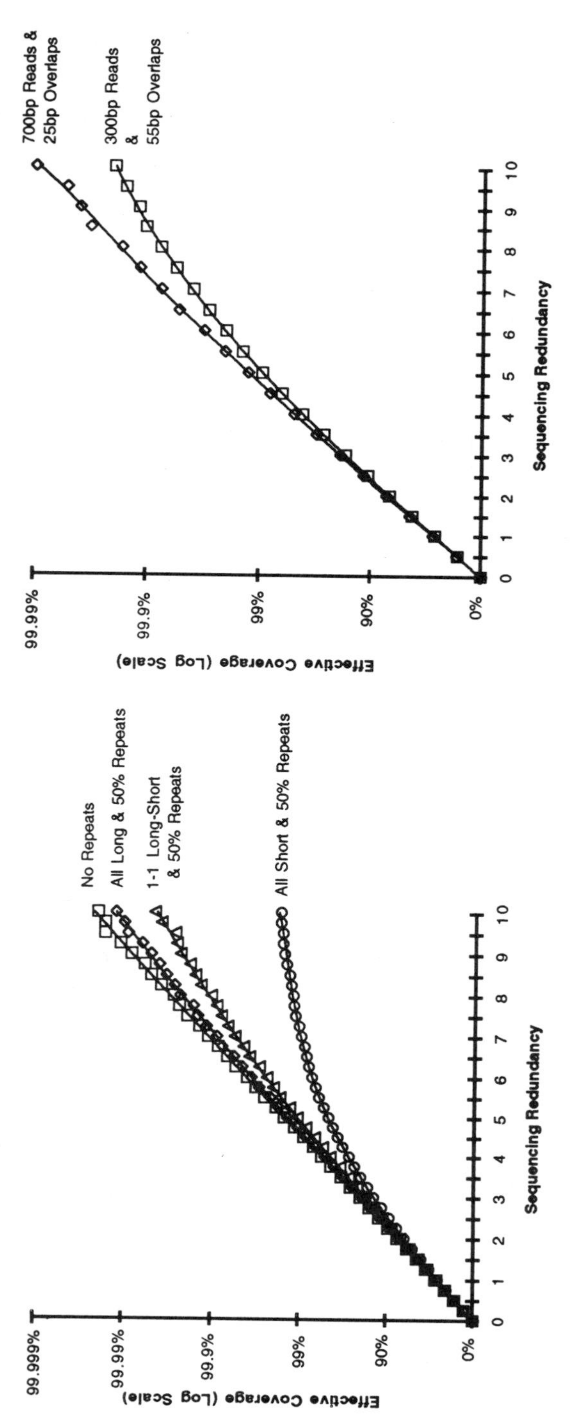

Figure 4. Effective Coverage as a function of (a) repetitiveness and long/short mix, and (b) read length and minimum overlap.

The total number of contigs and gaps in a whole genome data set depends on the average length of a contig which in turn depends on the mix of long to short inserts and the repetitiveness of the genome. The plots in Figure 5 depict curves for the average contig length as a function of LS_RATIO, one curve for a different degree of repetitiveness of the underlying genome. Figure 5a shows the situation assuming a 5x sequencing redundancy, and Figure 5b is for 10x sequencing. Assuming no repeats in the genome, short inserts are effectively 700*bp* reads and long inserts provide 400*bp* reads, the connection between pairs being of effectively no value in terms of contig building. Thus a greater ratio of shorts is produces longer contigs. However as soon as one begins to place repeats into the genome, the LINE elements become a barrier which short inserts cannot cross. Thus when all inserts are short, the average contig length quickly becomes clamped to the average length between LINE elements. At the other end of the curves, where all inserts are long, there is a slight degradation but it is limited only by the number of inserts that are rendered useless. Note that in Figure 5a where sequencing redundancy is 5x, a mix of shorts increases contig length up to about a 1 to 2 ratio, but in Figure 5b, where sequencing is 10x, there is only a steady loss as shorts are introduced into the mix, with a precipitous decline in contig length as the ratio becomes less than 1 to 1. One would conclude that the mix ratio should therefore be kept at least this high.

The other aspect of whole genome sequencing that we investigated via simulation, was the nature of the scaffolds linking contigs together. We found that in almost all cases, that at the default settings and for a ''super-chromosome'' of 500*Mbp*, that (1) there was one scaffold that spanned almost the entire chromosome, (2) there were 2–5 small scaffolds of 2–5 contigs near the ends of the chromosome, and (3) there were 10–20 isolated, and small contigs of one or two reads nestled between the contigs of the super-scaffold. Thus the scaffold information linked together is very strong, effectively giving a global linking of all contigs.

Recall that to assemble a whole genome data set, we will be assuming the availability of an STS map of some average resolution, hopefully on the order of 100*Kbp* between each marker. The aspect of scaffold connectivity that is thus the most relevant is the frequency with which a scaffold exists that links together a given pair of markers in such a map. To this end, the simulator was augmented to generate pairs of markers of length MARKER_LEN at a distance MARKER_SPACING apart. The location of the markers was chosen uniformly subject to the additional criterion that the markers were not in a repetitive element. The default length of markers was 300*bp* and for data point in Figure 6 was generated with 1000 trials on marker pairs. Each plot displays curves showing the effect of the long-to-short insert mix, the length of the long inserts, and the spacing between the markers. What is plotted is the percentage of the trials for which a scaffold containing both markers failed to exist, i.e., the ''STS spanning failure rate''. Figure 6a shows a series of curves for 5x sequencing redundancy and Figure 6b for 10x sequencing. The results shown are very encouraging: for long inserts of 5*Kbp* or more and a long-to-short ratio of 1 or more, a spanning scaffold exists over 99% of the time for redundancies as low as 5x. One should further consider that these plots assume that 5% of the genome between the markers is part of long repeats that are barriers to connectivity.

2.3 Inter-Marker Assembly

We now turn to an examination of the problem of determining the sequence between two markers, *A* and *B*, given a whole-genome data set. First consider the problem of finding a scaffold that connects the two markers. Imagine a graph in which each vertex represents an

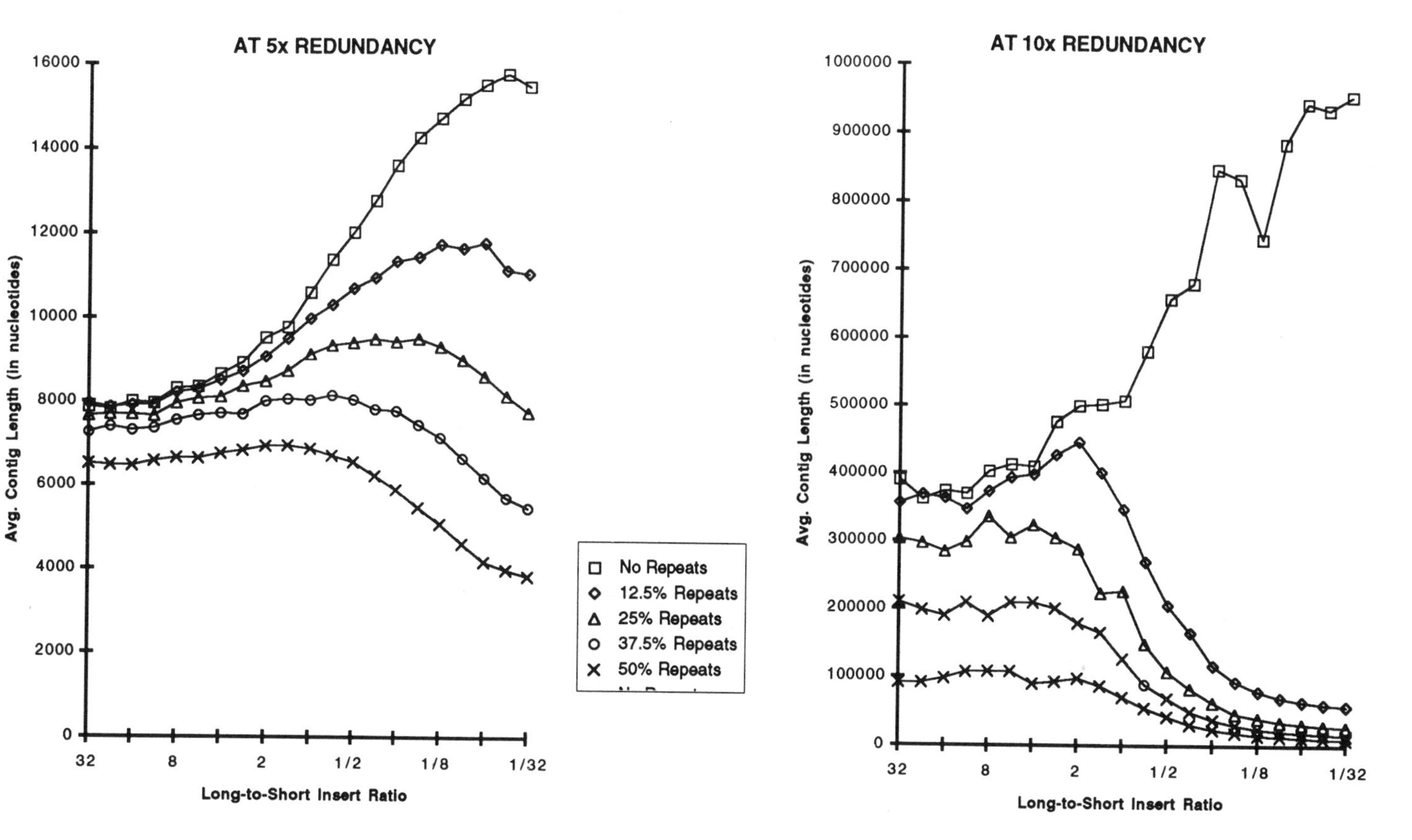

Figure 5. Average contig length as a function of long/short mix, repetitiveness, and sequencing redundancy.

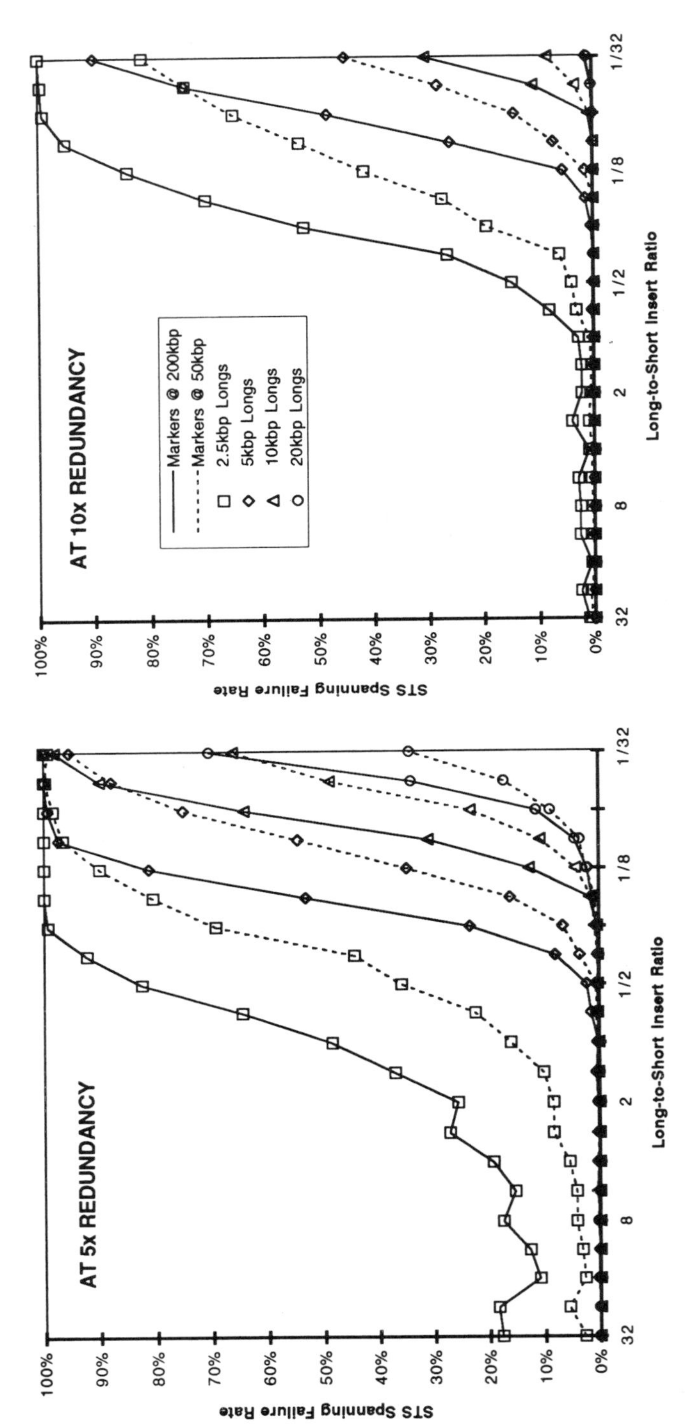

Figure 6. STS spanning failure rate as a function of long/short mix, long insert length, inter-marker spacing, and sequencing redundancy.

insert (not a read) and where there is an oriented edge between two vertices if their inserts have overlapping reads. Let O_A be the set of inserts whose reads have a significant overlap with marker *A*, and let O_B be the corresponding set for the marker *B*. It is easy to see that a scaffold between *A* and *B* must corresponds to a path from a vertex in O_A to a vertex in O_B, i.e., our problem reduces to a simple graph connectivity question. The difficulty is that while a ''true'' scaffold must correspond to such a path, there may be connecting paths that involve false overlaps and hence do not model a valid spanning scaffold.

In a redundancy $\bar{c}$ data set, one would expect that each read will on average have true overlaps with $2(\bar{c}-1)$ other reads. However, a read that contains part of a repetitive element will have many more overlaps due to the repeat overlaps induced by all the other copies of the repetitive element. In terms of the graph, reads that do not contain overlaps will typically have degree $O(\bar{c})$ whereas those containing repeats will have degree $O(\overline{cn})$ where *n* is the number of copies of the repeat in the genome. While there is statistical fluctuation in the degrees of vertices, note that vertices containing high copy repeats such as Alu's will easily be distinguished by their enormously large edge degree. But an even more useful observation is that vertices with degrees less than $2\bar{c}$ very likely do not contain repeats and hence all their edges/overlaps are very likely to be true.

It now follows from the simulation results of the previous section that a path from O_A to O_B using predominantly low degree vertices is very likely to exist. Thus while we have yet to simulate the following, we venture that the following algorithm will find a minimal scaffold with high probability. The algorithm first finds the vertices in O_A and O_B and then begins a depth first search traversal from these vertices towards those in the opposing set in parallel. Because the edges are oriented the algorithm knows in which direction it is extending. If extending from a point estimated to be less than half way between the two markers, then the algorithm explores long inserts before short inserts, otherwise it gives preference to low-degree vertices over high-degree vertices. The algorithm terminates when paths from opposite markers meet.

From our prior simulation experience we know that the average number of inserts in a spanning path is on the order of the marker spacing divided by the length of long inserts and that favoring long inserts up to the half-way point significantly reduces the number of edges that need to be explored. Because high degree vertices are avoided it will almost always be the case that the spanning scaffold found will not involve any false overlaps. The occasional exception will be for inter-marker stretches containing several copies of a low-copy number and very large repeat (such as a gene). Even in these cases, the repeats will be skirted unless they are tandem or chance works against the algorithm. However, provided that insert sizes are know to within plus or minus 50% this situation will be detected because the path found will be too short. Further evidence for this situation will be the presence of one or more distinct loops in the paths traversed, and possible dead-end paths induced by copies of the repeat in other parts of the genome Figure 7 provides an illustration.

Given a minimal and true scaffold between two markers the inter-marker assembly problem then becomes like any other shotgun assembly problem save that pulling in the relevant pieces involves much more time intensive overlap comparisons than in standard problems. We estimate that with current high-performance workstatsion, one inter-marker assembly between markers at 100*Kbp* spacing and a 10x data set over the human genome can be solved in one day. Ultimately, how well or easily the inter-marker assembly problem can be solved depends upon how tightly the lengths of inserts are known and the degree to which dual end reads are asserted to be mates but are actually not (because of chimerism or data tracking errors). It is at this time an open question to ascertain what can be expected but our preliminary experience suggests a positive result.

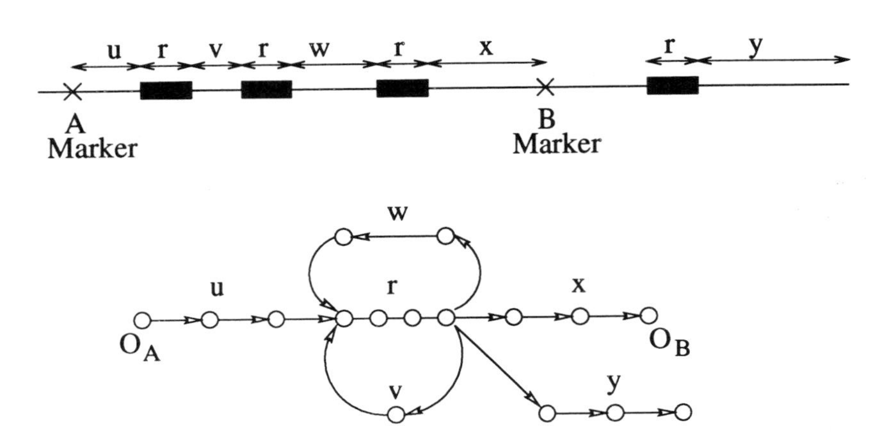

Figure 7. Illustration of inter-marker path structure as a function of the inter-marker sequence.

3. DISCUSSION

The map-based whole genome assembly problem may be succinctly stated as follows. One is given a 5–10x shotgun data set of end-sequenced inserts and an STS map of the target genome. The problem is to determine, for each adjacent pair of markers, the sequence between the markers. There are two things to note immediately. First, is that the proposition is applicable to not only whole genomes but to any assayable target. One might consider sequencing a large target by simultaneously collecting shotgun data while developing a 100*Kbp* resolution STS map. Second, is that the computational problem is divided into a collection of independent inter-marker assembly problems, one for each pair of adjacent markers. Thus a very large computation is nicely broken into manageable subproblems that may be solved in parallel.

From the simulation results it follows that, with a suitable choice of redundancy, long insert length, and long-to-short insert mix, one will find a scaffold spanning a pair of markers over 99% of the time. Moreover, one can then assemble the contigs of this scaffold between the marker. It follows from these simulation results that one can expect a number of contigs and average gap length between these contigs that is *identical* to that which would have been obtained had one shotgun sequenced the stretch between the two markers. The only drawback is that the extreme order statistic for gap length is larger than in clone-by-clone approaches so there will be a small percentage of cases where gaps can be PCR-filled in the clone-by-clone data sets but not in the whole genome case. Otherwise the amount of information is equivalent. At the default setting of the parameters a scaffold will exist 99.8% of the time, and in such cases 99.6% of the sequence between the markers will be delivered, implying that 99.4% of the genome will result from such a data set in regions that are not wildly repetitive (e.g. centromeres and telomeres).

The principal appeal of map-based whole genome sequencing is its simplicity. There is no physical library to construct or to maintain, greatly simplifying the number of steps, and, more importantly from the point of view of automation, the number of *types of steps* involved. There is no need to track reads, the only data organization required is list of the reads and their long insert mates. Clone-by-clone approaches such as whole genome OSS will involve resequencing 5–10% the genome as this much overlap is estimated in a = 20 minimum tiling set. Whole genome sequencing does not suffer this waste of effort. It is also possible to immediately derive comprehensive polymorphism information by collecting inserts from several different individuals. The clone-by-clone approaches would require completely sequencing each individual to obtain the same information.

On the other hand, there are drawbacks to map-based shotgun sequencing. First, the STS spanning failure rate does not become small until sequencing redundancy reaches 4x. Thus one must perform 40–80% of all sequencing, before one can begin to actually deliver sequence information. While we have given an indication that false repeat overlaps can effectively be inferred and that inter-marker assembly is very likely computationally feasible, we have not definitively demonstrated this. Therefore, the risk of collecting so much data without the certainty of successful prior experience is daunting.

For the future, we plan on further simulation aimed at exploring the reliability/feasibility of inter-marker assembly. In a first phase we will do so by simulations of the type presented. If these reveal a positive outcome, then we will proceed with the more expensive trials of building an inter-marker assembler and running it on data sets obtained by taking the largest available stretches of human DNA and building a simulated whole genome data set from an extrapolation of said sequence. In a final test-of-feasibility stage, a real data set could be generated for a 1–5*Mbp* segment of a genome.

REFERENCES

1. Sanger, F., Coulson, A.R., Hong, G.F., Hill, D.F., and Peterson, G.B 1982 Nucleotide sequence of bacteriophage λ DNA. *J. Mol. Biol.* 162:729.
2. Sanger, F., Nicklen, S. and Coulson, A.R 1977. DNA sequencing with chain-terminating inhibitors. *Proc. Natl. Acad. Sci.* USA 74:5463.
3. Peltola, H., Söderlund, H., and Ukkonen, E. 1984. SEQUAID: a DNA sequence assembly program based on a mathematical model. *Nucleic Acids Research* 12:307–321.
4. Kececioglu, J. and Myers, E. 1995. Exact and Approximate Algorithms for the Sequence Reconstruction Problem. *Algorithmica* 13:7.
5. Myers, E. 1995. Toward Simplifying and Accurately Formulating Fragment Assembly. *J. of Computational Biology* 2:275.
6. L. Rowen, and B.F. Koop. Zen and the art of large-scale genomic sequencing. *Automated DNA Sequencing and Analysis* (M.D Adams, C. Fields, & J.C Venter, eds.) Academic Press (London, 1994), 167–174.
7. Bell, G.I. 1992. Roles of repetitive sequences. *Computers Chem.* 16:135.
8. Iris, F.J.M. 1994. Optimized methods for large-scale shotgun sequencing in Alu-rich genomic regions. *Automated DNA Sequencing and Analysis* (M.D Adams, C. Fields, J.C. Venter, eds.) Academic Press, London, 199–210.
9. Lander, E.S. and Waterman, M.S. 1988. Genomic mapping by fingerprinting random clones: a mathematical analysis. *Genomics* 2:231.
10. Wilson, R.K., Chen, C., and Hood, L.E. 1990. *BioTechniques* 8:184.
11. Edwards, A., and Caskey, C.T. 1991. *Methods: Companion Methods Enzymol.* 3:41.
12. Berg, C.M., Wang, G., Strausbaugh, L.D., and Berg, D.E. 1993. *Methods Enzymology* 218:279.
13. Myers, E. Advances in Sequence Assembly. *Automated DNA Sequencing and Analysis* (M.D. Adams, C. Fields, & J.C. Venter, eds.) Academic Press (London, 1994), 231–238.
14. Larson, S., Jain, M., Anson, E., and Myers, E. An interface for a fragment assembly kernel. *Tech. Rep. TR96–04*, Dept. of Computer Science, U. of Arizona, Tucson, 85721.
15. Roach, J.C., Boysen, C., Wang, K., and Hood, L. 1995. Pairwise end sequencing: A unified approach to genomic mapping and sequencing. *Genomics* 26:345.
16. Chen, E.Y., Schlessinger, D., and Kere, J. 1993. Ordered shotgun sequencing, a strategy for integrated mapping and sequencing of YAC clones. *Genomics* 17:651.
17. Hudson, T.J. et al. 1995. An STS-based map of the human genome. *Science* 270: 1945–1954.
18. Downey, P.J., and Maier, R.S. 1988. Logarithmic moment bounds for extreme order statistics. *Technical Report TR 88–32*, Dept of Computer Science, University of Arizona, Tucson, AZ 85721–0077.

7

SEQUENCE PATTERNS DIAGNOSTIC OF STRUCTURE AND FUNCTION

Temple F. Smith,[1] R. Mark Adams,[1] Sudeshna Das,[1] Lihua Yu,[1] Loredana Lo Conte,[1] and James White[2]

[1]BioMolecular Engineering Research Center
College of Engineering
Boston University
36 Cummington Street, Boston, Massachusetts 02215
[2]TASC, Inc.
55 Walkers Brook Drive, Reading, Massachusetts 01867

1. INTRODUCTION

There currently exists a vast wealth of amino acid sequence data from a great many different genes and different organisms. It is anticipated that this wealth will continue to increase. These data represent much of our current raw knowledge of the biological systems on earth. Yet our ability to exploit these data is still limited: first, by our inability to predict a protein's biochemical functions or the structure encoding those functions directly from the knowledge of its amino acid sequence; and second, by our limited ability to carry out direct experimental determination of either the function or structure. Interestingly, it is currently easier to predict function rather than structure. This is true even though knowledge of the folded structure is essential to understanding the detailed steric constraints and side chain chemistry that determine the function(s). Only a small fraction of genes have had biochemical or cellular functions directly determined by experiment. Through the recognition of sequence similarities or shared patterns of sequence elements, we have been able to infer the function of tens of thousands of additional genes. This has proven a powerful approach. By suggesting a probable function for a newly determined sequence, only a few experimental tests need to be carried out for confirmation. In some cases the mere suggestion provided by a shared set of common elements generates obvious insight into past experimental observations making validation all but obvious. Thus recognition of biologically significant sequence similarity has become the Rosetta Stone of modern molecular biology. The tools used depend on methodologies drawn from mathematics, computer science, and linguistics as well as physics, chemistry and the rest of biology. Finally it is believed that analogous approaches will be developed for the accurate prediction of protein structure.

Theoretical and Computational Methods in Genome Research, edited by Suhai
Plenum Press, New York, 1997

The utility of common pattern recognition arises, as has been pointed out many times from Earth's evolutionary history. Evolutionary processes are both opportunistic and conservative. Firstly, any organism can use only what is currently at hand, independent of whether or not there might possibly be a much better solution. Secondly, once having solved a particular "problem", it will conserve those aspects essential for that solution, while "exploring" modifications there to. To some extent this all seems rather obvious: the more closely related in time any two sequences are, the more likely they will look similar, and the more likely they will encode nearly identical functions. For example, we find that if we simply cluster the mammalian Cytochrome-c's or alpha hemoglobins, based on the percentage of sequence identity, the standard evolutionary taxonomy is obtained rather accurately. Such similarities can be observed over much greater distances. For example, all globins contain similar overall patterns of hydrophobicity, implying similar helical structures [1] with similar distribution of buried and exposed amino acid positions. Also all globins contain the identical conserved Histidine required to coordinate the Iron found in the common Heme ring. In a like manner, there is a common structural and associated amino acid pattern among nearly all ATP binding proteins, consisting of a hydrophobic central beta strand and a Glycine N- terminal alpha helix[2] whose dipole moment is directly associated with the close proximal binding of the alpha phosphate.

The nature of the beast, however, is more complex and subtle. We know, for example, that only a few common amino acids, out of hundreds, can be diagnostic of identical biochemistry, while sequences apparently sharing twenty percent of their amino acids in common can be unrelated and encode very dissimilar functions. In fact, there may be only a hundred or so basic protein folds possible. In addition, nearly identical protein structures can be encoded by very different sequences, nature having apparently evolved similar structure from completely independent lines of descent. Again some of this is just the requirement of any evolving system. If any particular structure and/or function was not encodable by innumerable sequences, how would nature have discovered them in the first place, and how could "she" have continuously modified them, one random step at a time, to serve the subtle and ever-changing differences among the inter-, intra- and extra-cellular environments found across millions of species.

A protein's overall three-dimensional structure is very robust to most substitutions, and their encoded functions are thus also robust to substitution in all but a limited number of chemically critical surface sites. In addition, it appears that for soluble proteins a hundred or greater amino acids in length, there may only be a few tens of basic folding topologies available that satisfy the two basic structural constraints of minimizing the number of exposed hydrophobic side chains while maximizing the number of inter or solvent backbone Hydrogen bonds.

Finally, one needs to remember that there are a number of reasons why sequences appear to share a common pattern of sequence elements. Most obviously, this happens when they have been derived from a recent common ancestor, or they may have derived from a very distant common ancestor, but still carry out the original function and thus have conserved those amino acids essential to that function. The latter can also arise as a result of sequence elements essential to a structural component that has been conserved since the time of the last common ancestor, but not directly related to the currently divergent functions. Finally, two sequences can appear to share a pattern of sequence elements by chance rather than as the result of any common function or ancestor. The problem here is, what does one mean by "chance"? Note that for globular soluble proteins of similar size (or domain structure), there are a limited number of basic folds. Any central alpha-beta protein of similar size, for example, will have similar hydrophobicity profiles resulting from similar number and distribution of the basic structural repeat of amphipathic-helix-turn-buried-strand. One would guess that most extant proteins will look more similar to other extant proteins than to any random string of amino acids —even if of the same composition. It is this anticipated similarity of sequence elements, independent of common function and/or ancestry that has led to various statistical approaches for predicting protein structure by homologous extension modeling and its generalization.

It is the aim of this short paper to review the simplest of the protein functional pattern methods, and to show how these can be combined with those sequence pattern methods currently used to predict structure. Part of the motivation here arises from the fact that few of the structure prediction methods work very well alone, and that many important functional patterns, such as the His-Asp-Ser catalytic triad in the serine proteases, are too short to have useful specificity in any large database search.

In particular, we will look at combining functional sequence patterns representable as regular expressions or profiles, with two of the approaches currently employed in the so-called Protein Inverse Folding Problem. These are the threading and discrete state or Markov modeling methods. Both of these are conceptual extensions to the older homologous extension method. Recall that in homologous extension, one uses the recognition of a high degree of overall sequence similarity to align a sequence of unknown structure to one of determined structure, and thereby align the sequence of unknown structure to the one of determined structure. All of the various inverse folding methods begin with the construction of a set of structural models, and then attempt to optimally align a sequence to these models based on some scoring schema, identifying the highest scoring alignment as an indication of the most probable structure.

2. METHODS

2.1 Patterns Diagnostic of Structure

The most obvious and simple structural patterns are those associated with secondary structures. For example, a hydrophobicity plot along a protein sequence showing a clear period of 3.6, 2.0 or zero is a pattern indicative of surface amphipathic helices, surface strands or a completely buried region (including transmembrane segments). Such patterns can be expressed as regular expressions (the lowest rung of the Chomsky formal language hierarchy[3] as in Figure 1) or environmental profiles[4]. The latter is a sequence of amino acid preference vectors, one associated with each patterned structural position.

Structural patterns can be considerably more complex, however. They can be hierarchical, consisting of a pattern of secondary structures (indicative of a particular 3-D folding class) each of which is, in turn, composed of a pattern of positional environments (with associated amino acid and/or hydrophobicity preferences). For example, the class of simple four helix parallel bundles would have a pattern consisting of four nearly equal length amphipathic helices separated by three or more loop- or turn-compatible amino acids. Each helix in turn would be represented by a pattern of buried and exposed positions, each of which would be associated with a vector of helix-compatible amino acid preferences. Here one generally needs to move up the Chomsky hierarchy to "context free grammars" to be able to enforce the "non local" constraint of similar lengths among the four helices. This can be done using Discrete State Models [5] or Hidden Markov Chains[6]. For example see Figure 2.

Consensus patterns

Only the conserved residues are represented:

```
CLVKWVYPFLWIDSK
CLVRWMYPKLPIKSK
CLIKWAYPYLWIESK
CLIKWAYPFLWIDSK
---------------
CL--W-YP-L-I-SK
```

Frequency Profiles

Matrix containing frequencies of occurance for each amino acid at each residue:

```
CLVKWVYPFLWIDSK
CLVRWMYPKLPIKSK
CLIKWAYPYLWIESK
CLIKWAYPFLWIDSK
```

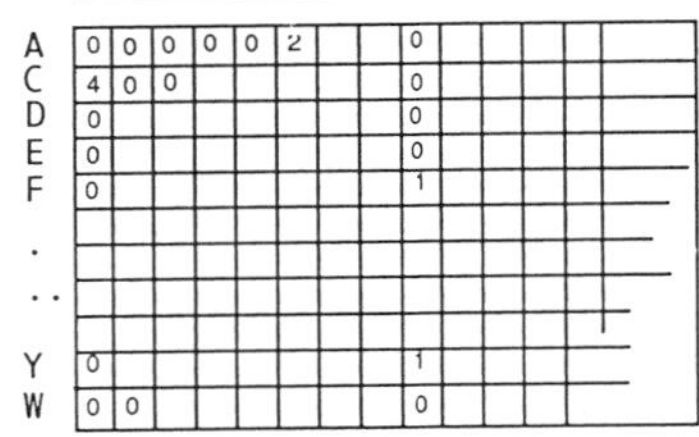

Normally converted to a matrix of log likelihood ratios.

Regular Expressions

Using minimal amino acid covering or probable substitution classes:

```
CLVKWVYPFLWIDSK
CLVRWMYPKLPIKSK
CLIKWAYPYLWIESK
CLIKWAYPFLWIDSK
---------------
CLanWfYPxLxIrSK
```

Converted to a UNIX *grep*-style regular expression statement

```
CL[ILV][KR]W[IVLMFWYCA]YP.L.I[DENQKRH]SK
```

Figure 1. An example of a simple pattern expressed as a consensus, a profile and a regular expression. Normally the weights or scores in the weight matrix are given as log likelihood ratios of the observed frequencies to the random expected, rather than the raw counts given here. The regular expression is expressed in the form of a sequence of minimal amino acid covering classes following Smith and Smith [27].

Models used for the "Threading" approach[7,8,9] to the inverse protein folding problem are often represented by even more complex patterns. There, in addition to containing patterns of sequence position environments and the associated amino acid preferences, pairwise constraints or contact preferences are added. These can be represented by a pattern of pairwise amino acid preferences associated with an adjacency matrix (or contact graph) among some subset of positions. Usually there is a score of pseudo energy associated with each pair of implied contacts for any particular threading or sequence alignment to structural model (see Figure 3). Here one is forced to a context-free representation such as the Random Markov Fields.[10]

2.2 Patterns Diagnostic of Function

As in the simplest case of a structural pattern, regular expressions[11] or profiles[12] are used to represent the patterns of conserved amino acids common to a given functional protein family. It should be obvious that such patterns normally include many structural elements, since identical functions are nearly always encoded within similar structures. In addition, there are common pattern elements included, only because the allowed variation has not yet been observed. Both of these are just the expected result of the conserva-

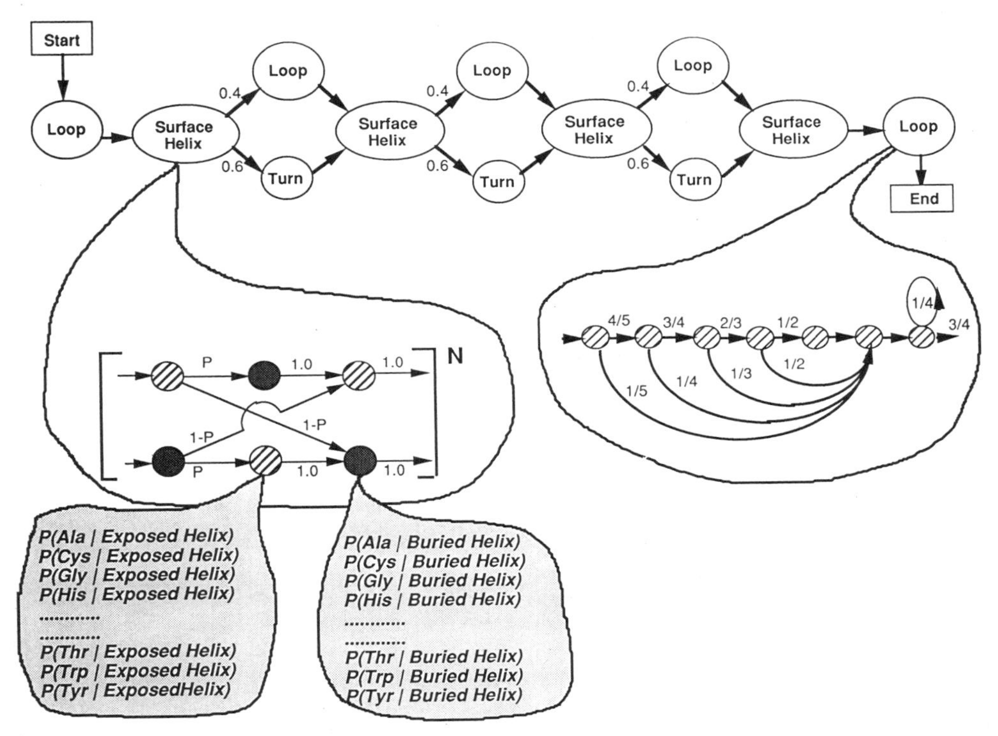

Figure 2. A hierarchical graphic representation of a Discrete State Model of a four helix parallel bundle. The arrows between modeled states represent allow transitions, and are labeled by their relative transition probabilities. At the lowest level of the hierarchy are the so-called hidden states or conditional probabilities associated with the likelihood of observing any given amino acid in that state. The value of "P" in the helical state repeat is normally set to generate a period in exposed verses buried positions of 3.6. Also note that the position state transition probabilities in the loop state have been designed to allow for a uniform probability distribution over the lengths of two to six with an exponentially decaying probability for longer loop lengths.

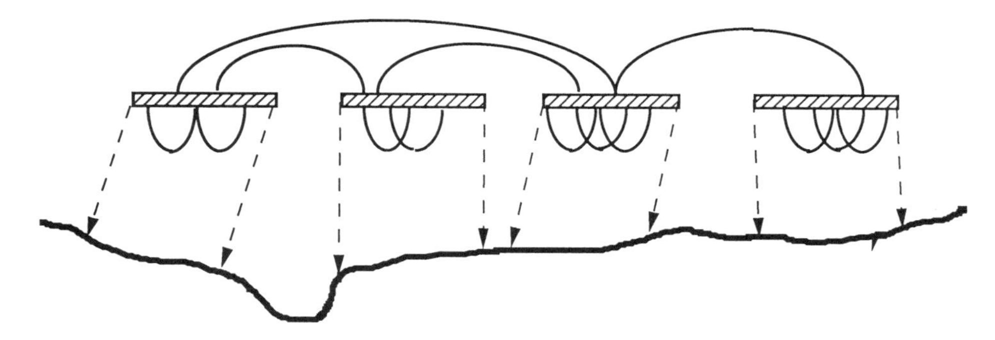

Figure 3. A graphical representation of a possible set of pairwise contacts to be included in a threading model. The rectangles represent four secondary structural elements with potential inter element contacting side chain positions shown as arcs drawn above, and intra potential contacting positions drawn as arcs below. In the most general case, the optimal threading or alignment of a sequence to such a model would be the one that placed the four secondary elements along the sequence in such a manner as to maximize the probability of observing the amino acid pairs aligned with the arch ends in actual contact.

tive/opportunistic nature of evolution. While it is in general difficult to separate these classes of elements, it is often possible to clearly identify experimentally those pattern elements that are essential to an "active site". As for the inclusion of chance common elements, one assumes that with sufficient sampling (taxonomic breadth) or biochemical modeling that these can be kept to a minimum.

All of the current methods for common sequence pattern discovery begin with a set of sequences believed to be functionally related. This is followed by one of two basic iterative procedures. The first constructs optimal alignment(s) from which a pattern of common elements is extracted. The pattern is then used in constructing subsequential alignments updating the pattern. A second general approach is to begin with a large set of "seed" patterns or short words. These may be generated either from the entire set of sequences or from any subset. Next their distribution (or close matches) among the sequences is obtained and used to construct preliminary patterns. The procedures can be iterated by generalizing the words until a set is found that is common to some desired fraction of the total. The two basic approaches differ primarily in that only the first directly involves optimal sequence to sequence alignment, while the search for the occurrence of any pattern in a new sequence always involves some type of pattern to sequence alignment procedure.

Pattern alignment is particularly useful since it not only suggests an overall function, but also information as to which parts of a new sequence are most likely to be critical to its function.[11,12,13,14,15,16] However, there are limitations to most current implementations. None of the current pattern identification methods deal convincingly with families of proteins sharing heterologous sets of functional and/or structural domains. Current methods produce patterns of common elements while eliminating heterologous sequence elements. This is their strength when dealing with single-domain proteins, for it is the highly variable regions among homologous sequences that can give rise to random and misleading full database search matches. However, in the heterologous multiple domain protein families, entire and important domains may also be discarded or be reduced to near complete degeneracy in the generation of a single pattern common to all.

We will concern ourselves here only with the two most common representations of common sequence patterns. These are the regular expressions and the profiles. While strict Boolean matches to the regular expression may be used, it is more common to use some form of dynamic programming[11] to search for optimal matches to either of these pattern representations.

2.3 Hidden Markov Model

Considerable research has taken place in the area of speech pattern recognition using Hidden Markov or "discrete-state" models. While there have been limited applications to protein sequence analysis, the problems are conceptually very similar. Speech is a linear sequence of elements. Its pattern interpretations are highly complex and very context dependent, not different from the protein sequence's encoding of structure and function. Such models can be viewed as signal generators, or amino acid sequence generators in our case. The models are often represented as graphs as in Figure 2. Each of the N nodes represents a modeled state. The connecting arcs represent the allowed transitions between the states and its probability. Each model state includes a subset of "hidden" output states. Each output state defines a distribution over some alphabet of output symbols, or in our case the amino acids. Hidden Markov Model represented patterns are thus composed of an N by N state-to-state transition matrix and a set of output state distribution functions.

While Hidden Markov or Discrete State models (DSM), like Markov chains, are often thought of as symbol sequence generators, they are, however, employable in the "reverse" manner, that is, given an observed sequence the model can return the probability that it would have generated such a string[17]. In the protein structure case, output states can represent a pattern of sequence position with the distribution of amino acids expected at that position, similar to the columns in the weight matrix or profile. White et al.[6] have developed such a set of Discrete State models for a number of general protein domain folds and have shown the limiting case equivalence to the environmental profiles of Bowie et al.[1] The output states may be more complex than just a sequence of pattern positions and their associated environments. They can represent a hierarchy of super secondary structure, secondary structure, hydrophobic patterns and discrete conserved functional amino acids. In such a hierarchical pattern, each model state can itself be an additional discrete state model. Each state may in fact be a discrete state model itself containing many other states (see Figure 2).

2.4 Combined Patterns

In those cases where a functional sequence pattern is thought to contain primarily only those sequence elements directly encoding the function rather than indirectly encoding the structure common to that functional family, the patterns are often too short or of insufficient specificity to be diagnostic. A classic example is the pattern of residues associated with the catalytic triad of the Serine proteases. Thus since all of the known Serine proteases are found to be all beta (crudely wound double) sheets, a pattern composed of the highly sensitive, but non specific catalytic triad "embedded" in a general all beta structural pattern might have greatly improved specificity without reduced sensitivity. This is what we have tested in two cases. Firstly, by combining a discrete state model of an appropriate all beta fold with the Serine catalytic triad regular expression; and, secondly, by investigating the sensitivity of cross threading of Leghemoglobin into a secondary core model constructed from Myoglobin[9], when the pattern of two conserved Histidines is included.

2.5 Pattern Evaluation

At a minimum, two quantities should be reported for any pattern: its sensitivity and specificity (reflecting type I and type II errors).

Sensitivity = **TP**(True Positives)/{**TP** + **FN**(False Negatives)}

and

Specificity (selectivity) = **TN**(True Negatives)/{**TN** + **FP**(False Positives)}

Here TP is the number of defining (or positive) set sequence matches and FP is the number of control (or negative) set matches. The sensitivity of a pattern (e.g., diagnostic of zinc fingers) reflects the likelihood that all occurrences (e.g., all "real" zinc fingers) have been and thus will be found. It reflects how sensitive the pattern is as a detector (e.g., of zinc fingers). The specificity is a measure of the probability of correctly identifying the negative instances. The diagnostic ability of a pattern can be conveniently summarized in a normal two-by-two truth or contingency table, given a positive and negative control set.

Both these quantities should be reported for any patterns when possible. Patterns can be useful even if they lack near 100% sensitivity and specificity. One approach is to use an initial filter or sequential pass approach for patterns designed to identify the same function, but with different specificities. If two patterns have nearly 100% sensitivity, but only 70% specificity, their sequential application could have a greatly improved specificity if their false positives are uncorrelated. This is the expected case for active site patterns in combination with general structural fold patterns. For such uncorrelated false positives, the combined specificity, sp(a,b), will behave on a sufficiently large data set as:

$$1.0 - [1 - sp(a)][1 - sp(b)]$$

Such sequential application of patterns (of different representations) using different match algorithms has been used by Presnell et al.[18] to improve the prediction of alpha-helical structures, for example.

The computer science and statistical pattern recognition literature is replete with discussions of estimating class membership using the Bayesian approach [19]. Thus it may be useful to identify the relationship between these discussions and sensitivity and specificity. Let p_i be a pattern and f_j the protein structure or functional class. One can then define the relationship between the conditional probabilities from the definition of joint probability. This is known as Bayes' relation:

$$\mathbf{P(f_j \mid p_i) = \frac{P(p_j \mid f_j) * P(f_j)}{P(p_i)}}$$

Here $P(f_j)$ and $P(p_i)$ are the probabilities of a protein having the function (f_j) and matching the pattern (P_i) respectively. The conditional probability, $P(f_j|p_i)$, of having the function (j), given that you have matched the pattern (i), is referred to as the "posteriori" probability. The conditional probability, $P(p_i|f_j)$, of having the pattern given the function, is referred to as the prior probability. If these probabilities can be taken from the observed frequencies in the two-by-two truth table, then sensitivity is $P(f_j|p_i)$ and specificity is $P(\wedge f_j|\wedge p_i)$. Here $\wedge f_j$ and $\wedge p_i$ indicates negation, not having the function and not matching the pattern. The Bayes' decision rule for posteriori (or joint) probabilities assumes that one must classify all cases, even if there are only two hypotheses (having or not having the function); that is there is no indeterminate classification. In our case, the decision rule is that a protein with pattern (i) is assumed to have the function or structure (j) if and only if:

$$\mathbf{g_j(p_i) > g_k(p_i)}$$

for all k not equal to j where $\mathbf{g_j(p_i)}$ is a monotonic function of $P(f_j|p_i)$

This famous decision rule is equivalent to requiring that:

$$\mathbf{TP/(TP+FP) > FP/(TP+FP)}$$

Note that one can use this rule to set a value for any pattern match score cutoff or threshold used to define a match, as one must in profile analyses. This rule is insensitive to the number of true negatives, TN, which is a strong function of database size. However the number of false positives is also a function of database size. The functions $g_j(p_i)$, used in Bayes' decision rule can include "costs" or penalties associated with the different decision outcomes, but in protein pattern analysis, no standard has been proposed for such penalties. Many of the above considerations are reviewed in two recent book chapters [20,21].

3. RESULTS

Our results are summarized in Figures 4 and 5. Figure 4 shows the result of investigating the inclusion of the conserved Histidines in the threading of Leghemoglobin into a myoglobin derived model. Figure 5 shows the result of the inclusion of the catalytic serine protease triad in a DSM.

Both figures show, firstly, that the predictive accuracy of which amino acids are to be found where is clearly improved by the inclusion of some minimal functional pattern elements in either the threading or discrete state structural models. Secondly the figures show that the specificity of the minimal pattern of functional sequence elements is greatly improved by the included context of the generally structural class.

4. CONCLUSIONS

As expected, increasing the specificity of either a structure pattern or model, or a simple sequence functional pattern by combining the two, greatly increases the overall specificity in any large database search. In particular, this means a reduction in false positives with little or no loss in true positives. Perhaps more importantly, in detailed studies of prototypic cases, the predictive power of detailed structural placement of key residues is improved. Thus not only can one expect to accurately identify new Serine proteases, but to accurately place the catalytic residues into their proper three-dimensional context along with majority of other residues. This is the case not only in threading rather restricted assumed homologous models, but in the case of general structural folding class models such as those used in the DSM hidden Markov approach.

Major limitations are still associated with our ability to construct patterns and/or models to represent key structural features sufficiently to match all very distant homologues, and still be able to contribute to a combined pattern's specificity. Our recent experiences and those of others[9,23] on current threading model environmental descriptors or scoring potential functions, in combination with the more successful sequence pattern alphabets[11,24,25] strongly suggests a search for a better structure alphabet.

Clearly secondary structure, the degree of solvent exposure, and the various types of pairwise "neighborness" proposed, reflect important structural characteristics expected to be common between distant homologues. However, either they are not very robust over the observed range of homologue variation, or we have not yet been able to fully exploit them in the design of sequence to structure alignment scoring schemes. We have therefore begun to investigate a new set of structural environment descriptors based on two "natural" protein properties. The first is the tetrahedral geometry of the beta carbons, reflected in the various libraries of side chain gamma carbon rotamers. The second is the normal close packing of internal side chains. The latter results in there being few internal empty spaces or even included waters in most proteins. This is true even when different pairs of amino acid side chains fill equivalent internal volumes in different close homologues.

Thus we have defined a vector of "visible volume" about each structural core or modeled beta carbon. The three components of this vector are defined as that volume visible (not including any that is occluded by other beta carbons or any backbone atoms) from the beta carbon formed on each face by a tetrahedron whose faces are normal to the preferred beta-gamma carbon rotamer vectors. These can be simply defined based on the

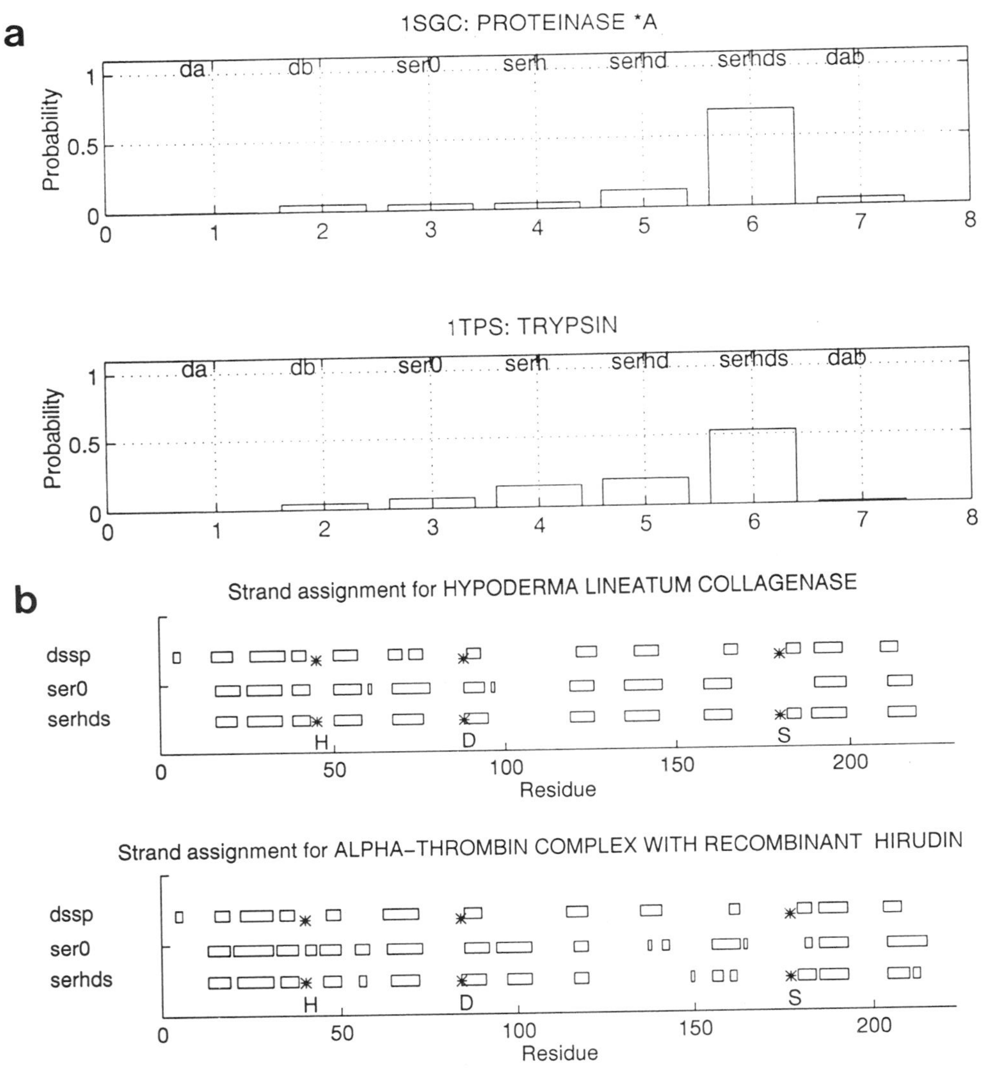

Figure 4. (a) Relative probabilities of seven different Discrete State Models having generated two different Serine protease amino acid sequences. The models [10] used were: "da", a diffuse or generalized all helical model; "dab", a generalized alpha-beta model; "db", a generalized all beta model; "ser0", an all beta model compatible with the number and length distribution of beta strands and loop observed in all known Serine protease determined structures; "serh", same as "ser0", but requiring a Histidine between strands three and four; "serhd", same as "serh" with the additional requirement of an Aspartic acid between strand five and six ; and finally "serhds" being the same as "serhd" with the additional requirement of a Serine between strands eight and nine or nine and ten. In the case of proteinase *A (Brookhaven ID, 1sgc), the preferred model is "db" unless either sequence specific "serhd" or "serhds" is included. While in the case of trypsin (Brookhaven ID, 1tps), the addition of each increment of sequence specificity improves the relative probability of that model and thus the correct folding class. **(b)** Two examples of increased secondary structure predictive accuracy with increasing sequence specificity combined with an all beta Discrete State Model. The first line for each protein shows in rectangles the DSSP [22] secondary beta strand assignments with asterisks indicating the relative positions of the catalytic triad H, D, and S. The second line shows the DSM smoothing algorithm's [10] secondary predictions from the "ser0" model. The third line shows those from the combined "serhds" model. In the case of Hypoderma lineatum collagenase (1hyl), two minor false positive strands are eliminated and the twelve strand is added with the addition of the HDS amino acid pattern requirement. In the case of Trypsin (1tps) there is an elimination of a short predicted strand in the neighborhood of the catalytic Histidine and an extension of the predicted strand just C-terminal of the catalytic Serine. However there is a loss of the partially predicted ninth strand.

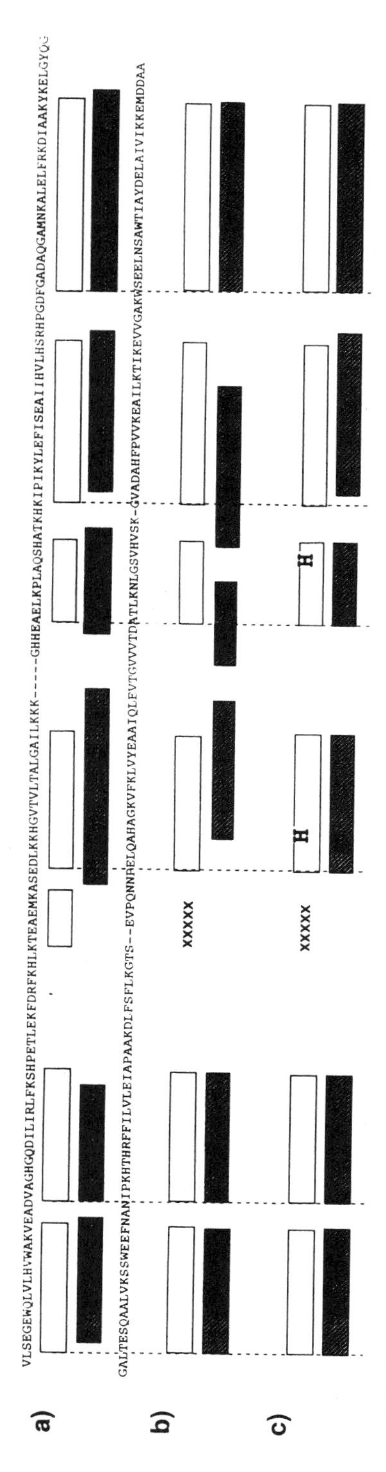

Figure 5. Leghemoglobin sequence threaded through the myoglobin structural model. (A) Reference alignment. (B) Short "D" helix removed, no active site constraints. (C) Short "D" helix removed, requiring model elements at positions E7 and F8 to be occupied by histidine, H, from the sequence.

geometry of the backbone coordinates or a model or real structure, independent of any particular set of observed side chains (including Glycine and Proline). The potential for such a local structural environmental descriptor is easily demonstrated. A simple sequential list of one or the sum of these visible volume components is capable of producing an accurate dynamic programming structural alignment between very distant homologues such as Leghemoglobin and Myoglobin without any direct inclusion of either the atomic coordinates or any amino acid information[26]. Note that such an alignment is very difficult [27] even when an optimal threading is found with a structural model that includes a couple of functionally essential residues as shown in Figure 4[9].

Current studies of the potential for characterizing the essential conserved features of any family of structures by sets of visible volumes have been quite suggestive. For example it has long been recognized that one of the key characteristics of any globular structure is the correlation between amino acid hydrophobicity and position exposure. As seen in Figure 6, one component of this visible volume clearly contains this information. In fact the Visible volume, which contains no direct specific side chain information, appears more sensitive in detecting the exposure/buried periodicity of an alpha helix than the side chain hydrophobicity and equal to that of full side chain exposure. Moreover the visible volume appears directly capable of recognizing whether the local backbone environment is extended or helical. This arises directly from the fact that the alpha beta vector has a very different angle relative to the helical and the beta sheet backbone-defined surfaces

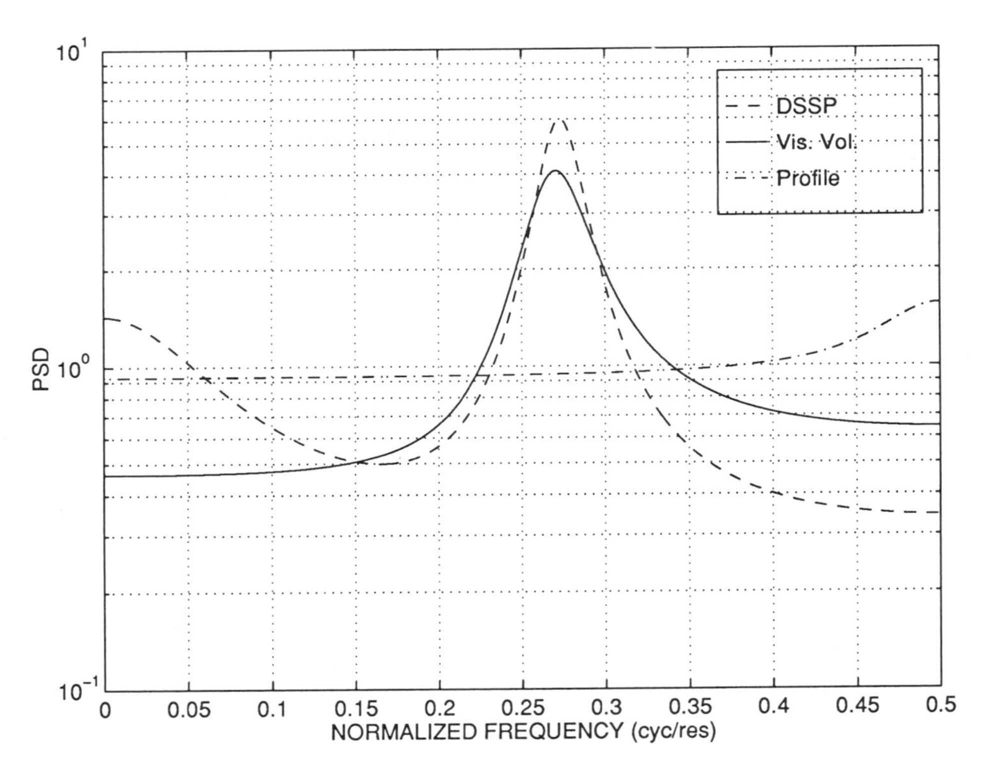

Figure 6. The relative sensitivity to detect helical periodicity in the globins. The three power spectra were generated on sperm whale myoglobin using respectively: the amino acid hydrophobicity [28], side chain exposure (using DSSP) and "visible volume" of the beta carbon tetrahedral face three.

(data not shown). Perhaps most importantly, this descriptor has the potential to constrain the allowed side chain packings and thus provide a better method of scoring potential pair-wise contacts in sequence to structure alignments.

In any case there will no doubt be many structural descriptors proposed that should allow new predictive structural patterns to be tested and combined with key functional or catalytic residue information. These in combination with the expanding pattern databases[24,25,29,30] will greatly increase our future ability to identify both a protein's function and its probable structure from sequence information alone. This is particularly important given that the anticipated near complete set of essential eukaryotic genes from yeast will soon be available.

REFERENCES

1. Bowie, F.U., Lüthy, R. and Eisenberg, D. (1991). A method to identify protein sequences that fold into a known three-dimensional structure. Science 253, 164–170.
2. Walker, J.E., Saraste, M. Runswick, M.J. and Gray, N.J. (1982). Distantly related sequences in the α- and β-subunits of ATP synthase, myosin, kinases and other ATO-requiring enzymes and a common nucleotide binding fold. EMBO J 1, 945–951.
3. Chomsky, Noam. (1957). *Syntactic structures*. Mouton: The Hague.
4. Bowie, F.U. and Eisenberg, D. (1993). Inverted protein structure prediction. Curr. Opin. Struct. Biol. 3, 437–444.
5. Stultz, Collin, White, James V. and Smith, Temple F. (1993). Structural analysis based on state-space modeling. Protein Science 2, 305–314.
6. White, James V., Muchnik, Ilya and Smith, Temple F. (1994). Modeling protein cores with Markov random fields. Mathematical Biosciences 124, 149–179.
7. Bryant, S.H. and Lawrence, C.E. (1993). An empirical energy function for threading protein sequence through the folding motif. Proteins: Struct. Funct. Genet. 16, 92–112.
8. Lathrop, Richard H. (1994). The protein threading problem with sequence amino acid interaction preferences is NP-complete. Protein Eng. 7, 1059–1068.
9. Lathrop, Richard H. and Smith, Temple F. (1996). Global optimum protein threading with gapped alignment and empirical pair score functions. Journal of Molecular Biology 255, 641–665.
10. White, James V., Muchnik, Ilya and Smith, Temple F. (1994). Modeling protein cores with Markov random fields. Mathematical Biosciences 124, 149–179.
11. Smith, Randall F. and Smith, Temple F. (1990). Automatic generation of primary sequence patterns from sets of related protein sequences. Proceedings of the National Academy of Sciences USA, 87, 118–122.
12. Henikoff, S. and Henikoff, J.G. (1991). Automated assembly of protein blocks for database searching. Nucleic Acids Research 19(23), 6565–6572.
13. Claverie, J.M. (1995). Progress in large scale sequence analysis. In: H. Villar, ed. *Advances in Computational Biology*. London: JAI Press.
14. Gribskov, M. McLachlan, A.D., Eisenberg, D. (1987). Profile analysis: detection of distantly related proteins. Proc. Natl. Acad. Sci. USA 84 (13), 4355–4358.
15. Staden, R. (1990). Searching for patterns in protein and nucleic acid sequences. Meth. Enzymol. 183, 193–211.
16. Taylor, W.R. (1986). Identification of protein sequence homology by consensus template alignment. J. Mol. Biol. 188 (2), 233–258.
17. Kitagawa, Genshiro. (1987). Non-Gaussian state-space modeling of nonstationary time series. Journal of the American Statistical Association 82 (400), 1032–1040.
18. Presnell, S. R., B. I. Cohen and F. E. Cohen (1992). A segment-based approach to protein secondary structure prediction. Biochemistry 31: 983–993.
19. Pao, Y.-H. (1989). *Adaptive pattern recognition and neural networks*. Reading, MA, Addison-Wesley.
20. Smith, Temple F., Lathrop, Richard and Cohen, Fred E. (1996). The identification of protein functional patterns. To appear in "Integrative Approaches to Molecular Biology", J. Collado-Vides, B. Magasanik &T. Smith eds. MIT Press, Cambridge, MA.

21. Stultz, C.M., Nambudripad, R., Lathrop, R.H., and White, J.V. (1996). "Predicting Protein Structure with Probabilistic Models." In: *Protein Structural Biology in Bio-Medical Research* (N. Allewell and C. Woodward, editors), JAI Press, Greenwich (in press).
22. Kabsch, Wolfgang and Sander, Christian. (1983). Dictionary of protein secondary structure: pattern recognition of hydrogen-bonded and geometrical features. Biopolymers 22, 2577–2637.
23. Lemer, Christian, M.-R., Rooman, Marianne J. and Wodak, Shoshana J. (1995). Protein structure prediction by threading methods: evaluation of current techniques. Proteins: Structure, Function, and Genetics 23 (3), 337–355.
24. Adams, R. Mark, Das, Sudeshna and Smith, Temple F. (1996). Multiple Domain Protein Diagnostic Patterns. Submitted to Protein Science.
25. Henikoff, S. and Henikoff, J.G. (1993). Performance evaluation of amino acid substitution matrices. Proteins 17, 49–61.
26. Lo Conte, Loredana and Smith, Temple F. Visible Volume: A robust measure for protein structure characterization. Manuscript in preparation.
27. Lesk, A.M and Chothia, C. (1980). How different amino acid sequences determine similar protein structures: the structure and evolutionary dynamics of the globins. J. Mol. Biol. 136, 225–270.
28. Eisenberg, D. and Weiss, R.M., Terwilliger, T.C. (1984). The hydrophobic moment detects periodicity in protein hydrophobicity. Proc. Natl. Acad. Sci. USA, 81, 140–144.
29. Smith, Randall F. and Smith, Temple F. (1992). Pattern-induced multi-sequence alignment (PIMA) algorithm employing secondary structure-dependent gap penalties for use in comparative protein modelling. Protein Engineering, 5 (1), 35–41.
30. Bairoch, A. (1994). The ENZYME Database. Nuccleic Acids Res. 22, 3626–3627.

8

RECOGNIZING FUNCTIONAL DOMAINS IN BIOLOGICAL SEQUENCES

Gary D. Stormo

Molecular, Cellular, and Developmental Biology
University of Colorado
Boulder, CO

ABSTRACT

Weight matrices are standard methods of representing the information contained in a set of related sequences, such as promoters, splice sites and other common sites in nucleic acids. They have also been used to represent common domains in protein families, such as in the BLOCKS database. Besides being useful for elucidating the sequence constraints associated with the related sites, they can be used effectively to identify new members of the family. Most often the weight matrix is determined solely from the information in the sequences of the family members, perhaps using a log-odds approach to maximize the difference between the family members and a random model of non-member sequences. In the case of protein motifs, much emphasis has also been placed on finding appropriate weightings for the family members to avoid problems due to the biased representations in the databases. We have tried an alternative approach, which is to utilize the actual sequences of both the family members and non-family members to find the weight matrix that gives optimum discrimination between the classes. Preliminary results for a few protein families listed in the BLOCKS database, and using the entire SwissProt database of proteins as the "background" against which we are trying to find optimal classifiers for the family members, show this approach can provide improved specificity.

1. INTRODUCTION TO WEIGHT MATRICES

Weight matrices are general representations of sequence specificity exhibited by common sites. They were originally developed to represent regulatory sites on DNA (or RNA), such as ribosome binding sites,[1] promoters,[2,3,4] splice sites[3] and others. They have also been extended to represent protein motifs,[5] and are used in the BLOCKS database of conserved regions in protein families.[6] Those approaches all used sites of constant size; that is the alignment of the sites did not require the addition of any gaps. Gribskov et al.[7] showed

Theoretical and Computational Methods in Genome Research, edited by Suhai
Plenum Press, New York, 1997

how the same basic idea can be extended to alignments that include gaps, in what are called "profiles." This extension requires dynamic programming to find the optimal alignments between the matrix and the sequence. More recently Hidden Markov Models (HMMs) have been used to define the patterns in conserved sequences.[8,9] Although classical weight matrices, profiles and HMMs differ in many details, they can also be viewed as variations of a common theme,[10] as will be described briefly below.

Before proceeding to describe current work, it is useful to provide a brief general overview of how weight matrices are used and some of their features. For more complete reviews see Ref. 1112. The concept is quite simple, although lots of approaches have been used to actually determine the appropriate values to be used in the matrix. The idea derives from thinking that each position within a site contributes something to its activity. Presumably the consensus base at a position would contribute the most to the activity, but any single change away from the consensus is likely to only diminish the activity somewhat, and not eliminate it altogether. This will, of course, depend on the type of site; some protein binding sites might be highly specific, like restriction sites, so that any changes from the consensus base will effectively reduce the activity beyond detection. But for sites like promoters this doesn't seem to be the case. Figure 1 shows how any sequence is scored using a weight matrix. The weight matrix has values for all possible bases at each position within the potential site, whereas the sequence has particular bases at each position. The weight matrix values for those bases are summed to get the total score for the particular site. In the case of a protein that simply binds to DNA, the values of the weight matrix can be thought of as partial binding energies contributed by each base interacting with the protein, and the sum of those is the total binding energy for the protein to the site. (Convention has it that lower energy corresponds to tighter binding, whereas higher scores are typically used for better sites, so that the score should really be considered a negative energy. Keeping track of the appropriate signs, especially when different papers may use different conventions, can be confusing.) In Figure 1 the matrix is shown at different positions of the sequence. The total score at each position depends on the sequence there; it is easy to imagine the protein sliding along the DNA and having different affinities for the different sequences that are encountered.

Consensus sequences can be considered a special case of weight matrices. Figure 2a shows a matrix for the consensus sequence of the -10 region of *E. coli* promoters, which is TATAAT. That sequence would get a score of 6, whereas any other sequence's score would be diminished by the number of mismatches to the consensus. Figure 2b modifies the relative importance of different positions. The consensus sequence gives a score of 60, but substituting an A for T at position 3 only reduces the score to 58. If we were to set a threshold of 51 for a site to be considered a promoter, the second and sixth positions would be forced to be the consensus, whereas almost any single change at the other positions would be permitted, and even a double change to TAAACT. Figures 2c and 2d show further refinements to the weight matrix which incorporate additional information about the specificity of -10 promoter regions.

An important feature of weight matrices is that they simply add together the elements for the corresponding sequence, which assumes that those elements contribute independently to the activity of the site. That need not be true, and in fact it is known to not be true for some sites. However, the basic weight matrix idea can be easily extended to allow some positions to contribute coordinately, rather than independently. For example, if adjacent dinucleotides do not contribute independently to the activity you can use a matrix with 16 rows, one for each dinucleotide, instead of 4. Nor do the non-independent positions have

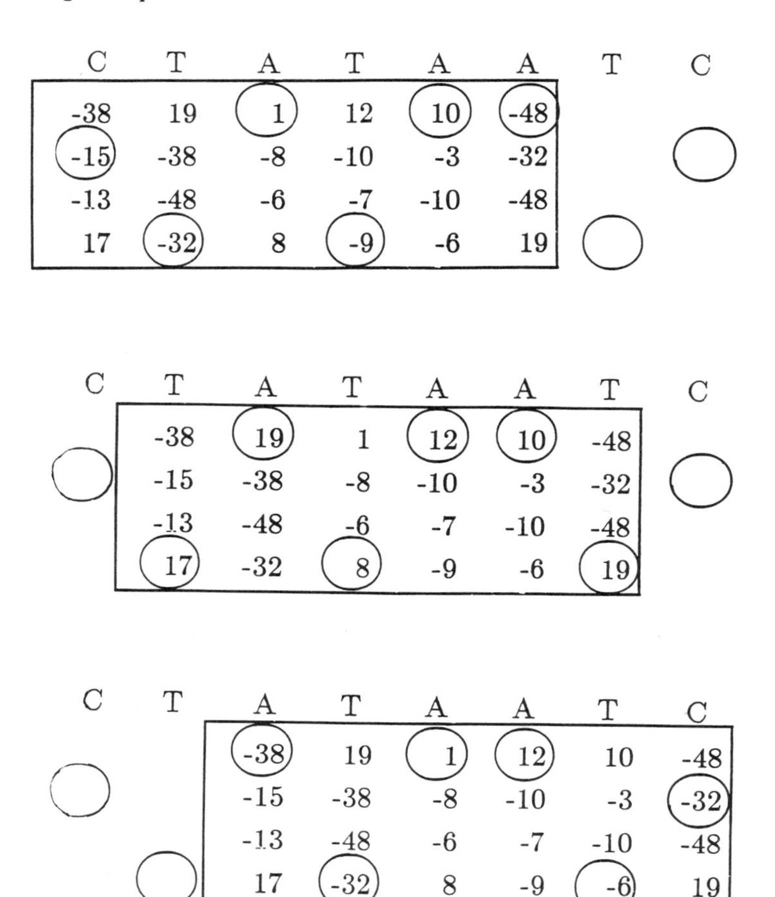

Figure 1. Using a matrix to score positions in a sequence. At each location of the matrix the score is the sum of the matrix elements corresponding to the sequence at that position (the circled elements).

to be adjacent. It is easy to imagine, and construct, matrices where different positions are taken into account together rather than separately. Such matrices are essential to representing sites on RNA sequences where important features are the base-paired secondary structures. Figure 3 makes a further point about the ability of weight matrices to account for non-additive contributions. On the left are shown the actual binding energies to all 4-long sequences for some hypothetical DNA binding protein. (Note that this figure displays energies, so the lower the value the higher the affinity; the highest affinity site is AGCT.) Because this hypothetical protein binds to DNA with additive interactions at each position, the simple weight matrix on the right contains exactly the same information as the long list (partially shown) on the left. That is, the correct binding energy for any 4-long sequence can be determined from the weight matrix. The hypothetical protein could have been designed to interact non-additively at adjacent positions, so that in order to accurately account for the binding data the weight matrix would have to be over dinucleotides. Or it might require using trinucleotides. However, even in the worst case, where there are no additive interactions with the DNA, a weight matrix representation still works because the list on

A	0	1	0	1	1	0
C	0	0	0	0	0	0
G	0	0	0	0	0	0
T	1	0	1	0	0	1

A	0	10	8	10	10	0
C	1	0	1	1	3	0
G	1	0	1	1	1	0
T	10	0	10	1	1	10

A	2	95	26	59	51	1
C	9	2	14	13	20	3
G	10	1	16	15	13	0
T	79	3	44	13	17	96

A	-38	19	1	12	10	-48
C	-15	-38	-8	-10	-3	-32
G	-13	-48	-6	-7	-10	-48
T	17	-32	8	-9	-6	19

Figure 2. Different matrices that represent the -10 region of *E. coli* promoters. The top matrix is a simple consensus sequence, whereas each following matrix adds additional information about the patterns observed at -10 regions.

the left is itself a weight matrix, having only one column and 256 rows. The main point is that a weight matrix of some sort must provide an accurate representation of the specificity of the protein, and the challenge is to find the appropriate matrix for the protein of interest. Of course it is much easier to obtain a smaller matrix, so we usually assume the additive units are mononucleotides and move to more complicated matrices only when forced to by the data. For many DNA binding proteins it appears that mononucleotide matrices provide good approximations to the actual binding activities.

If quantitative data are available for the binding to some of the possible sites (not necessarily the complete list shown in Figure 3), then regression methods can be used to find the weight matrix with the best fit to the data.[13,14,15] Furthermore, good statistical tests are available to ascertain whether or not the matrix provides an adequate fit to the data. If not, higher order matrices can be used, at the cost of requiring more quantitative data in order to fit the increased number of parameters. If quantitative data are not known, but the sequences of many example sites are, weight matrices can be made with statistical methods. Most current approaches are closely related to the method first described by Staden.[3] Schneider et al.[16] showed the relationship of this approach to information theory,

AAAA	2.63
AAAC	2.63
AAAG	2.63
AAAT	0.46
AACA	1.79
AACC	1.79
AACG	1.79
AACT	-0.38
.	
.	
.	
AGCG	-0.25
AGCT	-2.42
AGGA	0.59
.	
.	
.	
CTTT	-0.37
GAAA	4.01
.	
.	
.	
TTTG	2.63
TTTT	0.46

A	-0.55	+1.38	+0.42	+1.38
C	0	+0.55	-0.42	+1.38
G	+0.83	-0.66	+0.42	+1.38
T	+0.83	+0.42	0	-0.79

Figure 3. The left column represents the binding energy to all 256 4-long sequences (only some of them are shown). The matrix on the right provides the same information because every value in the list on the left can be calculated using this matrix.

and Berg and von Hippel[17] showed that, given some simplifying assumptions, the statistical method gives a matrix that is correlated to the binding energies. This approach is shown in Figure 4, again using −10 promoter regions as the example. Figure 4a shows the number of each base observed at each position in a collection of 242 aligned −10 region sequences, and Figure 4b converts those numbers to fractions. Figure 4c is the key step, where the fractions of Figure 4b are divided by the fraction expected by chance, and then the logarithm taken. For *E. coli*, the genome is about 25% of each base, so the fractions would each be divided by that same amount. For organisms with biased compositions the fractions need to be divided by the appropriate probability, which is the a priori expectation for observing that base. After taking the logarithm (this figure uses the base 2) the matrix elements are log-likelihood ratios of observing the particular base at the particular position in sequences that are sites (in this case −10 promoter sites) versus anywhere in the genome. Positive values are bases that occur more often in the sites than expected by chance, and negative values are for bases that are discriminated against in sites. When this matrix is used to search sequences, the score obtained is the total log-likelihood ratio of observing that sequence based on the −10 position frequencies versus the random base frequencies of the genome. Figure 4d shows the "information content" at each position in the sites. This is just the dot-product of the column vectors from the matrices in Figures 4b and 4c. This is also known as the relative entropy and the Kullback-Liebler distance. The formula for it is

$$I_{seq}(i) = \sum_{b=A}^{T} f(b,i) \log_2 \frac{f(b,i)}{p(b)} \tag{1}$$

where $f(b,i)$ is the frequency that each base b is observed at each position i, and $p(b)$ is the expected, or "genomic," frequency of base b. Taking the logarithm to base 2 gives information in *bits*.

2. ALIGNMENTS REQUIRING GAPS

As mentioned above, weight matrices can be extended to include alignments with gaps, as in profiles.[7] In this case the matrix contains extra rows to indicate the score associated with a gap in the alignment. Typically there will be scores for both the introduction of a gap and a different score for the extension of a gap, as it is well accepted that those two events have different probabilities. In HMM methods, there will be different scores for the introduction of insertions than for deletions,[8,9] but it has been shown how both HMMs and profiles can be described in the same syntactical framework.[10] The important point for this paper is that once a site in a sequence has been aligned with the profile, the score is calculated the same as for classical weight matrices, as the sum of the elements corresponding to the aligned sequence (including gaps) of the site. One advantage of the HMM methods is the training procedure that will align the sequences and generate the residue probabilities simultaneously. The training method is related to Expectation-Maximization (EM) methods used previously for ungapped alignments[18] and alignments with restricted gaps.[19] Other methods are also capable of finding gapped alignments and determining the matrix representations of those alignments simultaneously.[20]

A.

A	9	214	63	142	118	8
C	22	7	26	31	52	13
G	18	2	29	38	29	5
T	193	19	124	31	43	216

B.

A	0.04	0.88	0.26	0.59	0.49	0.03
C	0.09	0.03	0.11	0.13	0.22	0.05
G	0.07	0.01	0.12	0.16	0.12	0.02
T	0.80	0.08	0.51	0.13	0.18	0.89

C.

A	-2.76	1.82	0.06	1.23	0.96	-2.92
C	-1.46	-3.11	-1.22	-1.00	-0.22	-2.21
G	-1.76	-5.00	-1.06	-0.67	-1.06	-3.58
T	1.67	-1.66	1.04	-1.00	-0.49	1.84

D.

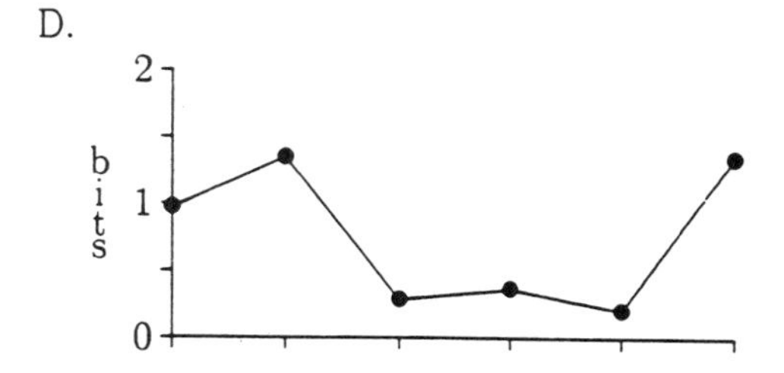

Figure 4. Determining the specificity matrix and information content of the −10 regions of promoters. The top matrix shows the occurrence of each base at each position in a collection of 242 promoters. The next matrix converts those numbers into the fraction of each base at each position. The third matrix is the log-likelihood ratio of the observed frequencies divided by the expected, using 0.25 as the expected occurrence for any base. (The logarithm is taken to base 2.) The bottom panel shows the information content at each position, using the formula from the text.

3. VECTOR INTERPRETATIONS

It is often useful to think of the weight matrices and the sequences in terms of a "sequence space." The weight matrix itself is a point, or vector to that point, in the space. And, of course, each sequence is also a vector in that space. However, we need to think of the space as defined by the weight matrix, have dimensions not only for all possible bases at each position, but also the possibilities of gaps, gap extensions and non-independent positions. In order to determine the vector for a particular sequence (or site) in this space, it is first necessary to align it with the matrix. The score for any sequence, given a matrix, is then just the dot-product of the vector for the matrix with the vector for the sequence. This is easily seen for simple ungapped matrices, as in Figure 1. The vector for the sequence just has a 1 for each base that occurs at each position (corresponding the circles below the sequence) and a 0 for all of the bases that do not occur. This is commonly used in statistical methods and is known as "dummy encoding," where the variables that exist for a particular example are given 1s, and the variables that do not exist are given 0s. Aligned sequences with gaps are still encoded the same way, but now the 1s may refer to positions of gaps, or gap extensions, instead of just to residues at positions. Then the score of any aligned sequence (called $Sc(S_i)$ for the sequence S_i) with the matrix is just the sum of the variables that occur, and is simply the dot-product of the sequence with the weight matrix:

$$Sc(S_i) = \vec{W} \cdot \vec{S_i} = |W||S_i| \cos \alpha_i \tag{2}$$

where $\vec{W}$ is the weight matrix. α_i is the angle between the weight matrix vector and the sequence vector. Since the weight matrix is fixed and every aligned sequence has the same length (i.e., a single 1 for each position in the alignment), the score is just proportional to the cosine of the angle between the weight matrix and sequence. If the weight matrix is a good representation of the sites it represents, then those sequences should all form a tight cluster around it, with small α, and other unrelated sequences should have larger angles.

When the elements of the weight matrix are logarithms of the probabilities of the residue at each position (including gaps as types of residues), as is commonly done in HMM methods, then the score is just the ln(Prob) for the whole site. That is, the probability of a particular site being generated by a model M (such as an HMM) is

$$\ln P(S_i|M) = \vec{W} \cdot \vec{S_i} \tag{3}$$

In more classical weight matrix methods the dot-product represents a log-likelihood ratio rather than a straight log-probability, but the same considerations hold. Then the problem of finding the the weight matrix that best represents a set of sites is that of finding the matrix that maximizes the product of the probabilities of all the sites, or equivalently that maximizes the sum of cosines of the angles between the weight matrix and the set of sites.

4. DISCRIMINANT MATRICES

One might want to use a weight matrix in a somewhat different manner. Rather than just take a collection of known sites and try to represent them effectively, or even try to make a matrix with good quantitative predictions of site activity, one might have the goal of finding a matrix that is optimally effective at discriminating true sites from other sequences, non-sites. For example, if one has a collection of sites that are known to be functional and another set of sequences that are known to not be functional sites, one could try to find a matrix that distinguished the two sets by giving every site a score higher than any non-site. In fact, this was the first application of weight matrices, which used a simple neural network called a "perceptron" to determine the matrix elements.[1]

When the goal is to identify the matrix that provides the best discrimination between the sites and other sequences, the objective to be maximized is not simply the probabilities of the sites themselves, but rather the relative probabilities of the sites compared to the other sequences. That is the purpose of the a priori base probabilities in the determination of the weight matrix of Figure 4c. This maximizes the Kullback-Liebler distance between the two distributions, those of the site sequences and those of the non-site sequences. The information content is just the average of the dot-products of all the sites with that maximally discriminant matrix.

Maximum information content has been used in several approaches at pattern, or motif, discovery. This is a problem where the exact positions of the sites are not known, nor is the weight matrix that represents them, and both must be found simultaneously. We developed a greedy algorithm to do that,[21,22] and have extended that method to work on alignments with gaps.[20] The same objective of maximizing information content, with the addition of small size corrections factors, was used to identify common motifs in proteins by a Gibbs' Sampling method.[23] Eddy et al.[24] used a related approach in a variation of the usual HMM method. Using a model of "background" as being random sequences of a given composition, they found the HMM to maximize the distinction between the proteins in the family being modeled and the background. Instead of maximizing the probability of the family alone, as in standard HMM methods, this approach maximizes the likelihood ratio between the family and the background.

We developed an alternative approach that allows for maximizing the discrimination between the site sequences and some specific set of non-site sequences.[25,26] As described previously, and shown in Figure 1, a given matrix will determine a score for any particular sequence. Equations 2 and 3 show that the is equal to the $\ln(Prob)$ of the sequence being generated by the model (or it may be a log-likelihood ratio if that is how the matrix is determined). Our goal was to find the matrix that maximized the product of the likelihood ratios for the set of sites, relative to those for all of the non-sites. If the sites are binding sites for some protein, this approach is directly related to finding the sites with maximum specific binding energy for the protein.[25] The method works by calculating the "partition function," or the total binding probability, over all of the possible sites in the collection (the entire genome for a DNA binding protein). Then the objective is to find the weight matrix that maximizes

$$U = \frac{1}{N} \sum_{i=1}^{N} \vec{W} \cdot \vec{S_i} - \ln Y \tag{4}$$

where the sum is over all N sites and Y is the average likelihood (i.e., $e^{\text{dot-product}}$) over the entire collection of sequences, both sites and non-sites. When the non-sites are modeled as random sequences with a given composition, the best weight matrix is exactly that as determined in Figure 4c, the one with maximum information content. However, if the no-sites sequences are not random this method can find a weight matrix that provides optimal discrimination between the two sets. The form of the matrix can be modified to allow complementary interactions so that it can identify RNA binding sites composed of both sequence and structure components.[25] It was also shown to work at finding common motifs in protein sequences, using as background random sequences of typical protein composition, as in the Gibbs' Sampling approach.[23,25]

More recently we have wondered whether this approach might provide improved searching methods for identifying new members of protein families. Current approaches use a variety of methods, from comparing a new sequence to individual sequences in the database, such as with BLAST or FASTA, to comparing the new sequence to family representations such as profiles, HMMs, BLOCKS or Prosite patterns. As an initial test we have compared the ability of our optimized matrices to distinguish family members to that of the BLOCKS database.[6] This database derives from the protein families described in Prosite, but represents the most conserved, ungapped regions as weight matrices that can be used to score new sequences and determine whether or not they are members of the family. Previously the weight matrices have been determined solely from the members of the family, not taking into account the sequences of non-family member proteins. Much emphasis has been placed on how to weight the sequences of the family members so as to compensate for the bias in the representation in the current database and to compensate for limited sample sizes (see Ref. 2728 for a recent example).

In our tests we have used BLOCKS version 5.0, which come from SwissProt version 22. We used the entire set of proteins in SwissProt 22 as the complete set of sequences. We used a gradient descent method to maximize our objective, U. We started with the weight matrices from the BLOCKS database to decrease the time to convergence and to see directly if we could improve the discrimination. At this time only a few of the BLOCKS have been tested, and only the one for homeodomains thoroughly. The increase in U over the BLOCKS matrix is fairly small, starting at 9.9 and increasing to 10.7 (these are the values of U in natural logarithm units). This represents a total increase of only about 8% in the average score for the family members compared to the background value of Y. This indicates that the current methods of generating the weight matrices from the family members only works quite well at discriminating them from other sequences. This is not too surprising since these methods have been shown to work well at identifying new family members. However, more interesting is the question of how the distribution of scores changes and whether the number of false positive and false negative sequences decreases. Both the ProSite Homeodomain listing and the BLOCKS Homeodomain listing indicate some incorrect predictions (although some of these may be the result of incomplete or inaccurate knowledge of the proteins' functions).

We used the optimized matrix to score all of the sequences in SwissProt 29. This newer database contains many sequences, both homeodomain proteins and others, that were not included in our training set, from version 22. The scores fall into two distinct peaks, one from about −4.5 to +2.9 representing the non-homeodomain proteins, and the other from +6.5 to +11.5, for the family members (Figure 5). No known homeodomain proteins fall into the lower peak. A few proteins that are not known to be homeodomain proteins fall into the upper peak, and our data would indicate suggest those really are members of the family. The separation into two peaks, with the same proteins in both peaks, is also obtained with the weights from the BLOCKS database, so our method does not change any false positives or false negatives. Indeed, there may not be any in either set, although some proteins may not be properly labeled. Again, this is not too surprising since this class of proteins is fairly easy to recognize by its sequence features. Further work will involve similar analyses on more difficult families. One encouraging result is that the distribution that results from our matrix is somewhat different from that using the BLOCKS matrix (data not shown).

```
-4|5555
-4|300
-3|987777776666666666665555555
-3|4444444443333333333322222222222221111111111111111111111111000000000000000000+
-2|9999999999999999999999999999999999999999999999999999999888888888888888888888+
-2|4444444444444444444444444444444444444444444444444444444444444444444444444444+
-1|9999999999999999999999999999999999999999999999999999999999999999999999999999+
-1|4444444444444444444444444444444444444444444444444444444444444444444444444444+
-0|9999999999999999999999999999999999999999999999999999999999999999999999999999+
-0|4444444444444444444444444444444444444444444444444444444444444444444444444444+
 0|0000000000000000000000000000000000000000000000000000000000000000000000000000+
 0|5555555555555555555555555555555555555555555555555555555555555555555555555555+
 1|000000000000000000111111111111111222222222222222222222333333333333444444444
 1|5555666799
 2|
 2|99
 3|
 3|
 4|
 4|
 5|
 5|
 6|
 6|58
 7|0113
 7|67778888999
 8|0224
 8|555788889999
 9|00000111111222333344444
 9|556666666677777778899999
10|001111111222222233333333333333333444444444444
10|55555555555555555666666666666666666667777777777777777777778888888888888888888+
11|000000000111111222
11|5
```

Figure 5. The distribution of scores for all sequences in SwissProt 29 using the optimized matrix for the Homeodomain BLOCK. The peaks of both distributions extend beyond the top of the figure and are not shown.

The main difference is that the peak corresponding to the family members is somewhat narrower and higher. This is consistent with idea that our method has found a weight matrix whose vector is more in the center of the cluster of the sequence vectors for the family members. That is, the average angle between our weight vector and all of the family vectors is smaller. At the same time the separation between the two peaks is somewhat larger, again indicating our weight matrix is providing slightly better discrimination between the two sets of sequences. Only tests of more difficult families will indicate how much is to be gained by this approach.

ACKNOWLEDGMENTS

John Heumann wrote most of the programs used in the recent analyses, and has contributed significantly to all the recent work. Alan Lapedes has also contributed significantly to this research, and Steve and Jorga Henikoff have helped with the comparison to the BLOCKS methods by facilitating access to the data and through very useful discussions. This work has been supported by grants from NIH (HG00249) and DOE (ER61606).

REFERENCES

1. Stormo, G. D., Schneider, T. D., Gold, L. and Ehrenfeucht, A. (1982) *Nucl. Acids Res.* 10: 2997-3011.
2. Harr, R., Haggstrom, M. and Gustafsson, P. (1983) *Nucl. Acids Res.* 11: 2943-57.
3. Staden, R. (1984) *Nucl. Acids Res.* 12: 505-19.
4. Mulligan, M. E., Hawley, D. K., Entriken, R. and McClure-W-R. (1984) *Nucl. Acids Res.* 12: 789-800.
5. Dodd, I. B. and Egan, J. B. (1987) *J. Mol. Biol.* 194: 557-64.
6. Henikoff, S. and Henikoff, J. G. (1994) *Genomics* 19:97-107.
7. Gribskov, M., McLachlan, A. D. and Eisenberg, D. (1987) *Proc. Natl. Acad. Sci. USA* 84:4355-4358.
8. Krogh, A., Brown, M., Mian, I. S., Sjolander, K. and Haussler, D. (1994) *J. Mol. Biol.* 235:1501-1531.
9. Baldi, P., Chauvin, Y., Hunkapiller, T. and McClure, M. A. (1994) *Proc. Natl. Acad. Sci. USA* 91:1059-63.
10. Bucher, P. and Bairoch, A. In: *Proceedings of the Second International Conference on Intelligent Systems in Molecular Biology*, pp. 53-61.
11. Stormo, G. D. (1988) *Ann. Rev. of Biophys. and Biophys. Chem.* 17:241-263.
12. Stormo, G. D. (1990) In: *Molecular Evolution: Computer Analysis of Protein and Nucleic Acid Sequences, Methods in Enzymology, Vol. 183,* (R. F. Doolittle, ed.), Academic Press, pp. 211-221.
13. Stormo, G. D., Schneider, T. D. and Gold, L. (1986) *Nucl. Acids Res.* 14:6661-6679.
14. Stormo, G. D. (1991) In: *Protein-DNA Interactions, Methods in Enzymology, Vol. 208* (R. T. Sauer, ed.), Academic Press, 458-468.
15. Barrick, D., Villanueba, K., Childs, J., Kalil, R., Schneider, T. D., Lawrence, C. E., Gold, L. and Stormo, G. D. (1994) *Nucl. Acids Res.* 22:1287-1295.
16. Schneider, T. D., Stormo, G. D., Gold, L. and Ehrenfeucht, A. (1986) *J. Mol. Biol.* 188:415-431.
17. Berg, O. G. and von Hippel, P. H. (1987) *J. Mol. Biol.* 193:723-750.
18. Lawrence, C. E. and Reilly, A. A. (1990) *Proteins* 7:41-51.
19. Cardon, L. R. and Stormo, G. D. (1992) *J. Mol. Biol.* 223:159-170.
20. Hertz, G. Z and Stormo, G. D. (1994) In: *Bioinformatics and Genome Research: Proceedings of the Third International Conference* (H. A. Lim and C. R. Cantor, eds.) World Scientific Publishing, Singapore. pp.199-214.
21. Stormo, G. D. and G. W. Hartzell III (1989) *Proc. Natl. Acad. Sci.* 86:1183-1187.
22. Hertz, G. Z., Hartzell, G. W. and Stormo, G. D. (1990) *Comput. Appl. Biosci.* 6:81-92.
23. Lawrence, C. E., Altschul, S. F., Boguski, M. S., Liu, J. S. Neuwald, A. F. and Wootton, J. C. (1993) *Science* 262: 208-14.
24. Eddy, S. R., Mitchison, G. and Durbin, R. (1995) *J. Comput. Biol.* 2: 9-23.
25. Heumann, J. M., Lapedes, A. S. and Stormo, G. D. (1994) In: *Proceedings of the Second International Conference on Intelligent Systems in Molecular Biology*, pp. 188-194.
26. Heumann, J. H., Lapedes, A. S. and Stormo, G. D. (1995) *Proceedings of the 1995 World Congress on Neural Networks* II: 771-775.
27. Henikoff, S. and Henikoff, J. G (1994) *J. Mol. Biol.* 243:574-578.
28. Henikoff, J. G. and Henikoff, S. (1996) *Comput. Appl. Biosci.* 12, in press.

9

THE INTEGRATED GENOMIC DATABASE (IGD)S

Enhancing the Productivity of Gene Mapping Projects

Stephen P. Bryant,* Anastassia Spiridou and Nigel K. Spurr

Human Genetic Resources Laboratory
Imperial Cancer Research Fund
Blanche Lane, South Mimms, Herts EN6 3LD, United Kingdom

ABSTRACT

Mapping genes using human pedigrees involves analysing the co-segregation of polymorphic markers with a trait phenotype. This technique exploits the pattern of meiotic recombination in the family to deduce the likely location of the unknown gene. In some cases, it is possible to visualize the transmission of haplotypes within a family and identify flanking loci, but more usually, genetic locations are estimated with statistical methods. Although genetic analysis in this way is supported by a variety of efficient and novel algorithms, implemented in a variety of programs, the rate limiting step continues to be the integration of data and software tools. The Integrated Genomic Database (IGD) approach offers an opportunity to remove the bottleneck and enhance the productivity of the linkage analyst. In this paper, we describe an approch to this problem using the IGD methods. We demonstrate the feasibility of IGD to increase the productivity of gene mapping projects and to facilitate communication between the clinic and the laboratory.

1. INTRODUCTION

Mapping genes using human pedigrees involves analysing the co-segregation of polymorphic markers with a trait phenotype (e.g. breast cancer). The technique exploits the

* Contact information. S.P.B.: Phone: (+44) 171 269 3850, Fax: (+44) 171 269 3801, mail to: s.bryant@icrf.icnet.uk; A.S.: Phone: (+44) 171 269 3849, Fax: (+44) 171 269 3801, mail to: a.spiridou@icrf.icnet.uk; N.K.S.: Phone: (+44) 171 269 3846, Fax: (+44) 171 269 3802, mail to: n.spurr@icrf.icnet.uk

Theoretical and Computational Methods in Genome Research, edited by Suhai
Plenum Press, New York, 1997

pattern of meiotic recombination in a family to deduce the likely location of the unknown gene. Markers that freely recombine with the trait gene are "unlinked" and are physically distant, whereas those that recombine at a lower frequency (or not at all) are "linked" and this result implies physical proximity of the trait and marker loci. In some cases, it is possible to visualise the transmission of haplotypes and to count recombinants but, in typical human pedigrees, missing data, reduced polymorphism and other complications would make this unfeasible. Instead, the statistical technique of Maximum Likelihood Estimation (Edwards, 1972) is usually routinely used in an analysis of this kind. The procedure is concerned largely with the estimation of the "recombination fraction" or genetic distance between loci (Ott, 1991) and requires a parameterised model for the trait, along with a set of marker and trait observations on a family or set of families. Specialised algorithms facilitate the estimation of the unknown parameters, e.g. LINKAGE (Lathrop and Lalouel, 1984; Lathrop et al., 1985) and CRI-Map (Green et al., 1989).

Although the linkage analyst is supported by a variety of analytical software, incorporating efficient (O'Connell and Weeks, 1995) and novel (Davis et al., 1996) algorithms, as well as programs for data display and management (Chapman, 1990), the bottleneck in an analysis continues to be the integration of data and software tools. The approach of the Integrated Genomic Database (IGD) project (Ritter, 1994) offers an opportunity to remove this rate-limiting step and enhance the productivity of the linkage analyst, as part of a collaborative gene mapping project.

IGD was originally conceived as a loose federation of existing molecular biological resources, including databases and algorithms, brought together into a single, integrated software framework, for exploration and analysis (Ritter, 1994). It is a collaborative project (Table 1) which has attracted substantial funding from the European Commission (EC Biomed 1 Grant GENE-CT93–0003).

IGD has developed the concepts of Resource End Databases (REDs) and Target End Databases (TEDs), communicating via "middle managers" (MIMs), frequently programs that apply transformations to data (Figure 1).

REDs are typically data collections that are autonomous, public entities such as GDB, GenBank or CEPH. The TEDs are a federated composite of the REDs, which are syntactically and semantically integrated. The TED is accessed by another component, the Front-End (FRED) , which may be a graphical browser such as ACEDB.

In all, IGD is a collection of interoperable data and program resources that can be downloaded into and merged with local, "private" collections. It also provides a means of managing local, experimental data, and offers laboratories the ability to mix-and-match components in order to solve a problem. IGD "adds value" to existing programs and database resources by providing integration tools. These tools may be programs that convert data from one format to another, display specialized objects, or support the use of programs for analysis.

Table 1. The IGD project: A collaboration of European and North American partners

DKFZ, Heidelberg, Germany
Imperial Cancer Research Fund (ICRF), London,UK
Max Planck Institute (MPI), Berlin, Germany
Medical Research Council Human Genome Mapping Project Resource Centre, Cambridge, UK
Weizmann Institute, Rehovot, Israel
CNRS, Montpellier, France
Lawrence Berkeley Laboratory, USA
Sanger Centre, Cambridge, UK

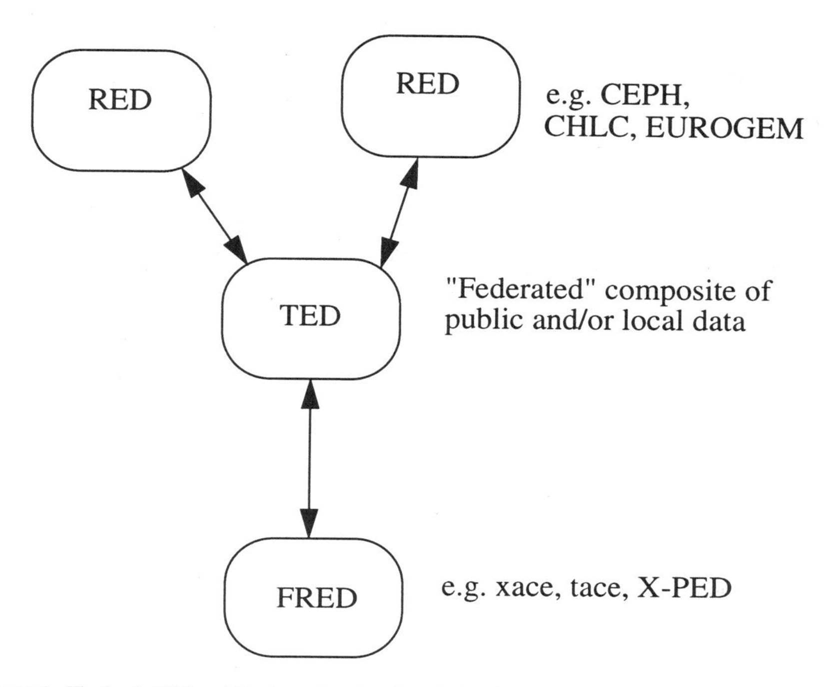

Figure 1. The basic IGD architecture, showing the relationship between the REDs, TEDs and FREDs.

The principal role of our laboratory within the IGD collaboration has been to consider the domain of genetic linkage analysis, with reference both to mapping disease genes in extended families, and to the construction, enhancement and use of reference maps from family resources such as the CEPH (Dausset, 1990). Our goals within the project can be stated as:

1. To model the domain sufficiently to enable a genetic mapping project to be supported at all stages.
2. To produce an algorithm for displaying families, including genetic and trait data, and implement it over X-Windows.
3. To integrate the LINKAGE and CRI-Map software for genetic linkage analysis.
4. To integrate the CEPH, EUROGEM and CHLC public map and genotype resources.
5. To integrate data from other, related techniques, e.g. physical mapping.

One of the main motivating factors of the work has been to enhance the communication between different components of a mapping project (Figure 2), particularly the clinic and the laboratory.

Genetic mapping projects require families segregating the relevant trait, which are typically sampled in clinics or in the field by pedigree workers, following the presentation of a proband individual. The genealogy is constructed by interview with key family members and the DNA extracted from blood samples sent to the laboratory, often followed by EBV transformation to form a cell line. Individuals from the families are genotyped, typically for microsatellite polymorphisms, and the genotypes analysed for co-segregration with the trait phenotype (see Figure 7 for an example).

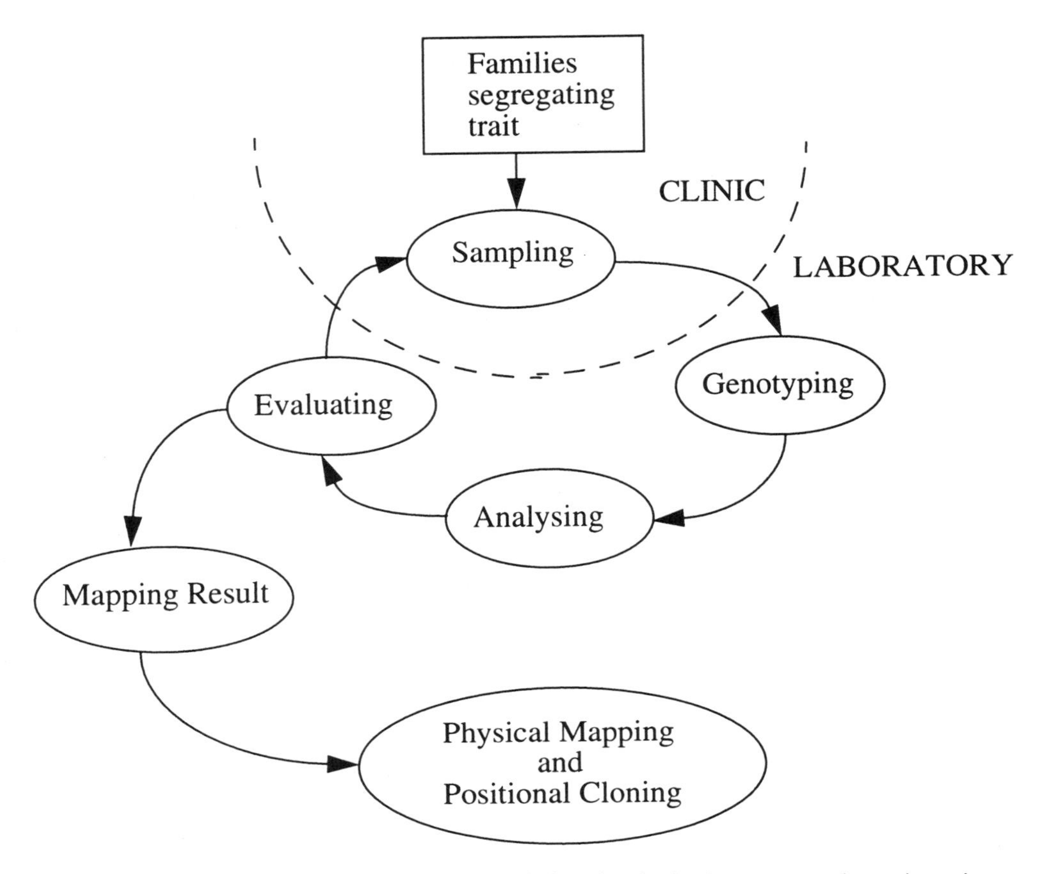

Figure 2. The cycle of data collection, genotyping, analysis and evaluation in a gene mapping project using genetic linkage analysis.

Linkage analysis is performed with one of a number of programs, e.g. LINKAGE (Lathrop and Lalouel, 1984) or CRI-Map (Green et al., 1989) and the results used to make inferences about the physical relationship of the marker and trait loci. Software that has been developed for managing pedigrees (e.g. LINKSYS, QDB/DOLINK) have elements in common, and we have used these as a starting point for our treatment. Our approach has been assisted by a close involvement with the IGD project. We have developed an effective way of modelling the components of a linkage analysis (e.g. Pedigrees, Polymorphisms, Traits and Trait Models) using the IGD method, which allows their reuse and minimizes redundancy. This model is incorporated into the global IGD model which automatically takes advantage of the relationship of genetic data to other areas of molecular biology (e.g. physical mapping, gene identification, sequence and function).

At all times, we have striven for biological clarity and recognition of current usage. Our data model allows the integration of new data and algorithmic resources by the mapping of an external data model onto the global IGD schema. Construction of the external data model, the translation of data and the handling of exceptions, which involves a detailed understanding of the semantics as well as the syntax of the resource, is very time-consuming. Using the technique, we have constructed a set of format translators and integrated the data and maps from CEPH (Dausset, 1990), CHLC (Buetow et al., 1994), the UK DNA Probe Bank and EUROGEM (Spurr et al., 1994). We have also integrated the LINKAGE (Lathrop and Lalouel, 1984) and CRI-Map (Green et al., 1989) software, albeit in a preliminary way.

The pedigree is the "unit of information exchange" in a gene mapping project and has been subject to long and careful scrutiny with regard to data management, display and analysis (Attwood and Bryant, 1988; Chapman, 1990; Lathrop and Lalouel, 1984; Ott, 1991; Mamelka et al., 1987; Bryant, 1996 for a review). The visual metaphor of the pedigree "tree" is very strong and the most common, but many differerent graphical nomenclatures exist. There have been recent attempts to promote a standard nomenclature for human pedigrees (Bennett et al., 1995) but the approach has been criticized (Curtis, 1995) as perhaps not taking account of existing display algorithms, which work with a data model that is not graphically based (e.g. PEDRAW, Pedigree/DRAW).

We have developed a pedigree display program (X-PED) that works over X-Windows and which is tightly integrated with IGD. X-PED displays pedigrees of arbitrary family structure, along with any marker genotypes and trait phenotypes. Although programs for displaying pedigrees were first developed in the Sixties, as computing technology began to be exploited in human genetics, and continue to be produced for personal computers such as the Apple Macintosh (MacCluer et al., 1986) and PC (Chapman, 1990), no program for the display of pedigrees over X-Windows exists in the public domain, which is a crucial part of the IGD architecture. Using a system such as X-PED increases the quality of data by providing visual verification, and indicates where problems exist in the data by examination of structure, traits and marker genotypes. It also serves as a basis for exchanging information between the clinic and laboratory. X-PED interoperates with IGD and behaves as a "helper" application analogous to those used by generic World Wide Web (WWW) browsers such as Netscape.

Following genetic mapping with families, physical maps are often constructed using radiation hybrids or YACs (Dunham et al., 1994), to reduce the region containing the suspected disease gene to a manageable size for sequencing and gene identification by positional cloning. We have also been keen to model these downstream activitiesas part of IGD, to move toward an integrated view of a mapping, cloning and sequencing project.

2. DATA MODELLING AND MANAGEMENT

The ACEDB system (Dunham et al., 1994) was used to construct the IGD data model. Some modifications have been made by us to the last production version of IGD (version 1.2), and these are in the process of being incorporated into the next release of the system. Data was managed directly by ACEDB. We also have developed scripts to transfer data from an Oracle database used in the laboratory to track material and experimental results (Bryant, 1994). Figure 3–Figure 5 shows some of the classes which have been developed, as they appear within ACEDB. The models are fully described on the IGD Web Site (`http://ftp.dkfz-heidelberg.de/igd/docs`).

Classes have been defined for the biological entities Polymorphisms, Pedigrees and Traits. Pedigrees are defined as a collection of Individual objects connected by the tags `father` and `mother`. Pedigrees and polymorphisms are collected into sets (`Pedigree_set` and `Polymorphism_set`). Traits are cross-referenced with a genetic transmission model (`Trait_model`) which defines a trait allele frequency and genotypic

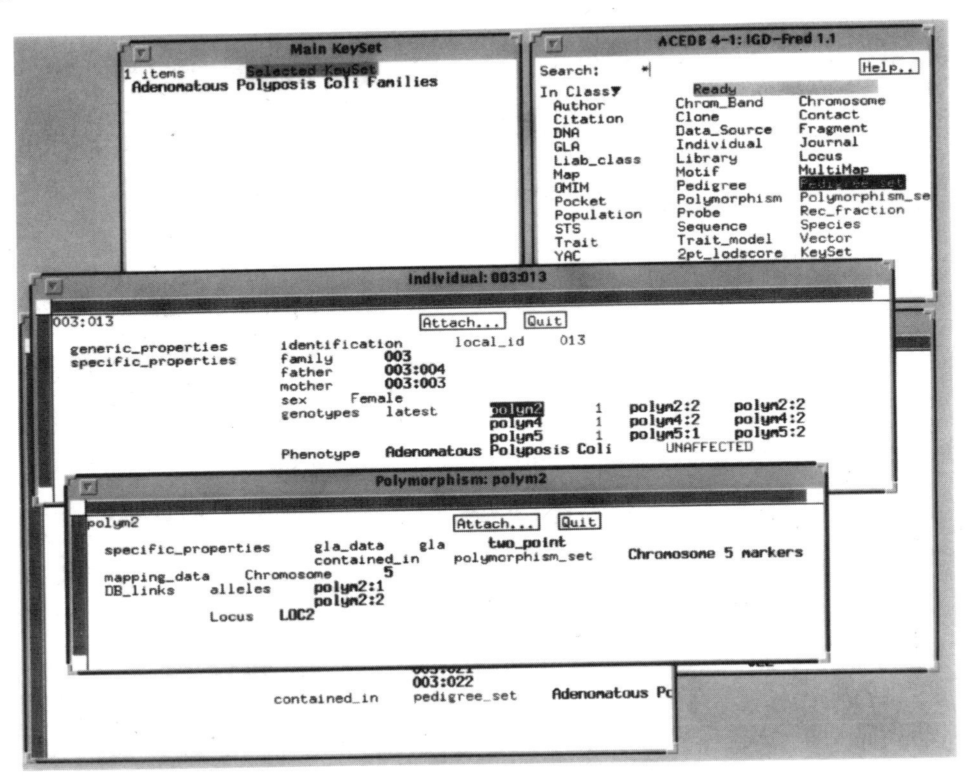

Figure 3. IIGD classes, as seen from ACEDB. Note that the individual 003:013 has tags that point to the father (003:004) and mother (003:003) and so define a triplet which, when repeated, defines the genealogy. Tags for genotypes and phenotypes are also included. The polymorphism polym2, highlighted as a tag on 003:013, is also displayed, showing the links to alleles and to locus.

penetrances, following the approach described by Ott (1991) and used in LINKAGE (Lathrop and Lalouel, 1984).

To support the linkage analyst, we developed a class called GLA (for Genetic Linkage Analysis). The GLA identifies the set of families (via `Pedigree_set` and `Pedigree`), the set of polymorphisms (via `Polymorphism_set` and `Polymorphism`), the trait (via `Trait`) and the genetic model (via `Trait_model`) that define the components of an analysis. In addition, there may be pointers to `Map` and `2pt_lodscore` objects for input and output. There are ACEDB "magic tags" (`Pick_me_to_call`) which are associated with both the `Pedigree` and `GLA` classes, and these enable the display of a single pedigree and the execution of an analysis, respectively.

3. PEDIGREE DISPLAY

A program (X-PED) was developed in the C and X/Motif languages to facilitate the visualisation of pedigree data (Spiridou et al., in prep. a). The program is integrated with

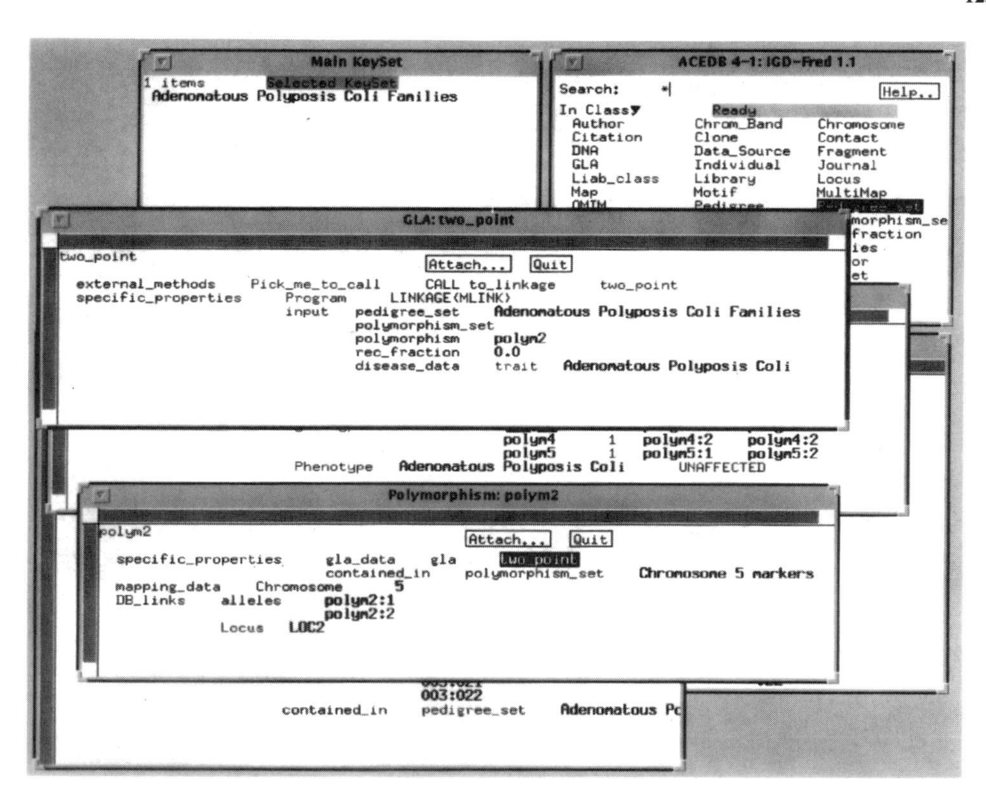

Figure 4. The GLA class, including tags to the set of pedigrees and polymorphisms defining an analysis, as well as the trait and the external program (LINKAGE) to be used.

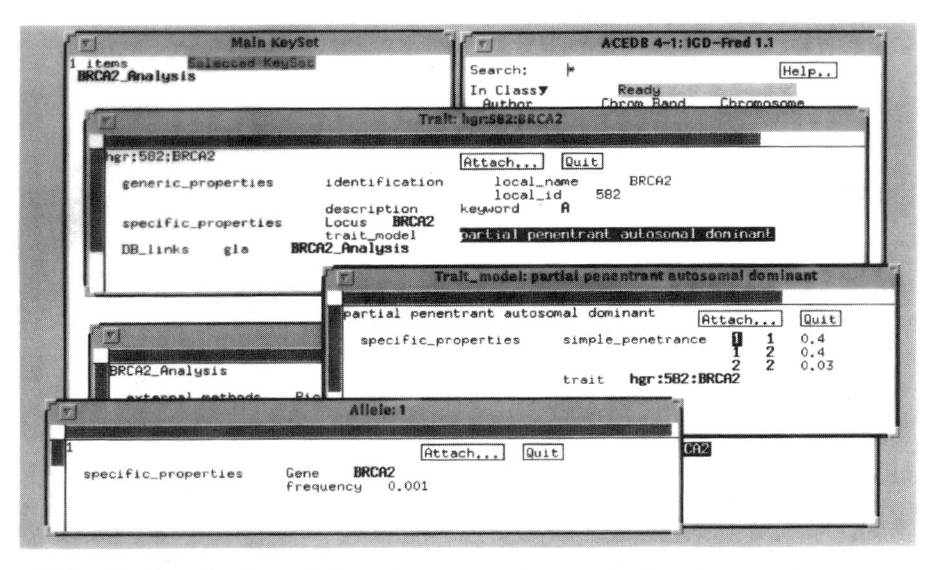

Figure 5. The `Trait`, `Trait_model` and `Allele` classes, showing their interrelationships. The `Trait_model` defines the penetrance matrix of the trait to be used in the analysis. This means that a pool of standard trait models (in this case partial penetrant autosomal dominant) can be created to be reused as and when required.

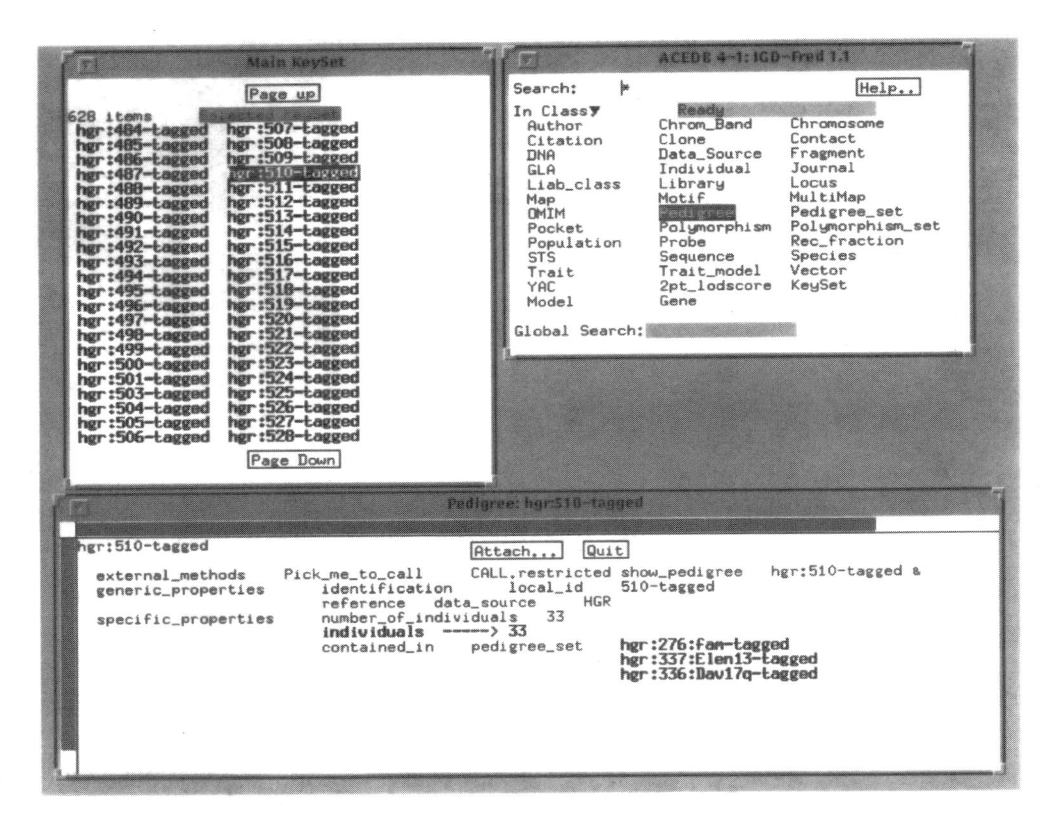

Figure 6. Displaying a pedigree using the `Pick_me_to_call show_pedigree` tag from the `Pedigree` class. By selecting this tag, a background process is invoked which extracts the appropriate data from ACEDB and passes it to X-PED, which appears as an icon on the screen.

IGD by the `Pick_me_to_call` tag of the Pedigree class . Invoking this tag calls X-PED to display the family (Figure 6).

X-PED displays pedigrees graphically (over X-Windows), along with any marker genotypes and trait phenotypes. It copes with consanguinous loops and multiple mates. X-PED can also display ancillary information about the pedigree, such as contact data. Generations can be numbered and an optional legend explaining the symbols used on the diagram is also supported (Figure 7).

The underlying drawing algorithm is split into two phases. First, individuals are mapped into a two dimensional array, and then the pedigree is drawn on the screen. During the mapping phase, inbreeding loops are broken by duplicating appropriate individuals. This simplifies the subsequent assignment of vertical and horizontal coordinates to each individual. Multiple matings are also resolved by duplicating individuals. Drawings

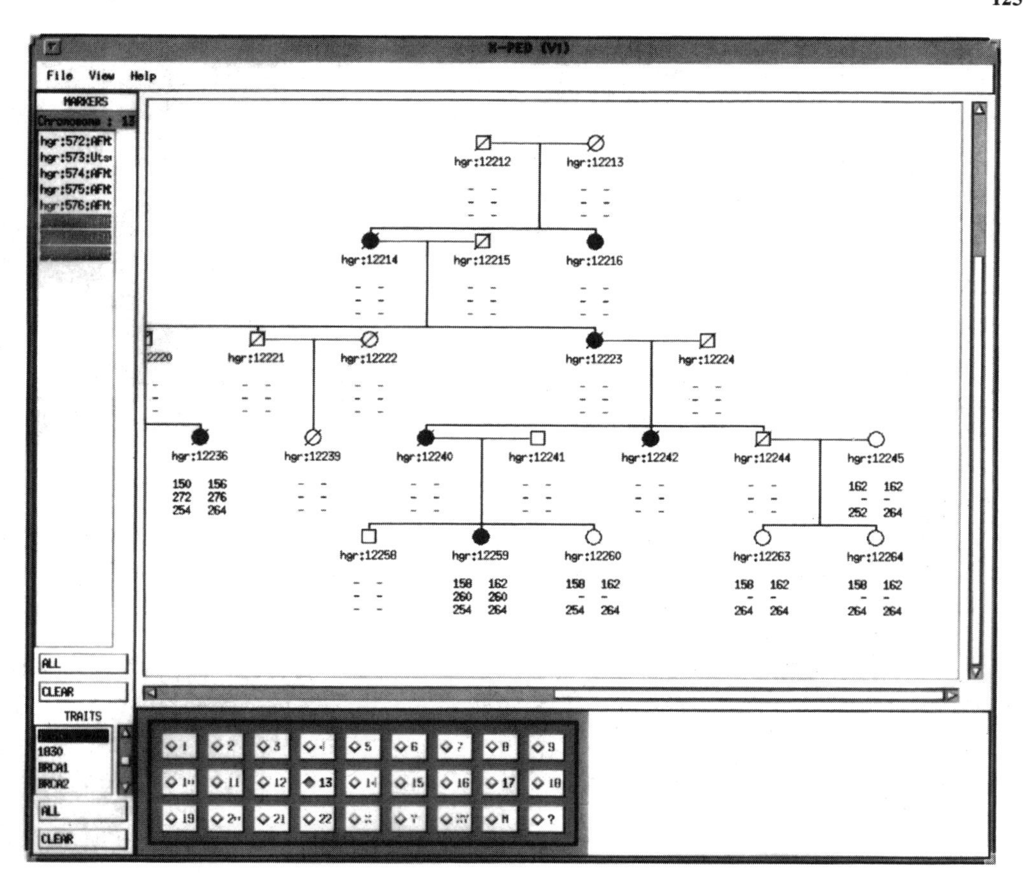

Figure 7. X-PED can display phenotypes and genotypes of family members, along with ancillary information about the pedigree. The grid at the bottom of the display shows the chromosomes available. Genotypes from markers on chromosome 13 (listed in the window at the left of the display) are being shown. The list of traits segregating in this family are shown at the bottom left of the display. The user has selected ICD9 code 1749. Inviduals shown in solid form are affected.

generated using this method are quite clear for the types of pedigrees usually encountered in mapping projects, and can cope with a remarkable degree of consanguinity and multiple mating (Figure 8).

4. INTEGRATION OF PROGRAMS FOR ANALYSIS

We identified two programs for pedigree analysis: LINKAGE (Lathrop and Lalouel, 1984) for disease gene mapping and general family analysis, and CRI-Map (Green et al., 1989) for reference map construction and enhancement.

The GLA class holds all the necessary information for a linkage analysis. This includes links to the trait being studied, the set of pedigrees and polymorphisms, as well as the genetic model for the trait. When an external program is called (using the

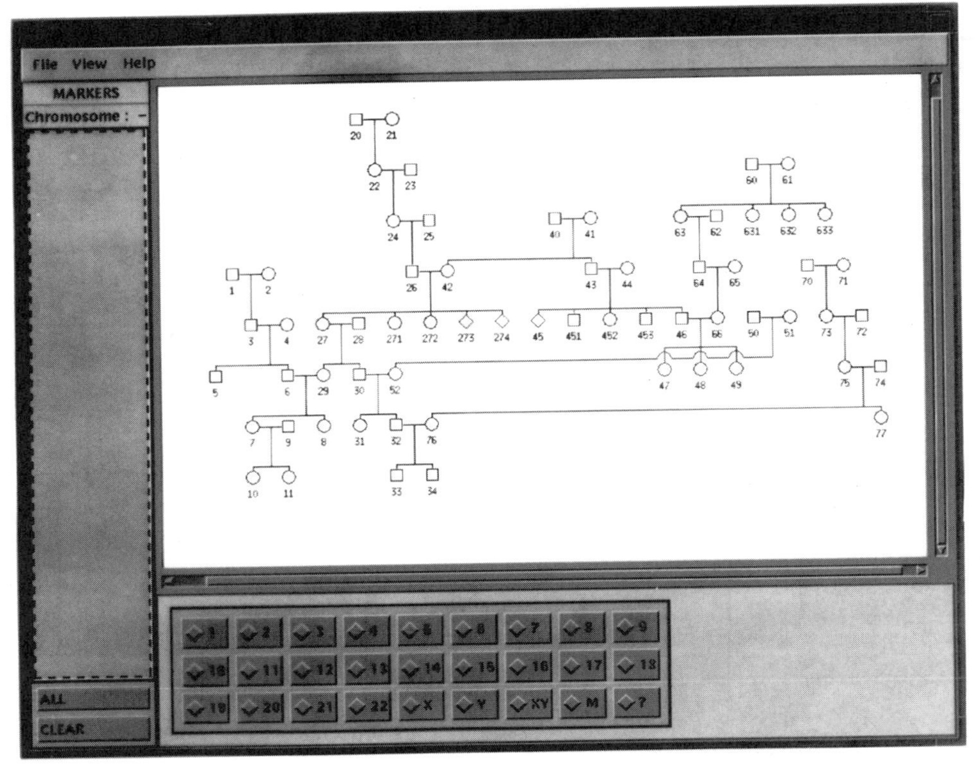

Figure 8. X-PED has no problem with complicated family structures. It copes with consanguinity and multiple mates by duplicating individuals.

`Pick_me_to_call` mechanism) all data referenced by the the GLA are extracted out of the ACEDB database and converted into the appropriate format (Figure 9).

The complete integration of LINKAGE and CRI-Map involves several steps:

1. *Extraction of ACE data from the database, as specified by the GLA.* A particular GLA definition establishes the queries sent to a textual version of ACEDB (`tace`) that provides query and file storage facilities. tace extracts the data in ACE format.
2. *Creation of input data files and shell scripts from the ACE data, by transformation.* The ACE data extracted by `tace` are then collected together and passed through a filter program (`ace2xped`) that produces input for X-PED. The ACE data are transformed into an external representation used by X-PED. X-PED is invoked with these data and an argument indicating the transformation (LINKAGE or CRI-Map) to be made. The end result is either a CRI-Map genotype file, or a pair of LINKAGE parameter and pedigree files (Figure 10).

   The shell script is created by using a template. At present, this is only implemented for 2-point analysis, but a generic script creation mechanism is being developed. In general, however, it is necessary to use a utility such as the Linkage Control Program (lcp) for this stage of the protocol.

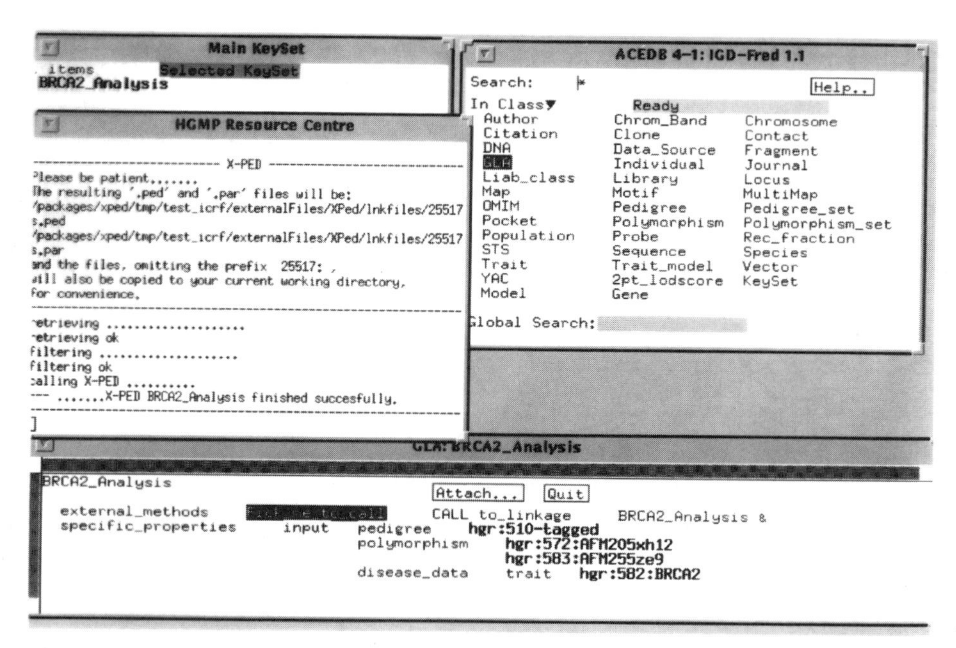

Figure 9. Extracting data into LINKAGE using the GLA `Pick_me_to_call to_linkage` tag. By selecting this tag, the problem specification defined by the GLA is translated into the equivalent LINKAGE or CRI-Map specification. The progress of the translation is shown in the window to the left of the figure.

3. *Execution of the shell script, to activate the program.* This is a straightforward step, applying the appropriate tool to the generated data in the way specified by the analysis.
4. *Parsing the results back into ACE classes.* This protocol step is at an early stage of development and involves mapping the LINKAGE or CRI-Map output back into appropriate IGD classes (`Map` or `2pt_lodscore`). This is not implemented in the current release.
5. *Loading the ACE data back into the running database.* The `tace` program is called to parse the ace result file. One problem, yet to be resolved, is that the user must quit `xace` at this point and restart, in order to see the new data. This is an ACEDB restriction.

5. INTEGRATION OF PUBLIC DATABASE RESOURCES

We identified four large public database resources, CEPH, EUROGEM, CHLC and the UK DNA Probe Bank, with the aim of integrating their data within the IGD model.

5.1 Centre d'Etude du Polymorphisme Humain (CEPH)

CEPH is a Foundation which collaborates with more than 100 scientists located in Europe, the USA, Canada, Japan, Australia and South Africa to facilitate the construction of a linkage map of the human genome. The databases contains genotypes from 65 nuclear

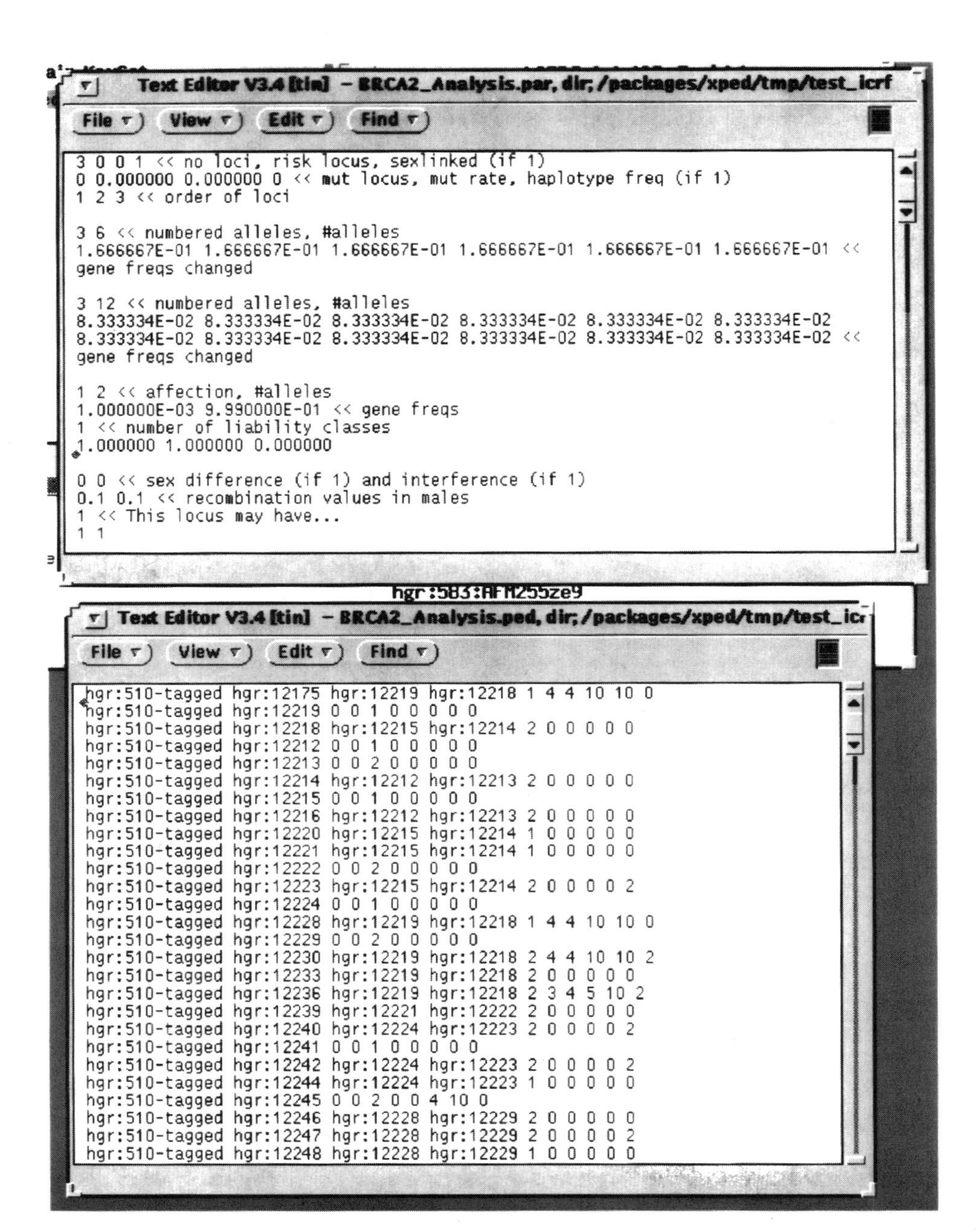

Text Editor V3.4 [tin] – BRCA2_Analysis.par, dir; /packages/xped/tmp/test_icrf

File View Edit Find

```
3 0 0 1 << no loci, risk locus, sexlinked (if 1)
0 0.000000 0.000000 0 << mut locus, mut rate, haplotype freq (if 1)
1 2 3 << order of loci

3 6 << numbered alleles, #alleles
1.666667E-01 1.666667E-01 1.666667E-01 1.666667E-01 1.666667E-01 1.666667E-01 <<
gene freqs changed

3 12 << numbered alleles, #alleles
8.333334E-02 8.333334E-02 8.333334E-02 8.333334E-02 8.333334E-02 8.333334E-02
8.333334E-02 8.333334E-02 8.333334E-02 8.333334E-02 8.333334E-02 8.333334E-02 <<
gene freqs changed

1 2 << affection, #alleles
1.000000E-03 9.990000E-01 << gene freqs
1 << number of liability classes
1.000000 1.000000 0.000000

0 0 << sex difference (if 1) and interference (if 1)
0.1 0.1 << recombination values in males
1 << This locus may have...
1 1
```

hgr:583:AFM255ze9

Text Editor V3.4 [tin] – BRCA2_Analysis.ped, dir; /packages/xped/tmp/test_ic

File View Edit Find

```
hgr:510-tagged hgr:12175 hgr:12219 hgr:12218 1 4 4 10 10 0
hgr:510-tagged hgr:12219 0 0 1 0 0 0 0 0
hgr:510-tagged hgr:12218 hgr:12215 hgr:12214 2 0 0 0 0 0
hgr:510-tagged hgr:12212 0 0 1 0 0 0 0 0
hgr:510-tagged hgr:12213 0 0 2 0 0 0 0 0
hgr:510-tagged hgr:12214 hgr:12212 hgr:12213 2 0 0 0 0 0
hgr:510-tagged hgr:12215 0 0 1 0 0 0 0 0
hgr:510-tagged hgr:12216 hgr:12212 hgr:12213 2 0 0 0 0 0
hgr:510-tagged hgr:12220 hgr:12215 hgr:12214 1 0 0 0 0 0
hgr:510-tagged hgr:12221 hgr:12215 hgr:12214 1 0 0 0 0 0
hgr:510-tagged hgr:12222 0 0 2 0 0 0 0 0
hgr:510-tagged hgr:12223 hgr:12215 hgr:12214 2 0 0 0 0 2
hgr:510-tagged hgr:12224 0 0 1 0 0 0 0 0
hgr:510-tagged hgr:12228 hgr:12219 hgr:12218 1 4 4 10 10 0
hgr:510-tagged hgr:12229 0 0 2 0 0 0 0 0
hgr:510-tagged hgr:12230 hgr:12219 hgr:12218 2 4 4 10 10 2
hgr:510-tagged hgr:12233 hgr:12219 hgr:12218 2 0 0 0 0 0
hgr:510-tagged hgr:12236 hgr:12219 hgr:12218 2 3 4 5 10 2
hgr:510-tagged hgr:12239 hgr:12221 hgr:12222 2 0 0 0 0 0
hgr:510-tagged hgr:12240 hgr:12224 hgr:12223 2 0 0 0 0 2
hgr:510-tagged hgr:12241 0 0 1 0 0 0 0 0
hgr:510-tagged hgr:12242 hgr:12224 hgr:12223 2 0 0 0 0 2
hgr:510-tagged hgr:12244 hgr:12224 hgr:12223 1 0 0 0 0 0
hgr:510-tagged hgr:12245 0 0 2 0 0 4 10 0
hgr:510-tagged hgr:12246 hgr:12228 hgr:12229 2 0 0 0 0 0
hgr:510-tagged hgr:12247 hgr:12228 hgr:12229 2 0 0 0 0 2
hgr:510-tagged hgr:12248 hgr:12228 hgr:12229 1 0 0 0 0 0
```

Figure 10. The Linkage input files produced from the GLA shown in Figure 9, which are part of a BRCA2 analysis on chromosome 13. The LINKAGE parameter file is shown at the top of the figure, and the pedigree file underneath.

families, many with grandparents. This kind of family structure is very efficient for genetic mapping. The Foundation sends DNAs from these reference families to collaborators who, with their own genetic markers, generate genotypes. The genotypic data are sent back to CEPH for inclusion in a database that is distributed to all collaborators and also made available on the Internet (`http://www.cephb.fr`).

The CEPH database (version 8.0) contains genotypes for over 10000 genetic markers located on all the human chromosomes. It also contains genotypes for over 2000 microsatellite markers including all the data from Genethon, most of which are highly polymorphic and very efficient for mapping. For this resource, we developed translation software (`ceph2igd`) that converted CEPH-formatted text files into IGD-formatted ACE classes. `ceph2igd` can either be used to convert data from the public CEPH database, or local lab data that are held in CEPH format, into a form that can easily be combined, for analysis.

5.2 European Gene Mapping Project (EUROGEM)

EUROGEM (EC grant Biomed 1 CT93–0077) was established in 1991 with financial support from the European Commission under the Human Genome Analysis programme. The primary aim of the project was to improve the resolution of the human genetic linkage map. The project has been structured around two resource centres and more than twenty laboratories carrying out genotypic analysis. The project has had a very close relationship with CEPH, as one of the resource centres, to the extent that all genotype data are transferred to CEPH and made available as part of their public database.

EUROGEM primary genotyping data are also freely available by anonymous ftp (`ftp://ftp.gene.ucl.ac.uk`) and on the Web (`http://www.ic-net.uk/axp/hgr/eurogem`) as well as being included as a subset of the CEPH database. They have the same format as CEPH data and are modelled within IGD in the same way. EUROGEM has also produced and published graphical genetic maps of the human genome, using CRI- Map (Green et al., 1989), which have been translated into ACE format using the IGD models. These are available from the EUROGEM Web site.

5.3 Co-operative Human Linkage Center (CHLC)

CHLC is a US federally funded, through NIH and DOE, genome center, with a goal to develop statistically rigorous, high heterozygosity genetic maps of the human genome that are greatly enriched for the presence of easy-to-use PCR-formatted microsatellite markers, with a particular emphasis on tri- and tetranucleotide repeats that are easy to genotype. The center synthesizes published genotypic data developed on the CEPH families by outside investigators, as well as genotypic information generated from marker development in CHLC core laboratories. The center is also open to assisting outside investigators who would like to incorporate their own genotypic information into these maps, as well. The CHLC data and maps are available on the World Wide Web (`http://www.chlc.org`). All CHLC data and maps are available in CRI-Map format and, in collaboration with John Attwood at University College London, we have produced translation software that converts CRI-Map genotypes and “fixed” CRI-Map output into IGD format.

5.4 UK DNA Probe Bank

This was established in 1989 with funding from the UK Human Genome Project (HGMP). The starting resource was based on a collection of over 400 DNA probes detecting restriction fragment length polymorphisms, obtained by our laboratory at the ICRF. The funding was given with the aim of developing the in-house facility into a nationally available collection of markers for genetic mapping. The work has been carried out by our laboratory and by the HGMP Resource Centre (now at Hinxton Hall, Cambridge, UK), to which the Bank has now been transferred.

For all these resources, the integration was achieved by developing translation software that mapped the syntactic structure and semantic content of the resources on to the IGD models (Spiridou et al., in prep. b). All these programs are available from the IGD Web site (`http://ftp.dkfz-heidelberg.de/igd`).

In addition, collections of data that have already been converted are available from the UK Human Genome Mapping Project Resource Centre (HGMP-RC), on the Web at `http://www.hgmp.mrc.ac.uk`. Registered users can obtain the data from the directory `/packages/xped/example-files`.

6. INTEGRATION OF DOWN-STREAM MAPPING DATA

As an example of the integration of physical mapping data with the data model presented here, Figure 11 shows an ACE-formatted file containing YAC/STS data for chromosome 13.

The Cross-reference between the `Locus` and `Locus_out` tags in the YAC class act as a bridge between maps, so that `Multimap` objects can be created from the different Maps, aligned by their common elements. The links between classes also make it easy to see which biological reagents have been used in which experiments (Figure 12).

7. DISCUSSION

As a pedigree display tool, X-PED is in the tradition of other programs which map the structure of a family onto a 2-dimensional graphical representation, e.g Pedigree/DRAW (MacCluer et al., 1986) rather than those that offer a palette with which to define the layout of the family, e.g. CYRILLIC (Chapman, 1990). It has the advantage of using an external data representation, easily mapped from IGD. It can connect to an alternative data store, as long as the mapping is defined. It will probably be prudent at some point to extract the data translation facility out of X-PED, so as to remove the dependence on X-Windows and increase the modularity of the system.

The use of IGD/X-PED to manage and visualise pedigree data has proven very useful in locating data errors. Common errors include impossible pedigree structures (e.g. an individual that is both female and defined to be a father); possible but unlikely pedigree structures (e.g. inbreeding loops), as well as errors on trait phenotypes and marker genotypes. X-PED displays helpful error messages indicating the problem. Erroneous data are also much easier to identify once they are displayed graphically. The overall result is improved data quality.

We have applied X-PED to several mapping problems, requiring both LINKAGE and CRI-Map analyses. The system was able to specify the problems in a satisfactory way

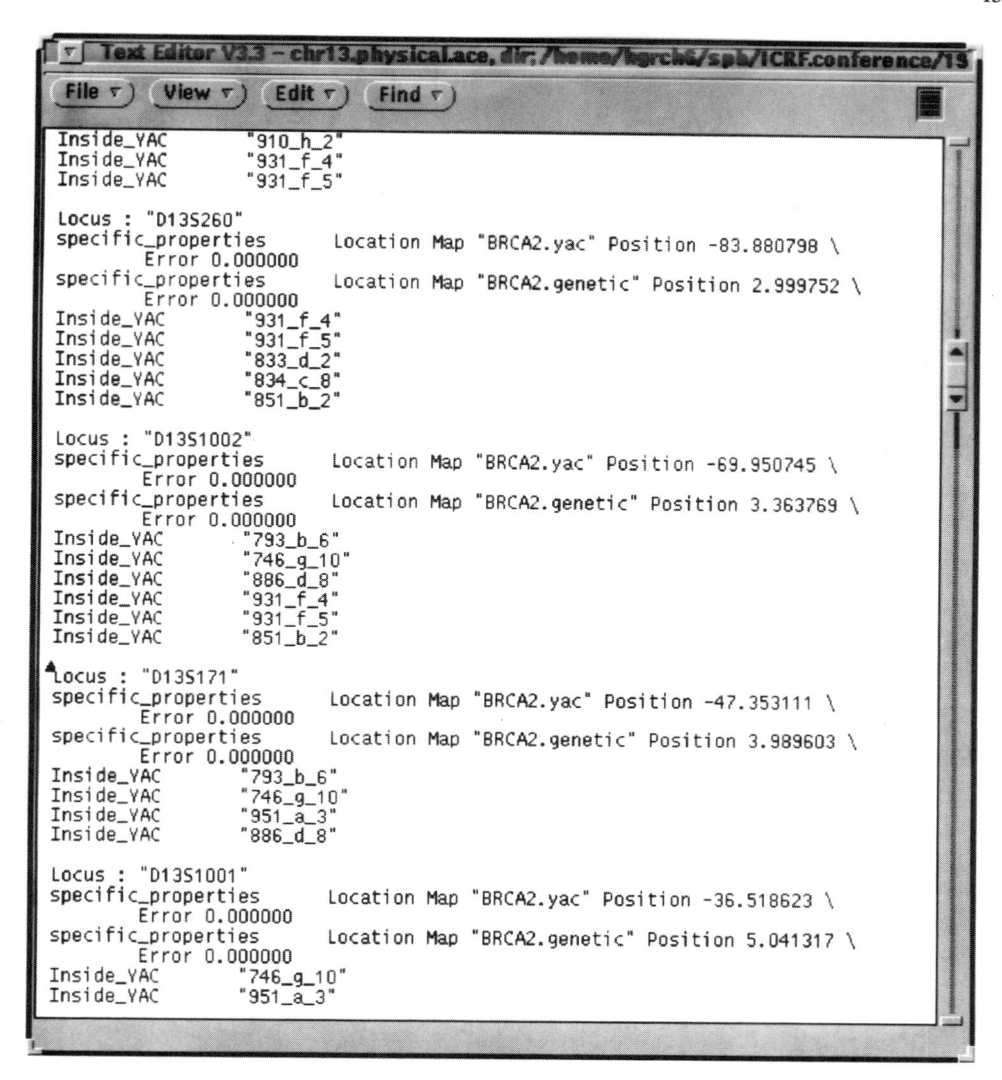

```
Inside_YAC        "910_h_2"
Inside_YAC        "931_f_4"
Inside_YAC        "931_f_5"

Locus : "D13S260"
specific_properties      Location Map "BRCA2.yac" Position -83.880798 \
        Error 0.000000
specific_properties      Location Map "BRCA2.genetic" Position 2.999752 \
        Error 0.000000
Inside_YAC        "931_f_4"
Inside_YAC        "931_f_5"
Inside_YAC        "833_d_2"
Inside_YAC        "834_c_8"
Inside_YAC        "851_b_2"

Locus : "D13S1002"
specific_properties      Location Map "BRCA2.yac" Position -69.950745 \
        Error 0.000000
specific_properties      Location Map "BRCA2.genetic" Position 3.363769 \
        Error 0.000000
Inside_YAC        "793_b_6"
Inside_YAC        "746_g_10"
Inside_YAC        "886_d_8"
Inside_YAC        "931_f_4"
Inside_YAC        "931_f_5"
Inside_YAC        "851_b_2"

Locus : "D13S171"
specific_properties      Location Map "BRCA2.yac" Position -47.353111 \
        Error 0.000000
specific_properties      Location Map "BRCA2.genetic" Position 3.989603 \
        Error 0.000000
Inside_YAC        "793_b_6"
Inside_YAC        "746_g_10"
Inside_YAC        "951_a_3"
Inside_YAC        "886_d_8"

Locus : "D13S1001"
specific_properties      Location Map "BRCA2.yac" Position -36.518623 \
        Error 0.000000
specific_properties      Location Map "BRCA2.genetic" Position 5.041317 \
        Error 0.000000
Inside_YAC        "746_g_10"
Inside_YAC        "951_a_3"
```

Figure 11. ACE-formatted YAC/STS content data from chromosome 13. All data collections can be maintained in an external format such as this, conforming to the IGD models. Where cross-referenced tags imply common elements, great care has to be taken to ensure that the relationship between the objects is meaningful.

and enabled rapid solutions to be obtained (Deckert et al., 1995; Dewald et al., in press; Smith et al., in press).

We have also road-tested the software as part of the training courses run by the UK Human Genome Mapping Project Resource Centre. Initial response from the students has largely been favourable and a pool of regular users has been established.

A major component of pedigree analysis which is currently missing from our implementation is the construction of haplotypes. We have recently extended the IGD models to include haplotypes, although these cannot as yet be displayed using X-PED. Integration of haplotyping algorithms (Weeks et al., 1995) is an important future goal.

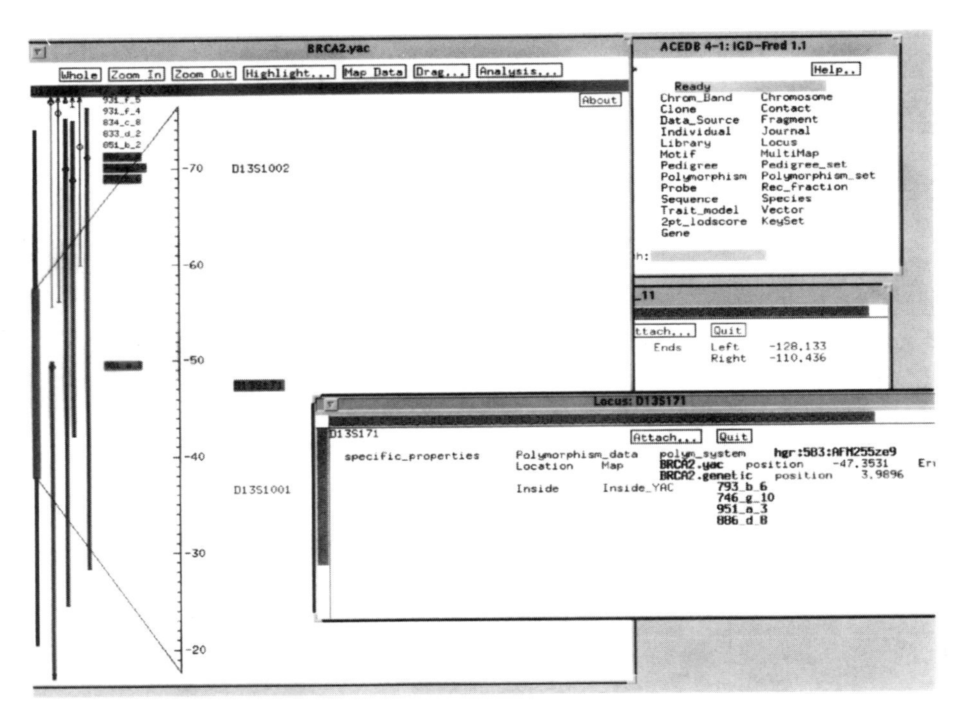

Figure 12. Integration of genetic and physical mapping data. The locus D13S171 has been used as a bridging object between the physical and genetic maps. The YAC map is shown, as well as a link to the genetic map, which could be displayed in association, as a `Multimap` object.

Dissatisfaction with the trait model-based approach to pedigree analysis has led in recent years to the emergence of tests for linkage which are not dependent on a parameterized trait model. These methods have their origins in the sib-pair method of Penrose (1935) and have been extended to sib-sets (Lange, 1986a,b) and extended families (Weeks and Lange, 1988, 1992). Early implementations of the approach lacked statistical power and confounded linkage and the estimation of allele frequencies, but recent improvements (Whittemore and Halpern, 1994; Davis et al., 1996) show great promise. Our implementation of the GLA and Trait_model classes are intimately bound up with the notion of a parameterized model of the trait, and a clear goal of future IGD work would be to embrace these new methodologies.

We also need to achieve a tighter integration of IGD with analytical programs such as LINKAGE, to complete the cycle of analysis and support the management of results, as discussed in the introduction.

This aim raises some problems which require novel solutions. We have gone some way toward achieving this level of integration, based on an IGD class we call GLA (Genetic Linkage Analysis), which links input data (Pedigrees, Polymorphisms and Traits) with output data (Lod Score tables and Maps). The co-existence of "raw" and "derived" data (whether inferred by logical or statistical estimation) introduces a problem of dependency control or truth maintenance. We are examining different approaches to the problem, including those similar to the version control mechanisms inherent in software development environments such as the well-known Source Code Control System (SCCS). Managing derived or computed results within the same software framework as the primary experimental data is an important goal of IGD, and one that will influence the success of the project.

Our conclusion is that, in design terms, "small is beautiful", but only if perfectly formed. The development of precise interface specifications between IGD components is needed to ensure that the architecture is extensible. The success of our approach in supporting gene mapping projects is an example of how increasing the modularity of IGD components, allowing a "Pick-and-Mix" solution to problems, can enhance the IGD concept, and provide creative ideas for the future.

ACKNOWLEDGMENTS

The authors wish to thank the other members of the IGD team, in particular Otto Ritter, for co-development of the system, and Elena Mavrakis, Vikky Murray, David Kelsell and Joanne Fallowfield in our laboratory for providing genotyping data. Sir Walter Bodmer and Tim Bishop are thanked for encouragement and discussion. The Imperial Cancer Research Fund provided support to SPB and NKS for this work. AS was supported by a grant from the European Commission (GENE-CT93–0003).

REFERENCES

Attwood J and Bryant S (1988). A computer program to make analysis with LIPED and LINKAGE easier to perform and less prone to input errors. *Ann. Hum. Genet.* 52, 259.

Bennett RL, Steinhaus KA, Ulrich SB, O'Sullivan CK, Resta RG, Lochner-Doyle D, Markel DS, Vincent V and Hamanishi J (1995) Recommendations for Standardized human pedigree nomenclature. *Am. J. Hum. Genet.* 56, 745–752.

Bryant SP (1994) Genetic Linkage Analysis. In: Bishop MJB (Ed.) *Guide to Human Genome Computing*. Academic Press, London: 59–110.

Bryant, SP (1996) Software for genetic linkage analysis - an update. *Molecular Biotechnology* 5(1), 49–61.

Buetow KH, Weber JL, Lundwigsen S, Scherpbier-Heddema T, Duyk GM, Sheffield VC, Wang Z , Murray JC (1994). Integrated human genome-wide maps constructed using the CEPH reference panel. *Nature Genetics* **6**, 391–393.

Chapman CJ (1990). A visual interface to computer programs for linkage analysis. *Am. J. Med. Genet.* 36, 155–160.

Curtis D (1995) Standardized pedigree nomenclature. *Am J Hum Genet* 57, 982–983

Davis S, Schroeder M, Goldin LR and Weeks DE (1996). Nonparametric simulation-based statistics for detecting linkage in general pedigrees. *Am. J. Hum. Genet.* 58, 867–880.

Dausset J, Cann H, Cohen D, Lathrop M, Lalouel J-M and White R (1990). Centre d'etude du Polymorphisme Humain (CEPH): collaborative genetic mapping of the human genome. *Genomics* 6, 575–577.

Deckert J, Nothen MM, Bryant SP, Ren H, Wolf HK, Stiles GL, Spurr NK and Propping P (1995) Human Adenosine A1 receptor gene: Systematic Screening for DNA sequence variation and linkage mapping on chromosome 1q31–32.1 using a silent polymorphism in the coding region. *Biochemical and Biophysical Research Communications* 214(2), 614–621.

Dewald G, Cichon S, Bryant SP, Hemmer S, Nothen MM and Spurr NK (in press) The Human Complement C8 gene, a member of the lipocalin gene family: polymorphisms and mapping to chromosome 9q34.3. *Annals of Human Genetics*.

Dunham I, Durbin R, Thierry-Mieg J and Bentley DR (1994). Physical mapping projects and AceDB. In: Bishop MJ (Ed) *Guide to Human Genome Computing*. Academic Press, London. 111–158.

Edwards AWF (1972). *Likelihood*. Cambridge University Press, Cambridge.

Green P, Falls K, Crooks S (1989). Documentation for CRI-MAP, version 2.4. Available from Phil Green.

Lange K (1986a). A test statistic for the affected-sib-set method. Ann. Hum. Genet. 50, 283–290.

Lange K (1986b). The affected sib-pair method using identity by state relations. Am. J. Hum. Genet. 39 (1), 148–150.

Lathrop GM and Lalouel JM (1984). Easy calculations of lod scores and genetic risks on small computers. *Am. J. Hum. Genet.* 36, 460–465.

Lathrop GM, Lalouel JM, Julier C and Ott J (1985). Multilocus linkage analysis in humans: detection of linkage and estimation of recombination. *Am. J. Hum. Genet.* 37, 482–498.

Mamelka PM, Dyke B and MacCluer JW (1987). *Pedigree/Draw for the Apple Macintosh.* Department of Genetics, Southwest Foundation for Biomedical Research, San Antonio.

O'Connell JR and Weeks DE (1995). The VITESSE algorithm for exact multilocus linkage analysis via genotype set-recoding and fuzzy inheritance. *Nature Genet.* 11, 402–408.

Ott J (1991). *Analysis of Human Genetic Linkage.* revised edition. Johns Hopkins University Press, Baltimore.

Penrose LS (1935). The detection of autosomal linkage in data which consist of brothers and sisters of unspecified parentage. *Ann. Eugen.* 6, 133–138.

Ritter O (1994) The integrated genomic database. In: Suhai S (ed) *Computational methods in genome research.* Plenum Press, pp 57–73

Smith FJD, Eady RAJ, McMillan JR, Leigh IM, Geddes JF, Kelsell DP, Bryant SP, Spurr NK, Kirtschig G, Milana G, de Bono AG, Owaribe K, Wiche G, Pulkinnen L, Uitto J, Rugg EL, McLean WHI and Lane EB. (in press). Plectin deficiency: Hereditary basis for Muscular Dystrophy with Epidermolysis Bullosa. *Nature Genetics.*

Spiridou, A., Spurr, N.K. and Bryant, S. (in prep. a). IGD/X-PED: A system for the management and graphical display of pedigree data. *Am. J. Hum. Genet.*

Spiridou, A., Spurr, N.K. and Bryant, S. (in prep. b). IGD: Facilitating genetic linkage analysis. *Genomics*

Spurr NK, Bryant SP, Attwood J, Nyberg K, Cox SA, Mills A, Bains R, Warne D, Cullin L, Povey S, Sebaoun J-M, Weissenbach J, Cann HM, Lathrop M, Dausset J, Marcadet-Troton, A and Cohen D (1994) European Gene Mapping Project (EUROGEM): Genetic Maps based on the CEPH Reference Families. *European Journal of Human Genetics* 2, 193–252.

Weeks DE and Lange K (1988). The affected-pedigree-member method of linkage analysis. *Am. J. Hum. Genet.* 42, 315–326.

Weeks DE and Lange K (1992). A multilocus extension of the affected-pedigree-member method of linkage analysis. *Am. J. Hum. Genet.* 50, 859–868.

Weeks DE, Sobel E, O'Connell JR, Lange K (1995) Computer programs for multilocus haplotyping of general pedigrees. *Am J Hum Genet* 56, 1506–1507

Whittemore, AS and Halpern, J. (1994) A class of tests for linkage using affected pedigree members. . *Biometrics* 50, 118–127.

10

ERROR ANALYSIS OF GENETIC LINKAGE DATA

Robert W. Cottingham Jr.,[1] Margaret Gelder Ehm,[2] and Marek Kimmel[3]

[1]Division of Biomedical Information Sciences
Johns Hopkins University School of Medicine
Baltimore, MD
bc@gdb.org
[2]Glaxo Wellcome Inc.
5 Moore Drive
Research Triangle Park, NC
meg_ehm@glaxo.com
[3]Department of Statistics
Rice University and Department of Cell Biology
Baylor College of Medicine
Houston, TX
kimmel@rice.edu

ABSTRACT

Any system of measurement has inherent sources of error. Research into the sources of error and how to compensate for them not only improves the accuracy of any analysis, but often leads to a better understanding of the underlying system and improved methods of measurement and analysis.

Like other systems of measurement, the mapping of the human genome at any level is subject to error. Here we will consider error in human genetic linkage mapping, however this is just one example of error to consider in the broader field of genome research and the approaches used in this study can serve as examples of how error analysis could be applied generally to other genomic measurements.

1. INTRODUCTION

Traditionally in genetic linkage mapping, errors were ignored. In early mapping studies, for experienced, efficient laboratories, ignoring errors was not a significant problem

Theoretical and Computational Methods in Genome Research, edited by Suhai
Plenum Press, New York, 1997

since the error rate was low relative to the recombination rate between markers in the low density maps of that time.

However an erroneous genotype usually results in a false recombination. In high density maps where recombinations rarely occur between markers an error is given much more weight and can substantially increase the estimate of the recombination fraction.

A method for detecting errors in genetic linkage data which is a variant of the likelihood ratio test statistic is described and then used to detect errors by testing the null hypothesis of no error for each individual at each locus versus the alternative hypothesis of error. High values of the index pinpoint individuals and loci with relatively unlikely genotypes. Power and significance studies of this method are reviewed to show its effectiveness in actual use.

1.1. Sources of Error in Genotype Data

Genetic linkage analysis considers pedigrees of individuals and their genotypes at two or more loci.[1] There are a variety of sources of error which include:

Pedigree errors:
- Misidentification of individuals
- Misidentification of relationships
- False paternity
- Unknown adoption

Typing errors:
- Sample switch
- Laboratory procedure error
- Misinterpretation of genotypes
- Keypunch data error

Algorithms can readily identify non-Mendelian inheritance, although the actual source of the error can be difficult to discern.[2,18] The present paper will concentrate on the relatively more difficult problem of finding genotype errors which are compatible with Mendelian inheritance.

It may seem counterintuitive to look for errors in genotypes which are consistent with Mendel's laws, but consider that competent laboratories will have usually detected and corrected the non-Mendelian errors in their data, so all of the remaining errors in a "good" dataset will be Mendelian. These errors do affect the results, and by knowing how the results are affected we can find ways to look for them.

1.2. What Are the Effects of Mendelian Genotype Errors?

When genetic maps were relatively low resolution (20 cM), the significance of errors normally was not great. Now that mapping is heading towards 0.1 cM resolution, erroneous genotypes can lead to very significant errors in distance. To see why, consider Ref. 17 which shows that

$$E(\hat{\theta}) = \theta_t(1 - \epsilon) + (1 - \theta_t)\epsilon = \theta_t + (1 - 2\theta_t)\epsilon.$$

where:

ϵ = the misclassification frequency or frequency with which recombinants are misclassified as nonrecombinants and nonrecombinants are misclassified as recombinants

θ_t = the true recombination fraction

When $\theta_t \to 0$, $E(\hat{\theta})$ approaches the error frequency, in the limit all observed recombination events will be erroneous. Also no true recombinants are classified as nonrecombinants since they do not exist. Therefore in dense maps any system which can detect the false recombinants will detect most of the errors. And since errors in dense maps are usually false recombinants, they increase the map distances and make locus order difficult to determine.

1.3. How Prevalent Are Errors in Practice?

A number of studies have been undertaken to estimate the magnitude of errors in genotypic data. These include:

Citation	Description	Est. Error Rate
Lathrop et al. (1983)	Tokelau	1.0%
Dracopoli et al. (1991)	CEPH Chromosome 1	0.6%
Buetow (1991)	CEPH Chromosome 4	1.4%
Brzustowicz et al. (1993)	CEPH Database	3.0%

These error rates were determined primarily by retyping the data. Of course this is an extremely expensive exercise which can not be carried out in general practice. So we want to look for a computational method which would allow us to find errors.

Note that these error rates are significantly larger than the recombination fraction for high resolution maps. So if these are typical rates, it will be virtually impossible to build high resolution genetic linkage maps without allowing for errors. As suggested in Ref. 3, typical error rates are probably about 3%, and this seems consistent with our experience.[6]

1.4. How Can Errors Be Dealt with?

In general there are two approaches —

compensate — develop techniques which estimate the actual error rate and adjust for its effects.

or

correct — identify actual errors which can then be retested and corrected.

Compensation approaches in genetic linkage analysis[15,14] are difficult to compute and therefore have not been developed to the extent required in large human pedigree studies. Simplified forms of the correction approach have been used by many labs, most commonly the CHROMPIC option of CRIMAP.[10] These identify regions or intervals within an individual's genotype which exhibit high recombination rates.

Only a few attempts have been made to develop computational methods for identifying specific individuals and loci likely to contain errors.[3,14,18] Because of the computational difficulty involved in large human pedigree studies, each of these resulted in an approximated solution. We were tempted by this challenge to develop a new method which would be a full likelihood ratio test starting with FASTLINK[12,11,13,8,19] as a basis[5] since this program significantly reduced the computation time in general linkage calculations. As it turned out we too were unable to produce a full likelihood ratio test however our approximation comes very close to this ideal. In addition, for the first time power and significance studies have been conducted to determine the capabilities of this method.[6] Here we review the method, its development, and how it can be applied.

2. METHODS

The advantage of using the likelihood method is that in its full implementation it captures maximum information. The likelihood of a pedigree is generally represented as[7]

$$L(\mathbf{X}; \theta_{jk}) = \sum_{m} \sum_{g} P(x \mid g) P(g|.)$$

where:

$\mathbf{x}_j = (x_{1j}, \ldots, x_{mj})$	is a vector of phenotypes at locus j, with one for each individual in the pedigree of size m
$\mathbf{g}_j = (g_{1j}, \ldots, g_{mj})$	is the corresponding vector of genotypes at locus j which in general are unobservable
θ_{jk}	is the recombination fraction between two loci j and k
$\mathbf{X} = (\mathbf{x}_j, \ldots, \mathbf{x}_k)$	is the joint phenotype between the two loci j and k

The likelihood as expressed above is for two loci j and k, but this can easily be extended for any number of loci by allowing θ to be a vector of recombinations for many loci. By analyzing many loci jointly the total information and therefore power increases. This is why we want to use multipoint analysis in general and specifically for error detection even though additional loci increase the computation time required.

With a fully penetrant locus such as the usual anonymous DNA markers used in mapping, for individual i at locus j, the penetrance is

$$P(x_{ij}|g_{ij}) = \begin{cases} 1 & \text{if } x_{ij} = g_{ij} \\ 0 & \text{if } x_{ij} \neq g_{ij} \end{cases}$$

Note that this form of the penetrance function describes what should happen, but does not take into account the possibility of error. As originally proposed by Ref. 9, error can be modeled be changing the penetrance function to account for the possibility that the measured phenotype is different from the actual genotype. In this case the penetrance function can be redefined as an error penetrance function

$$P(x_{ij}|g_{ij}) = \begin{cases} 1 - \varepsilon_{ij} + \frac{\varepsilon_{ij}}{h} & \text{if } x_{ij} = g_{ij} \\ \frac{\varepsilon_{ij}}{h} & \text{if } x_{ij} \neq g_{ij} \end{cases}$$

where:

ε_{ij} is the probability of error for individual i at locus j

and

$h = a(a+1)/2$ where a is the number of alleles at locus j.

This representation results in $h - 1$ small penetrance values for the erroneous phenotypes and as a result the likelihood is much more difficult to calculate. The disadvantage of a full likelihood model with variable penetrance is that it is much more computationally difficult.

2.1. Define the New Error Detection Method

The goal of the new method is to identify individuals whose measured genotype at a particular locus is in error. To do this we will compare the null hypothesis of no error against the alternative hypothesis of error for each individual i at each locus j, that is

$$H_0 : \varepsilon_{ij} = 0$$

versus

$$H_1 : \varepsilon_{ij} > 0$$

To make this comparison we do a likelihood ratio test for each individual at each locus where the likelihood is now expressed as $L_{ij}(\mathbf{X}; \theta, \varepsilon_{ij})$ which includes the new error parameter. The likelihood ratio test statistic then is

$$\lambda_{ij}(\mathbf{X}) = -2 \ln \left\{ \frac{L_{ij}(\mathbf{X}; \hat{\theta}_0, \varepsilon_{ij} = 0)}{L_{ij}(\mathbf{X}; \hat{\theta}_{\varepsilon^*}, \varepsilon_{ij} = \varepsilon^*)} \right\}$$

where $(\hat{\theta}_{\varepsilon^*}, \varepsilon^*)$ is the site of the global maximum of L_{ij}.

This is the exact form of the likelihood which we proposed to calculate. However even though we used FASTLINK, it turned out that the denominator was still uncomputable. This is due to the incomplete penetrance representing the probability of error which results in a dense array for every individual. Without the error model, the penetrance array is very sparse and easily calculated with the optimization techniques in FASTLINK. Therefore an approximation of the denominator had to be developed. The alternative chosen is

$$L_{ij}(\mathbf{X}; \hat{\theta}_0, \varepsilon_{ij} = \varepsilon_1)$$

where ε_1 represents a fixed small value of the estimate of the error rate. Typically we use $\varepsilon_1 = 0.01$. In specific simulations we have found that the resulting approximation

$$\tilde{\lambda}_{ij}(\mathbf{X}) = -2 \ln \left\{ \frac{L_{ij}(\mathbf{X}; \hat{\theta}_0, \varepsilon_{ij} = 0)}{L_{ij}(\mathbf{X}; \hat{\theta}_0, \varepsilon_{ij} = \varepsilon_1)} \right\}$$

varies insignificantly in power from λ_{ij} and can be shown to otherwise produce the same prediction of errors.

2.2. Procedure

To determine which individual's loci have likely errors in their measured genotype the following procedure is followed:

1. Begin by selecting 2 to 4 loci to be tested. There must be a minimum of 2 of course in order to detect recombination. Ideally we would use as many loci as available to maximize the power (see below). However there is an upper limit on the number of loci which current multipoint likelihood routines can process within available memory and reasonable time. Typically 4 is sufficient to insure reasonable power and not require excessive processing time however this depends heavily on the size of the pedigree.

2. The order of the loci and their distance must be known from previous mapping efforts such at those conducted by CEPH, Généthon and CHLC. The loci should be as close together as possible. The closer the loci are, the more powerful the error detection will be. Given the high resolution linkage maps now available from Généthon for instance Ref. 4 where the mean interval size between marker loci is 1.6 cM, it will usually be possible to find closely spaced markers.

3. $\tilde{\lambda}_{ij}(\mathbf{X})$ is calculated for each individual i at each locus j using a modified version of the ILINK program from FASTLINK 2.2 which is known as GENOCHECK.[6]

4. The calculated values of $\tilde{\lambda}_{ij}$ are then ordered. The highest value is the one most likely to contain an error and should be checked first. However before checking, we must should first understand the significance of the value so as to not waste time checking predicted errors which might be false positives and therefore of less interest. In the next section we describe a method for setting a cutoff which would have the false positive and false negative relationship desired by the lab. Then based on this cutoff, values which are higher than the cutoff will be retested to determine if they are actually in error.

2.3. Determining the Cutoff

To understand whether a particular value of $\tilde{\lambda}_{ij}$ is a significant indication of error, it is important to understand its distribution. As with any statistical test there will be a range of values. The higher values in the distribution are strong indicators of error. At some lower level there will be an increased probability of false positives. The user must decide what level of false positives or false negatives are acceptable and then adjust the cutoff accordingly. To understand this behavior one can generate simulations under H_0 which will give the appropriate cutoff usually determined as a particular p-value.

For instance the programs FASTLINK or SIMULATE[16,20,8] can be used to randomly generate large numbers of pedigrees with a specific structure (mimicking the real pedigrees under study), number of loci, recombination distances, and allele frequencies under the hypothesis of no error. Then $\tilde{\lambda}_{ij}$ can be calculated and an empirical cumulative distribution under H_0 can be formed. From this distribution the p-value can then be calculated as a percentage of the simulated values which are larger than the observed value under the null hypothesis. By increasing the cutoff, there will be fewer false positives but increased false negatives as with any test statistic.

2.4. Significance

The significance can be determined by simulating additional pedigrees which are error free. Then using the chosen p-value cutoff determined above, an estimate of the false positive error rate can be calculated by counting the number of test results greater than the cutoff which in this case are falsely indicating errors that do not exist. In practice we have found that for informative, closely spaced loci (θ¡0.02), 4-point calculations of $\tilde{\lambda}_{ij}$ yield false positive error rates with little inflation over selected nominal error rates between 0.00 and 0.04.

2.5. Power

Similarly, the power of the method can be determined by simulating additional pedigrees with particular errors and determining the proportion of these actual errors which are detected by the method. The power is not only a function of the method but also the intrinsic informativeness of the pedigree structure and genotypic data.

In practice we have found the power of the method to be greater than 0.80 for phase known individuals in pedigrees where again there are closely spaced loci (θ¡0.02), and 4-point calculations of $\tilde{\lambda}_{ij}$.

3. DISCUSSION

Researchers often consider their experimental results to be absolutely correct, even when they know that this is not true. The assumption of no errors can be a useful approximation, but of course when anything is measured there is error at some level. The study of errors and their cause provides a ready source of new detailed information about a system which can dramatically improve understanding and the associated analytic tools.

The error detection technique described here is designed to be used specifically in genetic disease mapping studies. This technique has now been applied in several studies.[5,6] When $\tilde{\lambda}_{ij}$ is found to be above the selected cutoff, labs typically initially reread the genotypes to see if there was a misread or data entry error. If further investigation is warranted, the genotyping experiments are repeated on occasion. The results in practice have been very useful in identifying actual errors.

Unlike some previous techniques[14] this one is completely general for any pedigree structure and able to compute results involving at least 4 loci in reasonable time. Another advantage is that this technique makes maximum use of the available information when compared with previous techniques which results in improved power and significance. In addition, this is the first, and to date the only technique which has published power and significance performance. This should be an integral part of any error analysis technique. In the present case the predicted performance has been shown to closely match actual performance in one detailed study.[6]

For the many experienced labs doing large scale genetic mapping, it has become common to check the maps for all double recombinants and to retype these. As noted, an erroneous genotype in the limit will produce a recombination which internal to a set of markers would be a double recombinant in the individual with the error if all of the other adjacent markers are typed correctly (since the phase has to switch back again). This raises the question, why not just look for these double recombinants and retype them rather than go to the trouble of running GENOCHECK.

The advantage of this program is that it will identify the most likely genotype to be in error so that it is unnecessary to hunt through other less likely errors. In addition it will identify single recombinants that are likely to be in error. And it uses the maximum information which necessarily will give more power.

In general one wants high power (true positive rate) and a low false positive rate in an ideal error detection test statistic. In using a full likelihood model GENOCHECK provides the optimum capability in this regard and allows the user to select the appropriate cutoff to balance power and significance as appropriate for a particular study and needs. The approximated test does not seem to have been less capable in tests or actual practice.

Because of the nature of genetic linkage, loci which are close together necessarily provide a form of redundancy testing within and between closely related individuals. Understanding this and the nature of genotyping errors has led to an effective method for detecting errors. As has been described elsewhere this method has not only detected genotyping errors but also a pedigree error which otherwise had been difficult to detect.

Since genome research now is supported by a variety of measurements it might be useful to consider how errors affect these measurements and the conclusions we derive from them. Here we have considered the detailed study of the effect of errors on genetic linkage analysis. Similar approaches toward finding test statistics which have high likelihood of identifying errors associated with other genomic measurements could yield useful new insights and analytic tools. Once a potentially useful test statistic is found, it is important to investigate its power and significance, and parameters affecting these properties to fully understand the tests utility.

3.1. GENOCHECK Distribution

GENOCHECK is available by anonymous FTP from softlib.cs.rice.edu.

REFERENCES

1. Jacques S. Beckman. Genetic mapping, an overview. In Sándor Suhai, editor, *Computational methods in genome research*. Plenum Press, 1994.
2. M. S. Boehnke and S.-W. Guo. Statistical approaches to identify marker typing error in linkage analysis. *American Journal of Human Genetics (Supplement)*, 51:A183, 1992.
3. L. M. Brzustowicz, C. Mérette, X. Xie, L. Townsend, T. C. Gilliam, and J. Ott. Molecular and statistical approaches to the detection and correction of errors in genotype databases. *American Journal of Human Genetics*, 53:1137–1145, 1991.
4. C. Dib, S. Fauré, C. Fizames, D. Samson, N. Drouot, A. Vignal, P. Millasseau, S. Marc, J. Hazan, E. Seboun, M. Lathrop, G. Gyapay, J. Morissette, and J. Weissenbach. A comprehensive genetic map of the human genome based on 5,264 microsatellites. *Nature*, 380:152–154, 1996.
5. Margaret Gelder Ehm, Marek Kimmel, and Jr. Robert W. Cottingham. Error detection in genetic linkage data for human pedigrees using likelihood ratio methods. *Journal of Biological Systems*, 3(1):13–25, 1995.
6. Margaret Gelder Ehm, Marek Kimmel, and Jr. Robert W. Cottingham. Error detection for genetic data using likelihood methods. *American Journal of Human Genetics*, 58(1):225–234, 1996.
7. R. C. Elston and J. Stewart. A general model for the analysis of pedigree data. *Human Heredity*, 21:523–542, 1971.
8. Robert W. Cottingham Jr., Ramana M. Idury, and Alejandro A. Schäffer. Faster sequential genetic linkage computations. *American Journal of Human Genetics*, 53:252–263, 1993.

9. B. J. B. Keats, L. Sherman, and J. Ott. Report of the committee on linkage and gene order. *Cytogenet. Cell Genet.*, 55:387–94, 1990.
10. E. S. Lander and P. Green. Construction of multilocus genetic linkage maps in humans. *Proc. Nat. Acad. Sci. USA*, 84:2363–2367, 1987.
11. G. M. Lathrop and J. M. Lalouel. Easy calculations of lod scores and genetic risks on small computers. *American Journal of Human Genetics*, 36:460–465, 1984.
12. G. M. Lathrop, J. M. Lalouel, C. Julier, and J. Ott. Strategies for multilocus linkage analysis in humans. *Proc. Natl. Acad. Sci. USA*, 81:3443–3446, June 1984.
13. G. M. Lathrop, J. M. Lalouel, and R. L. White. Construction of human genetic linkage maps: likelihood calculations for multilocus linkage analysis. *Genet Epidemiol*, 3:39–52, 1986.
14. S. E. Lincoln and E. S. Lander. Systematic detection of errors in genetic linkage data. *Genomics*, 14:604–10, 1992.
15. N. E. Morton and A. Collins. Standard maps of chromosome 10. *Ann. Hum. Genet.*, 54:235–51, 1990.
16. J. Ott. Computer-simulation methods in human linkage analysis. *Proc. Nat. Acad. Sci. USA*, 86:4175–4178, 1989.
17. J. Ott. *Analysis of Human Genetic Linkage*. The Johns Hopkins University Press, Baltimore and London, 1991. Revised edition.
18. J. Ott. Detecting marker inconsistencies in human gene mapping. *Human Heredity*, 43:25–30, 1993.
19. Alejandro A. Schäffer, Sandeep K. Gupta, K. Shriram, and Robert W. Cottingham Jr. Avoiding recomputation in linkage analysis. *Human Heredity*, 44:225–237, 1994.
20. D. E. Weeks, J. Ott, and G. M. Lathrop. SLINK: a general simulation program for linkage analysis. *American Journal of Human Genetics (Supplement)*, 47:A204, 1990.

11

MANAGING ACCELERATING DATA GROWTH IN THE GENOME DATABASE

Kenneth H. Fasman

Genome Database
Division of Biomedical Information Sciences
Johns Hopkins University School of Medicine
Baltimore, Maryland 21205-2236

ABSTRACT

The international Human Genome Project has entered a phase marked by phenomenal data growth. Rapid increases in the size and complexity of genomic maps, sequences, and macromolecular structures threaten to overwhelm the existing scientific information infrastructure. The traditional organization of public biological databases is inadequate to meet the problems arising from this increased flow. The Genome Database, an international repository of human genetic and physical mapping information, has been extensively revised to address these issues. Features of the new GDB architecture that were designed specifically to address expanding data volume include: direct public curation of the database, electronic data submission, efficient querying of specific genomic regions, and a modular database design. These features are described, with emphasis on their contribution to the query, display, and analysis of large data sets.

1. INTRODUCTION

The Genome Database, an international collaboration hosted at the Johns Hopkins University, is the major public repository for human gene mapping information[1,2,3,4]. Established in 1989, the project has evolved significantly through five previous revisions in order to keep pace with the scientific advancements of the Human Genome Project. Chief among the relevant advancements is the dramatic growth in information related to human genomic mapping.

The Genome Database version 6 (GDB v6) architecture[5] was designed specifically to address problems associated with its rapidly expanding data volume. Features of the new system include:

- **Direct public editing and curation of the database contents**, including a mechanism for third-party annotation of submissions and the creation of ad hoc

Theoretical and Computational Methods in Genome Research, edited by Suhai
Plenum Press, New York, 1997

links between data in GDB and information in other databases. No other method of data handling scales as well.

- **Electronic data submission system for bulk inserts *and* updates**. This is of critical importance to high-volume data sources such as genome centers.
- **Querying and display of map "slices" rather than entire genomic maps**. Slices are a single continuous region of the genome, usually defined by a pair of flanking markers or a cytogenetic location. Researchers must be able to pinpoint a region of interest in multiple genetic and physical maps easily, regardless of the overall size and complexity of those maps. This reduces query, download, and display times.
- **Separation of a complex monolithic database into component modules**. The revised GDB is managed as a federated database in miniature, a model that will be extended to include external biological databases in the near future. Attempts to gather vast areas of biology in a single database will not scale given current database management technology.

2. FEATURES OF GDB VERSION 6

The important new features of the Genome Database can be divided into two categories: those characteristics that are directly visible to users as new or enhanced functionality, and technical improvements that enhance performance and extensibility, or simplify maintenance of the system. User-oriented features include direct public curation and manipulation of map sections ("slices"). Technical features of GDB v6 include electronic data submission and the separation of a unitary database into modular components.

2.1. Direct Public Curation

The single most dramatic change in the Genome Database is a revision of its editorial model. Prior to the new release, all information was entered into the database by either the GDB staff or the HUGO editorial committees. While members of the genome community frequently submitted their data to GDB in electronic or paper form, it was not possible for them to enter the information directly into the database.

The same limitation exists for every other major public biological database, with the exception of the Genome Sequence DataBase (GSDB)[6], and its predecessor, Los Alamos' implementation of GenBank[7]. These systems pioneered the concept of "electronic data publishing"[8] in the biological research community by providing the first interfaces for direct user editing and the first protocol for bulk electronic submission of data. GDB v6 builds upon the GSDB/GenBank electronic publishing paradigm in a number of ways.

GDB's new editorial model is built on the premise that every data object has an owner, either an individual or an "edit group" that shares responsibility for a given set of data (such as the members of a laboratory or a larger-scale collaboration). Data may be submitted in confidence for a period up to six months in length, during which only the owner and GDB staff may view it. The security of unreleased and public data objects is insured by requiring individual contributors and edit group members to have a login account and password. Legitimate members of the genome research community can obtain these accounts easily from the Genome Database staff, but they will not be available to the general public in order to minimize the potential for abuse. However, the database remains freely queryable by anyone with access to the Internet.

Although the concept of object-level ownership is not new to community databases, true public curation required a novel extension of this concept. GDB v6 allows third parties to annotate the contributions of others. This capability is critically important to maintaining the accuracy and timeliness of the database contents. It enables the responsibility for correcting and enhancing database entries to be shared by the entire research community. Third-party annotation is implemented by having a separate "Annotation Object" in GDB that can be linked to any "primary" object in the database. While the proper attributes of the primary object may only be changed by its owner, the linked annotations are always displayed with the data object as associated ("derived") attributes. These attributes always appear when the main object is displayed, but they are clearly distinguished as having come from another contributor.

The mechanism of derived attributes is used frequently in the Genome Database to implement two very specific sorts of annotation. First, ad hoc links can be established from any object in GDB to corresponding information in another database, or anywhere at all on the World Wide Web. The external link is simply a structured annotation bearing the information required to access the related object from the external source. Second, derived attributes are also used to add properties to objects that are likely to be determined by other investigators, or even to involve multiple conflicting observations. For example, GDB v6 features the concept of a publicly editable map. While the maps themselves are owned by their initial contributor, others may place additional markers on these maps, even with multiple conflicting localizations.

Direct community contribution and curation is the only editorial model for public databases that scales well with increasing data and limited resources. Responsibility for placing information in the database and keeping it current lies with those who generate it. At the same time, third party annotation provides for continuous peer review of the full database content by the entire genome community.

Critics of this approach frequently raise concerns about the potential loss of database quality and consistency. They point to the problems that the sequence databases have had with erroneous or incomplete entries, or information that is simply out of date. (e.g, ref. 9,10). However, it can be argued that these problems were exacerbated by the historic lack of a direct update mechanism and third-party annotation capability in these databases. GDB is relying on the presence of these features, the continuing peer review of the HUGO editorial committees and the GDB staff, in addition to a reasonable limit on who can obtain an editing account to minimize these concerns.

2.2. Electronic Data Submission

The majority of the primary mapping and sequencing data generated for the Human Genome Project are coming from a limited number of centers throughout the world. Most of these groups maintain sophisticated laboratory information systems because of the complexity and sheer volume of experimental data and analysis that must be managed. The natural extension of these efforts is the bulk electronic submission of information to the relevant public databases.

Most community databases do not yet have a published interface for accepting bulk electronic submissions. Among the major biological databases, only the Genome Sequence DataBase and ACEDB have officially supported mechanisms for bulk submission. GSDB uses a transaction protocol originally developed during the management of the GenBank project at Los Alamos. The C. elegans database and other genomic databases built with the ACEDB software can process inserts and updates in ".ace" file format[11]. On an ad hoc basis, the National Center for Biotechnology Information will accept large submissions to GenBank using their ASN.1 format, and the EMBL Nucleotide Sequence Database will likewise accept bulk submissions based on their flatfile format[12]. A number of these approaches are neither formally defined nor well documented, and only the ".ace" format accommodates database updates as well as inserts.

The Genome Database now includes an electronic data submission (EDS) system that supports bulk inserts and updates. It has a simple, formal syntax designed for rapid, reentrant parsing, a semantic specification derived automatically from the database schema definition, and metacommands for specifying inserts, updates, symbol substitution, account information, etc.[13]. The EDS system supports very large data sets, such as maps or clone libraries with thousands of elements, comparable in size and complexity to multi-megabase sequence submissions. The submission system uses the same syntax and semantics as all communications between GDB client and server applications.

Unfortunately, the EDS system introduces yet another data exchange format to the genome community. This step was taken only after it was determined that none of the formats already in use met all of the design requirements. The Genome Database encourages the bioinformatics community to consider this format for their future data exchange implementations — the tools will be made freely available on the World Wide Web. In addition, GDB promises to support any other format adopted as a standard in the genome community that possesses equivalent functionality.

2.3. Map Slices

Recently reported genomic maps (e.g., ref. 14) and sequences (e.g., ref. 15) are denser, longer, and more complex than ever before. It is therefore incumbent on the genome informatics community to develop software systems for building, viewing, querying, and analyzing these very large data sets on computer platforms readily accessible to the average biologist. This requires changes to existing software systems that are based on manipulating an entire map or sequence at once, or those that can only display an arbitrarily defined section at one time.

The Genome Database version 6 makes significant use of the concept of a map interval or "slice." This is a single continuous genomic region defined by a pair of flanking landmarks or a single landmark and a defined distance to either side of it. Additionally, the user may request that only certain classes of marker be included in the returned slices: genes only, just clones and breakpoints, etc. Slices can be thought of as both spatial and categorical cuts through large genomic maps. The concept of a map slice is a natural one, since all genomic mapping ultimately involves the placement of new markers into intervals defined by an existing, well-ordered framework[16]. Users may query the database for multiple maps containing the designated interval, and the slices retrieved can be aligned for comparison based on shared markers.

Trafficking in slices reduces database query, network download, and screen display times for the user. Concerns about the loss of the complete picture are largely unwarranted. Whole-chromosome slices can still be considered when reduced to a reasonable number of framework markers or other landmarks of interest to the user. Attempts to display an entire integrated chromosome map at once, including many thousands of markers, are only useful for wall hangings at genome conferences. However, slice retrieval and manipulation must still be as rapid as possible to support interactive navigation through a reasonable section of the genome.

2.4. Modular Database Components

Each public biological database must handle an ever growing volume of data, reflecting the success of the Human Genome Project. However, there is another type of growth that database organizers find strongly compelling—the expansion of the scope of the database. This compulsion arises from the scientific community's demand for a more integrated view of biological data without the arbitrary separation imposed by the existing ad hoc partition of databases.

A fully integrated view of human biology is a worthwhile goal, but current database technology simply will not scale to provide it from a single source. Individual database schemas that attempt to capture many different kinds of information are often too complex to be maintained with the available resources—several recent attempts have had difficulty moving beyond the prototype stage. Inheritance, one of the great strengths of object-oriented data models, becomes the Achilles heel of a single integrated schema when class hierarchies need to be modified in conflicting ways to accommodate increasingly detailed and divergent biological information in their subclasses.

Federated databases provide a practical alternative by providing modular systems that can still be loosely coupled to provide an integrated view. Data models and other software technology can still be shared across these systems, but they can also be enhanced or revised somewhat independently of each other. Just as layered software architectures enabled systems of greater complexity to be developed more rapidly, modular databases can also be established in parallel as long as a common interface between them is defined.

Prior to version 6, the Genome Database schema had grown into a complex monolith that was increasingly difficult to maintain. GDB v6 was implemented as a "federation in miniature," a collection of related but separate relational databases storing mapping information, literature citations, and contact information for members of the genome community. An "object broker" was developed with limited capability to retrieve information across these database modules[17]. A system of accession identifiers was implemented for the federated database that could easily be extended across all public biological databases[18]. The citation and contact databases were developed in consultation with other database projects with the expectation that they would have widespread utility. It is hoped that other public databases will adopt similar practices, so that this in-house federation can be extended more easily to many categories of genomic data.

Software that can perform fully distributed queries and edits across a database federation is still not widely available. However, several commercial systems do exist for this purpose, as well as a number of prototype systems developed by the genome informatics community (e.g., ref. 19,20). These systems perform reasonably well for many straightforward distributed queries, but may have trouble with complex queries and edits of any sort.

3. IMPLICATIONS FOR OTHER BIOLOGICAL DATABASES

The new Genome Database release implements a number of features that anticipate requirements of all public biological databases in the near future. These include public curation, bulk data submission and exchange, limited views of the data for rapid querying and display, and modular schemas designed for a federated database environment.

For example, the genome community is beginning to balk at the effort required to submit their data to even a single database. Genome centers typically collect data that must be submitted to multiple repositories. There is a desperate need for joint submission tools to simplify this effort e.g., a protocol for jointly submitting clone mapping and sequencing data, or annotated peptide sequences combined with crystallographic coordinates. A unified format for bulk submission and exchange would make such tools much easier to develop. More importantly, it would result in greater integration among the biological databases, as jointly submitted data could be cross-referenced at submission time.

The concept of map slices can be readily extended to other biological databases as well. User interfaces should allow an arbitrary section of nucleotide or peptide sequence data to be retrieved with accompanying feature annotations. While molecular viewer software already allows structures to be viewed at varying levels of detail, the data filtering happens in the client program, and is not an inherent feature of the database itself. Similarly, the user may wish to view a particular subunit of a molecular complex. Researchers should be able to query a database of metabolic pathway information by retrieving any or all paths between two designated compounds, rather than having to browse a wall chart-style network of hopelessly interwoven reaction paths.

When implementing data submission or extraction mechanisms, it is important that the "slices" be defined by the user at run-time. They should not be artifactual, imposed by the database or interface software's arbitrary design limitations. For example, the approach being taken by the International Nucleotide Sequence Database Collaborators, limiting sequence entries to 350 kilobases[21], seems short-sighted now that megabase sequencing efforts are a reality.

The extensions to the Genome Database described here are only the first steps towards a more efficient biological information infrastructure. Only further experience will reveal which need to be further refined or replaced altogether with alternative approaches. It is the goal of the GDB project that the best of these and other shared information management technologies will lead to improved interoperability among the public biological databases. Enhanced sharing of infrastructure is crucial to the success of all, particularly in a time of growing responsibility and fixed resources.

ACKNOWLEDGMENTS

The Genome Database is the result of a rewarding collaboration among computer scientists and biologists, all of whom care deeply about the success of the Human Genome Project. It is a team effort, and credit for any success GDB has had should be shared by the entire staff. However, two individuals stand out as prime forces behind the development of the current release: Stan Letovsky and Peter Li. I am grateful to both of them for their wisdom and support. In addition, I would like to thank Sue Borchardt, John Campbell, and Bob Cottingham, without whom GDB v6 would remain unimplemented.

The Genome Database is supported by the U.S. Department of Energy (DE-FC02–9ER6130), the U.S. National Institutes of Health, and the Science and Technology Agency of Japan, with additional support from the Medical Research Council of the United Kingdom, the INSERM of France, and the European Union.

REFERENCES

1. Pearson, P.L. (1991) The Genome Data Base (GDB) - a human gene mapping repository. Nucl. Acids Res. 19: 2237–2239.

2. Pearson, P.L., Matheson, N.W., Flescher D.C., and Robbins R.J. (1992) The GDB Human Genome Data Base Anno 1992. Nucl. Acids Res. 20: 2201–2206.
3. Cuticchia, A.J., Fasman, K.H., Kingsbury, D.T., Robbins, R.J., and Pearson, P.L. (1993) The GDB Human Genome Data Base Anno 1993. Nucl. Acids Res. 21: 3003–3006.
4. Fasman, K.H., Cuticchia, A.J., and Kingsbury, D.T. (1994) The GDB Human Genome Data Base Anno 1994. Nucl. Acids Res. 22: 3462–3469.
5. Fasman, K.H., Letovsky, S.I., Cottingham, B.W., and Kingsbury, D.T. (1996) Improvements to the GDB Human Genome Data Base. Nucl. Acids Res. 24: 57–63.
6. Keen, G., Burton, J., Crowley, D., et al. (1996) The Genome Sequence DataBase (GSDB): meeting the challenge of genomic sequencing. Nucl. Acids Res. 24: 13–16.
7. Burks, C., Cassidy, M., Cinkosky, M.J., et al. (1991) GenBank. Nucl. Acids Res. 19: 2221–2225.
8. Cinkosky, M.J., Fickett, J.W., Gilna, P., and Burks, C. (1991) Electronic data publishing and GenBank. Science 252: 1273–1277.
9. Kristensen, T., Lopez, R., and Prydz, H. (1992) An estimate of the sequencing error frequency in the DNA sequence databases. DNA Seq. 2: 343–346. (Erratum appears in DNA Seq. 3: 337, 1993.)
10. White, O., Dunning T., Sutton G., Adams M., Venter J.C., Fields C. (1993) A quality control algorithm for DNA sequencing projects. Nucl. Acids Res. 21:3829–38.
11. Durbin, R., and Thierry-Mieg, J. (1994) The ACEDB genome database. In: Computational methods in genome research. S. Suhai, ed. Plenum Press: New York, 45–55.
12. Graham Cameron, personal communication.
13. Documentation for the electronic data submission system can be found on GDB's World Wide Web server at *http://gdbwww.gdb.org/*.
14. Hudson, T.J., Stein, L.D., Gerety, S.S., et al. (1995) An STS-based map of the human genome. Science 270: 1945–1954.
15. Fleischmann, R.D., Adams, M.D., White, O., et al. (1995) Whole-genome random sequencing and assembly of Haemophilus influenzae Rd. Science 269: 496–512.
16. Cox, D.R., Green, E.D., Lander, E.S., Cohen, D., and Myers, R.M. (1994) Assessing mapping progress in the Human Genome Project. Science 265: 2031–2032.
17. Li, P., Waldo, D., Pineo, S.V., and Campbell, J.M. (1996) An extensible object broker for the Genome Data Base V6. Department of Energy Human Genome Program Contractor-Grantee Workshop V. Santa Fe, NM.
18. Fasman, K.H. (1994) Restructuring the Genome Data Base: a model for a federation of biological databases. J. Comp. Bio. 1: 165–171.
19. Chen, I.A., Kosky, A., Markowitz, V.M., and Szeto, E. (1996) OPM*QS: The Object Protocol Model Multidatabase Query System. Lawrence Berkeley National Laboratory Technical Report LBL-38181 (available at *http://gizmo.lbl.gov/DM_TOOLS/OPM/OPM_QS/DOC/OPM_QS_DOC.html*).
20. Leone, J. And Shin, D.-G. (1996) A graphical ad hoc query interface capable of accessing heterogeneous public genome databases. Department of Energy Human Genome Program Contractor-Grantee Workshop V. Santa Fe, NM.
21. GenBank enters the megabase sequence era. NCBI News, September 1995, p. 1.

12

ADVANCES IN STATISTICAL METHODS FOR LINKAGE ANALYSIS

Jeffrey R. O'Connell[2] and Daniel E. Weeks[1,2]

[1]The Wellcome Trust Centre for Human Genetics
University of Oxford
Oxford, United Kingdom
[2]Department of Human Genetics
University of Pittsburgh
Pittsburgh, Pennsylvania

ABSTRACT

The past few years have seen a quantum leap in the amount of genetic marker information available in a typical disease-gene mapping study. Whereas before we were dealing with tens of markers, now we are dealing with hundreds of markers spanning the genome. New computational tools are needed to handle multiple marker analyses efficiently. We describe here a recent advance in statistical methods for linkage analysis that involves the merger of statistics, numerical analysis, computer science, and mathematical genetics: The VITESSE algorithm for rapid exact computation of multipoint lod scores in the presence of untyped individuals. The development of this algorithm involved understanding what parts of the computation were redundant and which parts were essential, and then the creation of a very efficient way to dynamically identify and eliminate the redundant parts. The VITESSE algorithm is much more memory efficient and much faster than previous linkage analysis programs, and significantly extends the size of the problems that can be solved.

VITESSE has already proven useful in practice, enabling us to compute better multipoint lod score curves that provide more precise localization of the disease gene. In addition to explaining this algorithm, we discuss briefly our plans for the future, which include parallelization of these algorithms to enable optimal use of modern computational resources.

INTRODUCTION

Rapid progress is being made in mapping disease genes, since advances in laboratory technology and reagents have made it much easier to type large numbers of genetic

Theoretical and Computational Methods in Genome Research, edited by Suhai
Plenum Press, New York, 1997

markers rapidly. Thus, while in the past, the amount of genetic linkage data was the limiting factor, now the limiting factor has become the capabilities of available software for *linkage analysis*, which is a statistical method for assessing the evidence that a putative disease gene is near a genetic marker (Ott 1991). It is based on the key idea that the further apart genetic loci are, the more likely it is that recombination will occur between them. Thus, we can assess the distance between two genes by examining their tendency to pass together one generation to the next. The strength of evidence in favor of linkage is often reported as a *lod score*, which is essentially the log of the odds in favor of linkage (Morton 1955). Linkage analysis is extremely important in terms of constructing genetic maps and for localizing disease genes. Ideally we would like to localize the disease gene to the smallest interval possible, to provide the best possible starting point for physical mapping. To do this, we would like to be able to carry out multipoint computations, where we simultaneously analyze marker data from several closely linked marker loci spanning the region of interest. However, computational bottlenecks often preclude the desired analyses. We will describe here one recent advance in computational algorithms for linkage analysis (O'Connell and Weeks 1995).

THE ELSTON-STEWART ALGORITHM

Multipoint lod score computations can be very time-consuming, with runs of relatively simple problems easily taking 3 to 4 weeks on a very fast workstation. These analyses are difficult because they involve computing a sum of products where the sum is carried out over all possible multilocus genotypes of all the members of the pedigree. The algorithm used is known as the Elston-Stewart algorithm (Elston and Stewart 1971), and it can be represented in the abstract as:

$$\sum \text{Penetrance} * \text{Prior} * \text{Transmission}$$

where the summation is over all possible multilocus genotype vectors for the pedigree, and the three terms (Figure 1) are:

1. The *penetrance* is the probability $P(x \mid g)$ of having phenotype x given the genotype g. This is usually thought of as the probability of a person being affected given they have a copy of the disease allele.
2. The *prior* is only defined for founders, which are individuals whose parents are not members of the pedigree. The prior probability $P(g_k)$ for a founder k is a function of the population gene frequencies.
3. The *transmission* probability $P(g_{i_1} \mid g_{i_2}, g_{i_3})$ is the probability that parents with genotypes g_{i_2} and g_{i_3} produce a child with genotype g_{i_1}.

The global summation may be made more efficient by splitting it into separate summations over individuals and then moving the summations as far into the products as possible (this is known as peeling the pedigree). However, the algorithm is still slow. The amount of time that the Elston-Stewart algorithm takes is a function of the number of markers and the number of multilocus genotypes in the pedigree. Complexity is greatly increased when there are:

- Large numbers of untyped people
- Highly polymorphic markers with many alleles
- Inbreeding loops

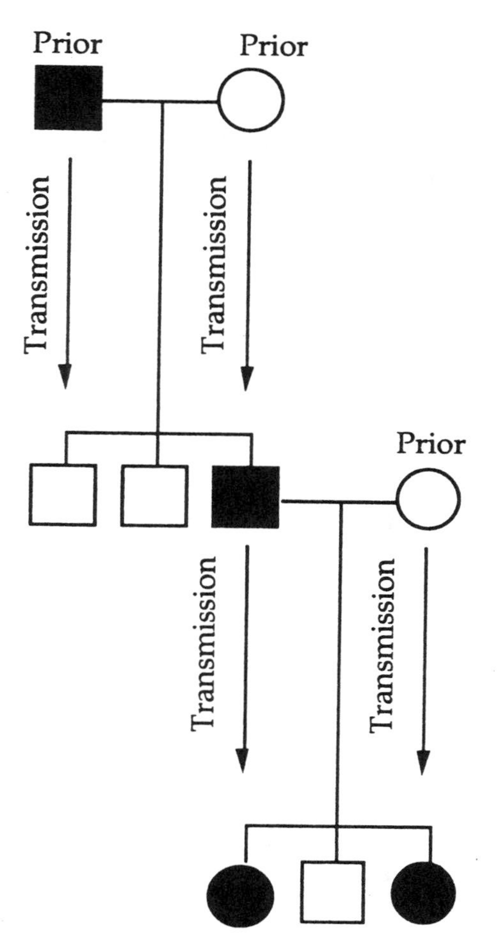

Figure 1. In the Elston-Stewart algorithm, every founder has a prior term, while every parent-child triple has a transmission term. Every person has a penetrance term.

For example, if there are 6 alleles at a marker locus, then each untyped person initially has 21 unordered genotypes:

1/1	2/1	3/1	4/1	5/1	6/1
2/2	3/2	4/2	5/2	6/2	
3/3	4/3	5/3	6/3		
4/4	5/4	6/4			
5/5	6/5				
6/6					

If the computation involves 5 such markers, then *each* untyped person has 21^5 Å 4 million multilocus genotypes.

COMPUTATIONAL STRATEGIES

Thus, the fundamental challenge is to *reduce* the number of genotypes needed in the calculation without altering the likelihood at all. In other words, the goal is to reduce the number of terms in the summation index; *e.g.:*

$$\sum_{j=1}^{100} A_j * B_j * C_j = \sum_{k=1}^{50} D_k * E_k * F_k$$

Note that reduction at even a single locus can be significant. For example, if there are 4 markers with 10 genotypes each, then an untyped person has 10,000 possible multilocus genotypes, while if we reduce each marker by one genotype, then an untyped person would only have 6,561 possible multilocus genotypes.

There are two basic strategies for reducing the number of summations:

1. Eliminate terms that have a known factor of zero, since these terms contribute nothing to the sum. In the Elston-Stewart algorithm, this occurs in two basic ways:

   a. The transmission probability is zero if and only if the child's haplotype and the parental genotype are incompatible (A maternal *haplotype* represents the "half" of the genotype that is presumed to have come from the mother).

   b. The penetrance term is zero for any genotypes that are incompatible with the observed phenotype. For example, for a fully-penetrant dominant disease, if a person is phenotypically normal, then the penetrance probability is zero if the genotype under consideration contains at least one disease allele. Genotype elimination algorithms (Lange and Goradia 1987; Lange and Weeks 1989) have been developed which efficiently identify all genotypes that are incompatible with known phenotypic information.

2. Combine terms with common factors (distributive law). In the Elston-Stewart algorithm, this primarily occurs when the transmission probabilities are the *same* for different genotypes. In addition, it is sometimes possible to combine like terms by applying the technique of global lumping, where all the alleles that do not appear in the pedigree are lumped into a single dummy allele, whose frequency is the sum of the missing allele (Ott 1991).

While genotype elimination algorithms were developed many years ago, they do not help much when there are several untyped generations in the pedigree. For example, consider the pedigree in Figure 2. Assuming the marker has 6 alleles (which is the minimum required even if we applied global lumping), then each untyped person starts out with 21 possible genotypes. Genotype elimination, which is based on considering the consistency of each possible genotype with the genotypes of "close" relatives, reduces the genotypes somewhat (Table 1). However, spouses 1 and 2 have 11 possible genotypes each at this single locus. As this small pedigree illustrates, genotype elimination is often not sufficient to reduce the number of possible genotypes, particularly on large pedigrees where there are many untyped generations (The farther a given individual is from a genotyped relative, the less genotype elimination will work).

It was clear to us that the crux of the problem of how to accelerate linkage analyses lay in the answer to the question: How can we reduce the genotype lists of untyped people

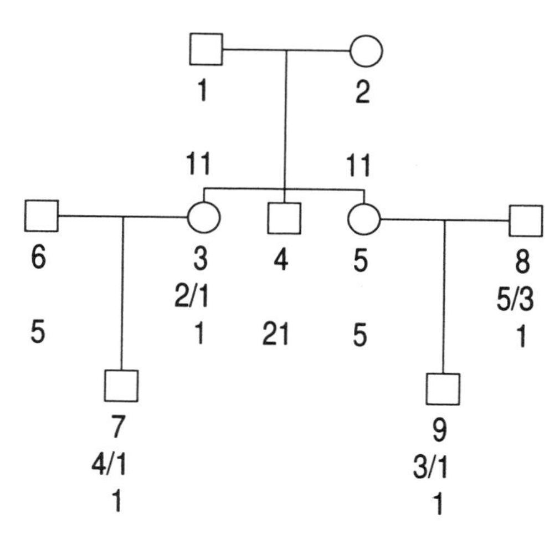

Figure 2. Four members of the pedigree, persons 3, 7, 8 and 9, have been genotyped at a 6-allele marker. The number indicates the number of genotypes after genotype elimination.

without losing information? We have recently presented our new and novel solution in the journal Nature Genetics (O'Connell and Weeks 1995). Our solution basically involved creating a very efficient way to determine which genotypes are the "same" in terms of their role in the Elston-Stewart algorithm. It is based on the following observations regarding the two distinct roles that alleles have:

Role 1. *Alleles determine recombination events.* Recombination events in a child's haplotype are identified by examining the origin of the child's alleles in an ordered parental genotype. The specific label of the allele *per se* does not matter. For example, suppose we wish to compute the transmission probability that the child's haplotype arose from the parent's genotype, where we are using ordered genotypes (in an "ordered" genotype, the order of the alleles indicates the maternal or paternal origin of the allele):

Parental Genotype	1|2	5|6
	3|4	7|8
Child Haplotype	1|	5|
	4|	8|

Here both haplotypes represent the *same* recombination event, but the parental genotypes have *different* prior probabilities. If we can develop an efficient way

Table 1. The possible genotypes after genotype elimination

Person	Unordered genotypes						Number of genotypes
1, 2	1/1	2/1	3/1	4/1	5/1	6/1	
	2/2	3/2	4/2	5/2	6/2		11
5	1/1	2/1	3/1	4/1	5/1	6/1	5
6	4/1	4/2	4/3	4/4	5/4	6/4	5

to tell which transmission probabilities are the same, then we need only compute one transmission probability for each group of identical probabilities.

2. *Alleles determine prior probabilities of observing a particular genotype in the founders.* Here allele labels must be known, as each label is associated with a particular population allele frequency.

TRANSMITTED AND NON-TRANSMITTED ALLELES: SET-RECODING AND FUZZY INHERITANCE

In our paper (O'Connell and Weeks 1995), we developed a very efficient method for identifying common terms in the likelihood computation. This method involves classifying alleles of an untyped person into two classes:

1. Transmitted Alleles: Any allele that appears in a typed descendant that may have arisen from the person.
2. Non-transmitted Alleles: Any allele appearing in the person's genotype list that is not transmitted.

For example, on our example pedigree in Figure 2, we would have the following classification of alleles for each untyped person:

Person	**Transmitted alleles**	**Non-transmitted alleles**
1, 2	1, 2, 3	4, 5, 6
4	None	1, 2, 3, 4, 5, 6
5	1, 3	2, 4, 5, 6
6	1, 4	2, 3, 5, 6

This classification of alleles into "transmitted" and "non-transmitted" alleles on a person-by-person basis turns out to be extremely useful because, in terms of recombination indicators, a person's non-transmitted alleles are indistinguishable. In other words, a person's non-transmitted alleles play an identical role in the Elston-Stewart algorithm, and so we can combine non-transmitted alleles into sets and drastically reduce the number of multilocus genotypes needed to compute the likelihood.

We implemented this approach by first "set-recoding" all the non-transmitted alleles into a single set, and each transmitted allele into its own singleton set. For example, we would use the following recoding of alleles for the pedigree displayed in Figure 2:

Person	**Transmitted alleles**	**Single non-transmitted allele**
1, 2	{1}, {2}, {3}	{4, 5, 6}
4	None	{1, 2, 3, 4, 5, 6}
5	{1}, {3}	{2, 4, 5, 6}
6	{1},{4}	{2, 3, 5, 6}

After the alleles have been changed into sets, then we use these genotype sets to compute the likelihood rather than the original genotypes (Figure 3).

Then, based on an analogy to fuzzy logic, we use "fuzzy inheritance" to correctly keep track of the recombination events when the alleles are set-recoded. Normally, inferences about inheritance are based on allele equality. However, in set-recoding, instead of allele equality we use set inclusion (Note that we set-recode the genotypes of typed indi-

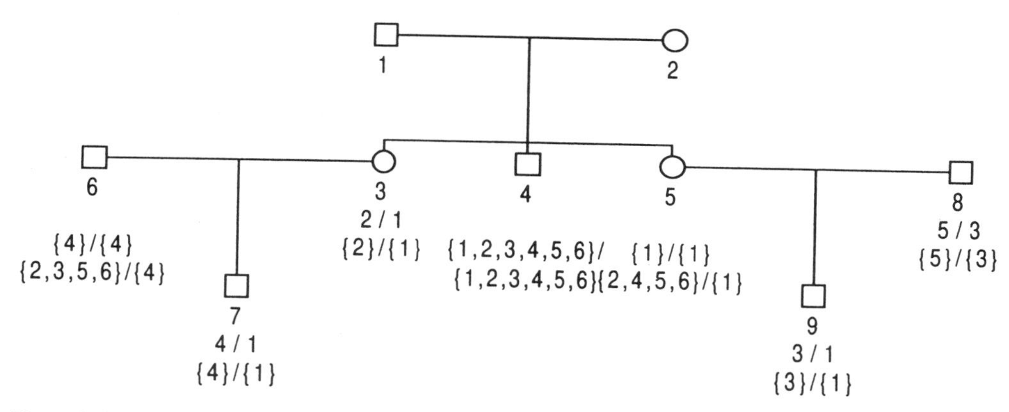

Figure 3. The original genotypes and the set-recoded genotypes. Note that the 11 possible genotypes for persons 1 and 2 are not shown.

viduals too; this permits the use of fuzzy inheritance throughout). For example, if we use fuzzy inheritance logic on the parent-child pair in Figure 4, then the parental genotype {4, 5, 6}/{2} appears 'homozygous' to person 5 with respect to the haplotype transmission since alleles {4, 5, 6} and {2} are subsets of {2, 4, 5, 6}; the set {2, 4, 5, 6} encodes the fact that, for example, both alleles 4 and 2 occur in pairs of haplotypes of person 5 with opposite phase.

The combination of set-recoding and fuzzy inheritance permits us to calculate the likelihood exactly while summing over many fewer genotypes. For example, on our example pedigree from Figure 2, the reduction in number of genotypes ranges from 36% to 95%:

Person	Genotypes	Before	After	Reduction
1,2	[Not shown]	11	7	36%
4	{1, 2, 3, 4, 5, 6}/{1, 2, 3, 4, 5, 6}	21	1	95%
5	{1}/{1}			
	{2, 4, 5, 6}/{1}	5	2	60%
6	{2, 3, 5, 6}/{4}			
	{4}/{4}	5	2	60%

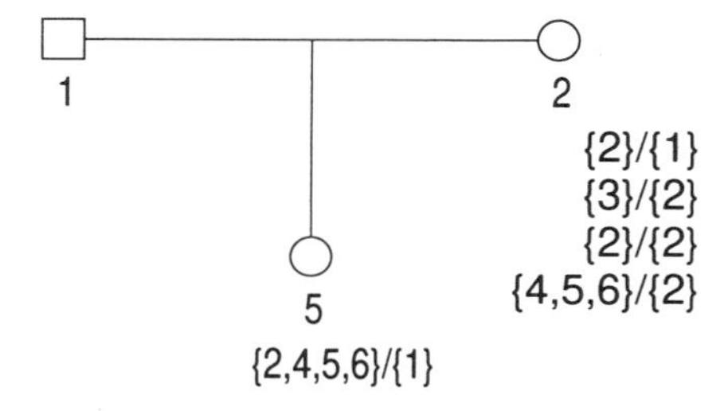

Figure 4. The possible genotype of person 2 that could have given rise to the {2,4,5,6}/{1} genotype of her child 5.

This novel approach to computing likelihoods has been implemented in the computer program VITESSE (O'Connell and Weeks 1995), which, on realistic problems, is much much faster than the previous generation of general linkage analysis programs. In addition, reduction of the number of genotypes means that not only is VITESSE faster, but it can handle much larger problems than the previous programs because it uses much less memory. In fact, VITESSE can do problems that were completely impossible before because of memory problems (For details, please see our paper (O'Connell and Weeks 1995).

CONCLUSION

Our VITESSE algorithm represents a major step forward in the rapid computation of multilocus likelihoods on pedigrees. However, now that VITESSE is available, investigators have simply increased the size of the data set they wish to analyze, and so are already clamoring for more speed. In order to provide more speed, we have recently parallelized VITESSE using PVM (Geist et al. 1993), BSP (Miller and Reed 1994), and threads. We have not yet extensively evaluated our parallel versions, but preliminary results indicate that their performance scales well with the number of processors available. While the parallel versions should help meet the demand for faster analyses, we are now exploring approximation techniques that will permit us to generate answers to the desired accuracy in a fraction of the time required to do a full computation. By working at the interface of computer science, human genetics, and mathematical statistics, we hope to continue to provide state-of-the-art computational tools to the international human genetics community.

ACKNOWLEDGMENT

This work was supported in part by funds from NIH grant HG00932, Association Française Contre Les Myopathies (AFM), the Wellcome Trust Centre for Human Genetics at the University of Oxford, the University of Pittsburgh, and the W.M. Keck Center for Advanced Training in Computational Biology at the University of Pittsburgh, Carnegie Mellon University, and the Pittsburgh Supercomputing Center.

REFERENCES

Elston RC, Stewart J (1971) A general model for the genetic analysis of pedigree data. Hum Hered **21**:523–542

Geist GA, Beguelin AL, Dongarra JJ, Mancheck RJ, Sunderam VS (1993) *PVM 3 User's Guide and Reference Manual*. Oak Ridge National Laboratory: Oak Ridge, Tennessee

Lange K, Goradia TM (1987) An algorithm for automatic genotype elimination. Am J Hum Genet **40**:250–256

Lange K, Weeks DE (1989) Efficient computation of lod scores: genotype elimination, genotype redefinition, and hybrid maximum likelihood algorithms. Ann Hum Genet **53**:67–83

Miller R, Reed J (1994) The Oxford BSP library users' guide. Oxford Parallel, Oxford

Morton NE (1955) Sequential tests for the detection of linkage. Am J Hum Genet **7**:277–318

O'Connell JR, Weeks DE (1995) The VITESSE algorithm for rapid exact multilocus linkage analysis via genotype set-recoding and fuzzy inheritance. Nat Genet **11**:402–408

Ott J (1991) Analysis of Human Genetic Linkage (Revised Edition). Johns Hopkins University Press, Baltimore

13

EXPLORING HETEROGENEOUS MOLECULAR BIOLOGY DATABASES IN THE CONTEXT OF THE OBJECT-PROTOCOL MODEL

Victor M. Markowitz,* I-Min A. Chen, and Anthony S. Kosky

Information and Computing Sciences Division
Lawrence Berkeley National Laboratory
Berkeley, CA

ABSTRACT

Solutions currently promoted for exploring heterogeneous molecular biology databases (MBDs) include providing Web links between MBDs or constructing MBDs consisting of links, physically integrating MBDs into data warehouses, and accessing MBDs using multidatabase query systems. Arguably the most difficult tasks in exploring heterogeneous MBDs are understanding the semantics of component MBDs and their connections, and specifying and interpreting queries expressed over MBDs. However, most existing solutions address only superficially these problems.

We propose a tool-based strategy for exploring heterogeneous MBDs in the context of the Object-Protocol Model (OPM). Our strategy involves developing tools that provide facilities for examining the semantics of MBDs; constructing and maintaining OPM views for MBDs; assembling MBDs into an OPM-based multidatabase system, while documenting MBD schemas and known schema links between MBDs; supporting multidatabase queries via uniform OPM interfaces; and assisting scientists in specifying and interpreting multidatabase queries. Each of these tools can be used independently and therefore represents a valuable resource in its own right. We discuss the status of implementing our strategy and our plans in pursuing further this strategy.

*To whom all correspondence should be sent: VMMarkowitz@lbl.gov (510) 486-4004 (Fax)

Theoretical and Computational Methods in Genome Research, edited by Suhai
Plenum Press, New York, 1997

1. INTRODUCTION

Data of interest to molecular biologists are distributed over numerous heterogeneous *molecular biology databases* (MBDs). These MBDs display heterogeneity at various levels: they are implemented using different systems, such as structured files or database management systems (DBMSs), are based on different views of the molecular biology domain, and contain different and possibly conflicting data. Furthermore, each MBD represents some part of the molecular biology domain and is often designed to address certain queries or applications. The data in an MBD are structured according to a *schema* specified in a *data definition language* (DDL) and are manipulated using operations specified in a *data manipulation language* (DML), where these languages are based on a *data model* that defines the semantics of their constructs and operations. Exploring multiple MBDs entails coping with the distribution of data among MBDs, the heterogeneity of the systems underlying these MBDs, and the semantic (schema representation) heterogeneity of these MBDs.

Strategies for managing heterogeneous MBDs can be grouped into two main categories(see Ref. 21421 for related classifications of heterogeneous database systems):

1. *Consolidation* strategies that entail replacing heterogeneous MBDs with a single homogeneous MBD formed by physically integrating the component MBDs, or by requiring MBDs to be reorganized using a common DDL or DBMS.
2. *Federation* strategies that allow access to multiple heterogeneous MBDs, while the component MBDs preserve their autonomy, that is, their local definitions, applications, and policy of exchanging data with other MBDs. Federation strategies include:
   (a) incorporating in MBDs references (links) to elements in other MBDs, or constructing MBDs consisting of such links;
   (b) organizing MBDs into loosely-coupled multidatabase systems; and
   (c) constructing data warehouses.

Heterogeneous MBDs can be connected via hypertext links on the Web at the level of individual data items. Data retrieval in such systems is limited to selecting a starting data item within one MBD and then following hyperlinks between data items within or across MBDs. Note that data item links (e.g., hypertext links) between MBDs do not require or comply with schema correlations across MBDs. Numerous MBDs are currently providing such links. However, missing and inconsistent links between MBDs prompted some archival MBDs to propose coordinating the management of links between their MBDs.[1] Systems such as SRS[10,11] and LinkDB[13] extract existing link information from (usually flat file) MBDs, and construct indexes for both direct and reverse links allowing fast access to these MBDs. These systems resolve heterogeneity issues such as duplicate or incompatible identifiers and provide only simple index and key match retrieval, but lack the ability of supporting full query facilities.

Multidatabase systems are collections of loosely coupled MBDs which are not integrated using a global schema. Querying multidatabase systems involves constructing queries over component MBDs, where a query explicitly refers to the elements of each MBD involved. Component MBDs of multidatabase systems can be queried using a common query language, as is done in Kleisli,[3] or can be both described and queried using a common data model, as entailed by the Object-Protocol Model tool-based strategy described in this paper. The common query language approach does not require the component MBDs to be represented using a common DDL or data model; however, the users are required to have some knowledge regarding the structure of the MBDs they query. On the other hand, the common data model approach requires all participating MBDs to have a view defined in a common DDL so that users can examine and query component MBDs in the context of the same data model. Unlike link-based MBD systems, multidatabase systems support query languages that allow specifying complex query conditions across MBDs. A query translator is needed for translating queries expressed in the multidatabase query language to subqueries targeting component MBDs, and for optimizing these queries.

Data warehouses entail developing a *global* schema (view) of the component MBDs, where definitions of these MBDs are expressed in a common DDL and discrepancies between these definitions are resolved before they are integrated into the global schema. Data from component MBDs are transformed in order to comply with this global schema, and loaded into a central data repository. The Integrated Genomic Database (IGD),[19] Genome Topographer (GT),[9] and Entrez[22] are examples of data warehouses, where IGD is developed with the ACeDB database system, GT is developed with the Gemstone commercial object-oriented DBMS, and Entrez is based on ASN.1 structured files. The query facilities of data warehouses are provided by the underlying system (e.g., ACeDB), and query processing is local to the warehouse. However, constructing data warehouses requires costly initial integration of component MBDs, followed by frequent synchronization with these MBDs in order to capture the evolution of their schemas. Moreover, data warehouses need to be updated on a regular basis in order to reflect updates of component MBDs.

In this paper, we propose a tool-based strategy for exploring heterogeneous MBDs in the context of the Object-Protocol Model (OPM). We will argue that possibly the most difficult tasks in exploring heterogeneous MBDs is understanding the semantics of component MBDs and their connections, and specifying and interpreting queries expressed over multiple MBDs. Our strategy involves developing tools that provide facilities for constructing and maintaining OPM views for MBDs implemented using a variety of DBMSs, assembling MBDs into an OPM-based multidatabase system, while documenting MBD schemas and known schema links between MBDs, examining the semantics of MBDs, supporting multidatabase queries via uniform OPM interfaces, and assisting scientists in specifying and interpreting queries. Each of these tools can be used independently and therefore represents a valuable resource in its own right.

The rest of this document is organized as follows. The main semantic problems of exploring heterogeneous MBDs are discussed in section 2. Our tool-based strategy is described in section 3. An example of applying our strategy for querying GDB 6.0 and GSDB 2.0 is described in section 4. In section 5, we review the status of implementing our strategy and our plans for continuing this work.

2. SEMANTIC PROBLEMS OF EXPLORING MOLECULAR BIOLOGY DATABASES

2.1. Semantics of Global Schemas and Views

Examining data within and across MBDs is currently hampered by the lack of information on MBDs and their semantics. The need for comprehensive documentation of MBD schemas was discussed extensively at the last two Meetings on Interconnection of

Molecular Biology Databases.[15,20] It was observed at these meetings that MBD schemas capture domain knowledge about biology and therefore the goal of schema design is not only achieving an efficient implementation but also supporting biological exploration.

Existing systems for exploring heterogeneous MBDs do not address the problem of understanding the semantics of component MBDs. For example, systems that support links between MBDs do not provide any information regarding the structure or semantics of the linked MBDs. Some multidatabase systems require users to know the structure (schemas) of component MBDs, without providing them with any support for this purpose. The highest expectations are promoted by data warehouses which present a single unified view while insulating users from the component MBDs. Unfortunately, systems such as GT, IGD, and Entrez are no better documented than their component MBDs, if at all. IGD, GT, and Entrez are based on global schemas (views) of their component MBDs, expressed in ACeDB DDL for IGD, Gemstone DDL for GT, and ASN.1 for Entrez. As of March 1996, no documentation was available on the structure of the GT data warehouse; IGD's global schema specified in ACeDB DDL is scarcely discussed (see Ref. 19) and questions, such as the relationship between the IGD schema and the schemas of its target MBD components, are left unanswered; the structure of the ASN.1 files underlying Entrez's is also scarcely documented (see Ref. 22).

The global schemas of systems such as GT, IGD, and Entrez are not based on schema integration techniques, but are the result of independent schema design processes based on the domain knowledge underlying the component MBDs. The global schemas for GT, IGD, and Entrez were developed locally by small groups, and therefore do not represent 'consensus' schemas. In order to reduce the complexity of schema design and to make global schemas general enough, developers usually design these schemas using 'generic' classes and/or attributes, which may not be applicable to individual component MBDs and may not fully capture the semantics of the data. Little or no information regarding the relationships between global schemas and participating MBD schemas is provided.

Constructing global schemas or local views for exploring heterogeneous MBDs usually requires detecting semantic conflicts between schemas of component MBDs, ranging from naming conflicts and inconsistencies to detecting identical entities of interest that are represented differently. The same concept can be represented in different schemas by using synonyms, alternative terminology, or different data structures. For example one database could use the term *primer* to represent a class of primer sequences, while another could use the term *oligo*, and yet another could simply represent primers directly using their sequence data. Homonyms can cause naming conflicts in a heterogeneous MBD environment. Domain conflicts can be caused by storing similar values using different units or formats in different MBDs, or from conflicting data arising from different experiments or experimental techniques. Entities of interest can be represented using various data structures in different MBDs, where the diversity of representations stems from different views of the data (e.g., an `Author` can be represented only as an attribute of a `Citation` or as an independent object) and on the underlying DDL (e.g., a `Citation` can be represented as an object of a class within an object data model, but needs to be represented with one or several tables using a relational DDL). Other causes of conflicts include different ways of representing incomplete information (e.g., the meaning of nulls), and different ways of identifying objects in MBDs.

Resolving schema conflicts is a very complex task and may involve various methods ranging from simple renamings in order to resolve naming conflicts to schema restructurings in order to resolve structural dissimilarities. Systems such as IGD and SRS detect and resolve only some schema and data conflicts, such as name and object identification conflicts. Alternatively, a heterogeneous MBD system could leave conflict resolution to users. If users are responsible for conflict resolution, such a system could provide a mechanism for recording such resolutions and making them available to other users.

2.2. Semantics of Data Exploration

MBDs are usually explored via specially constructed schemas or views. These views may not necessarily preserve the *information capacity*[17,18] of component MBDs. The tool-based strategy described in the next section, for example, involves constructing OPM views of component MBDs, where an OPM view may entail constraints (e.g., referential integrity constraints) that are not enforced in the underlying MBD. Consequently access to an underlying MBD through an OPM view is restricted to those data which comply with these constraints, while other data are discarded. Discrepancies between the information capacities of the views employed for exploring heterogeneous MBDs and the underlying component MBDs can be a source of confusion if not properly documented and explained. This is especially critical for data warehouses where data are converted from the format of component MBDs into that of the data warehouse and are subsequently physically loaded into the data warehouse. Anecdotal evidence suggests that some data from target MBDs are not converted into IGD, but the causes and extent of the information gap between IGD and its component MBDs have not been examined or documented.

Often the power of the facilities provided for exploring heterogeneous MBDs is not properly characterized. Query capabilities provided by MBDs vary between two extremes. Systems such as SRS[10] support only queries with limited keyword matching capabilities. Entrez[22] provides two query interfaces, NetEntrez and WebEntrez, both supporting the expression of form-based queries. The query language of IGD consists of, and is limited to, the ACeDB query language. On the other hand, systems such as Kleisli allow users to query databases using powerful programming languages such as CPL. Although users of these systems can submit very complex queries, it is difficult to imagine a biologist mastering such languages.

Often interfaces for exploring heterogeneous MBDs do not provide any help in clarifying the semantics of the queries users specify or in interpreting the semantics of the query results. For example the Entrez interface provides a number of query forms and various choices of attributes on which to search, but does not offer any description of the extents over which these queries search, or the semantics of the individual attributes. These problems are compounded by the extremely large and complex molecular biology nomenclature and by various differences in interpretations of this nomenclature within the molecular biology community.

Users of many MBD systems, such as Entrez and ENQUire (see `http://csl.ncsa.uiuc.edu:80/ENQUire/`), interact directly with a Web query interface, and may not be aware of the existence of a global schema. It is difficult or even impossible for users of these systems to detect whether there are conflicts in component MBDs and to realize how the conflicts are resolved. Therefore, when a user receives uninterpreted answers to a query involving conflicting MBDs, information regarding conflicts and how the conflicts are resolved are all hidden from the user.

3. EXPLORING HETEROGENEOUS MOLECULAR BIOLOGY DATABASES IN THE CONTEXT OF THE OBJECT-PROTOCOL MODEL

In this section we briefly describe our tool-based strategy for exploring heterogeneous MBDs, and the tools we are developing in order to pursue this strategy. We will start by presenting the Object-Protocol Model (OPM) and the existing suite of OPM tools which form the backbone for this strategy. Then we will describe how our strategy addresses the semantic problems mentioned in the previous section. An overview of OPM and OPM tools can be found in Ref. 4.

3.1. The Object-Protocol Model and Tools

Objects in OPM are uniquely identified by object identifiers, are qualified by attributes, and are classified into classes. Classes can be organized in subclass-superclass hierarchies, where a subclass *inherits* the attributes of its superclasses.

Attributes can be *simple* or consist of a *tuple* of simple attributes. A simple attribute can have a single value, a set of values, or a list of values, and can be *primitive*, if it is associated with a system-provided data type or a controlled value class with controlled values or ranges, or *abstract*, if it takes values from other classes. The attributes of an object class can be partitioned into *non-versioned* and *versioned* attributes, where the former represent stable properties, while the latter represent evolving properties.

OPM provides protocol classes for modeling laboratory experiments. Protocols can be recursively expanded, where a protocol can be specified (expanded) in terms of alternative subprotocols, sequences of subprotocols, and optional protocols. A protocol class can be associated with regular as well as input and output attributes that are used for specifying input and output connections between protocols.

OPM supports the specification of derived attributes using derivation rules involving arithmetic expressions, aggregate functions, and attribute compositions (or path expressions). OPM also supports derived subclasses and derived superclasses. A derived subclass is defined as a subclass of another derived or non-derived class with an optional derivation condition. A derived superclass is defined as a union of two or more derived or non-derived classes.

An OPM query consists of a select, insert, delete, or update statement; only select queries are considered in this paper. These statements can involve conditions consisting of and-or compositions of atomic comparisons. An OPM select query on target class O_i can involve local, inherited, and derived attributes associated with O_i, as well as path expressions starting with these attributes. Although each OPM query is associated with a single target class, this limitation can be offset in part by using abstract attributes (referencing other object or protocol classes) in select and condition statements. Furthermore, multi-target class queries can be constructed using derived classes and derived attributes.

OPM data management tools provide facilities for developing databases using commercial relational DBMSs. OPM schemas can be specified using an OPM Schema Editor or a regular text editor, and can be published in various formats, such as LaTeX and Html. An OPM Schema Translator can be used for mapping OPM schemas into DBMS-specific relational schema definitions and SQL stored procedures.[5] The OPM Schema Translator also generates a *mapping dictionary* with information regarding the mapping of OPM elements into DBMS elements. The OPM Query Translator processes OPM queries and, using the mapping dictionary mentioned above, translates them into SQL queries.[7]

3.2. The OPM Tool-Based Strategy

Our strategy for exploring heterogeneous MBDs involves additional OPM tools that provide facilities for (1) constructing OPM views for MBDs developed with or without OPM; (2) assembling MBDs within an OPM-based multidatabase system, while documenting MBD schemas and known schema links between MBDs; and (3) expressing, processing, and interpreting multidatabase queries in this multidatabase system.

Constructing OPM views for MBDs. OPM views for MBDs can be constructed using an OPM Retrofitting tool.[6] This tool allows constructing one or more OPM views for existing MBDs developed without the OPM data management tools, or constructing multiple OPM views for MBDs developed using the OPM tools.

The OPM Retrofitting tool follows an iterative strategy of constructing OPM views for MBDs. First, a canonical (default) OPM view is generated automatically from the underlying MBD schema. Then this canonical OPM view can be refined using schema restructuring operations, such as renaming and/or removing classes and attributes, merging and splitting classes, adding or removing subclass relationships, defining derived classes and attributes, and so on.

A *mapping dictionary* contains information on the DBMS representations of the view (OPM) constructs. This mapping dictionary is used for generating appropriate retrieval and update methods for the view attributes and classes, and underlies browsing and querying MBD via OPM views.

The multidatabase directory. Incorporating an MBD into an OPM multidatabase system involves constructing one or more OPM views of the MBD, and entering information about the MBD and its views into a *Multidatabase Directory*. The multidatabase directory stores information necessary for accessing and formulating queries over the component MBDs, including:

1. *General information* describing each MBD accessible in the system, and the information necessary in order to access that MBD. In particular, for each MBD, the MBD Directory will contain: (i) the *MBD name* and a *brief description* of the purpose of the MBD; (ii) the *physical location* and *history* of the MBD, including the original charter of the MBD, pointers to related MBDs, etc; (iii) information on the *DDL* and *implementation details* for the MBD, including either precise definitions and examples for the DDL or references to where such information can be found, information on the underlying DBMSs, and implementation strategy for the MBD; (iv) *contact information* for the MBD; (v) *access information*, such as the type (e.g., browsing, querying, update) of data manipulation supported, internet addresses, URLs, and subscription information, as well as references and/or links to more detailed documentation; and (vi) *keywords* (e.g., `sequence` and `human genome`) for high level searches of the schema library described below.

2. An *MBD Schema Library* containing the schemas and related information for component MBDs. Each component MBD can have multiple schemas, including OPM schemas, schemas in the native DDL of the MBD and schemas in other DDLs of interest. The MBD Schema Library contains information on each major schema component (class, attribute), including: (i) *semantic descriptions* describing the physical or real world concepts represented by the schema component, and possibly constraints on the values of instances of the component that cannot be expressed in the underlying DDL; (ii) *design motivation* for the use of a particular construct in representing application data; (iii) *synonyms* and *keywords* for identifying differences in terminology and for establishing potential correspondences between the components of MBD schemas; and (iv) *sample data* providing examples of how the schema constructs are used and how typical data may appear. In addition general information on each schema is stored, including explanations and motivation for the particular view of the MBD provided by the schema. For OPM schemas, the mapping dictionary containing the OPM–DBMS correspondence is also recorded.

3. An *MBD Link Library* containing information about known links between classes in different MBDs. Information on each link includes a description of its semantics, the nature of the correspondence (one-to-one, surjective and so on), and any data-manipulations, such as reformatting of accession numbers, that need to be performed in order to traverse the link.

The Multidatabase Directory is maintained and can be examined using various tools, such as tools for keyword searches that allow identifying MBDs and schemas relevant to a particular query. The Multidatabase Directory is an essential part of processing multidatabase queries. Note that the 'general information' part of our multidatabase directory is similar to, but much more restricted in scope than the MBD dictionary proposed by Peter Karp in Ref. 16, which he called the *Knowledge Base of Databases*.

Supporting multidatabase queries. Queries in an OPM-based multidatabase system are expressed in the OPM multidatabase query language (OPM*QL).[8] OPM*QL extends the single-database OPM query language, OPM-QL, with constructs needed for querying multiple databases. These extensions include the ability to query multiple classes, possibly from distinct databases; constructs that allow navigation between the classes of multiple databases following inter-database links; and the ability to rename fields of a query in order to resolve potential naming conflicts between multiple databases.

An example of a simple OPM*QL query over databases GSDB and GDB is:

```
SELECT Name = GSDB:Gene.name, Annotation = GDB:Gene.annotation
FROM GSDB:Gene, GDB:Gene
WHERE GDB:Gene.accessionID = GSDB:Gene.gdb_xref AND GSDB:Gene.name = "ACHE" ;
```

In the query above, the term "`GSDB:Gene`" refers to class `Gene` of database schema `GSDB` (which must be recorded in the MBD directory) while term "`GSDB:Gene.name`" refers to attribute `name` of class `Gene`. If a class name is unique among all the classes listed in the multidatabase directory, the database name can be omitted from a term, for example "`Gene`" could be used instead of "`GSDB:Gene`".

The `WHERE` statement consists of and-or compositions of atomic comparisons. Conditions can involve multiple classes, possibly from different databases.

Processing OPM multidatabase queries involves generating OPM-QL queries over individual databases in the multidatabase system, and combining the results of these queries using a local query processor. The stages of generating OPM-QL queries and manipulating data locally may be interleaved depending on the particular query evaluation strategy being pursued.

Formulating and interpreting multidatabase queries. The most difficult problems of querying multiple heterogeneous MBDs are (1) formulating a query, which involves determining the MBDs contain relevant data, understanding how data are represented in each of these MBDs, and how data in these MBDs relate to one another; and (2) interpreting the result of a query. Addressing these problems requires comprehensive information on the MBDs that are explored, and unfortunately such information is seldom available. While it cannot fill existing gaps in the documentation of MBDs, the Multidatabase Directory can help by making existing documentation available through a single resource and in a uniform representation. Browsing and keyword-search tools can be used to identify MBDs potentially of interest. Documentation on MBDs and their schemas can be then examined in order to determine whether they do indeed contain relevant data. When the MBD schema and documentation are not sufficient to clarify certain semantic issues, sample data can provide additional insight by allowing comparisons of data representations for the same or similar data in different MBDs. The MBD Link Library can be consulted in order to determine known correspondences between relevant data in heterogeneous MBDs. Furthermore, using inter-database links in multidatabase queries simplifies their formulation by resolving representational incompatibilities, such as different formats for accession numbers.

The information in the Multidatabase Directory together with the semantics of the operations underlying multidatabase query processing can be used for interpreting query results. For example, information on the semantics of objects in a given class can be used for annotating query results, information about inconsistent inter-database links can be used for explaining null query results, and so on. Consider, for instance, class `Citation` in database Map_X containing only citations published between April 1990 and March 1996; the result of a query requesting all the citations in `Citation` can then state that the results refer to citations published in this time range.

4. AN EXAMPLE

In this section we illustrate how our strategy can be used for exploring heterogeneous MBDs, by describing an application involving the Genome Data Base (GDB) and the Genome Sequence Database (GSDB).

4.1. The GDB-GSDB Multidatabase System

The Genome Data Base (GDB) is an archival MBD of genomic mapping data maintained at Johns Hopkins School of Medicine, Baltimore.[12] The new version of GDB, GDB 6.0 (see `http://wwwtest.gdb.org/gdb/`), was developed with the Sybase DBMS using the OPM toolkit.[4] GDB contains objects identified by *accession numbers* and are classified in classes organized in a class hierarchy. The main classes of this class hierarchy contain objects representing genomic data, literature references, and information on people and organizations.

The Genome Sequence Database (GSDB) is an archival MBD of genome sequence data maintained at the National Center for Genome Resources, Santa Fe. The current version of GSDB, GSDB 2.0 (see `http://www.ncgr.org/gsdb/gsdb.html`), has also been developed with Sybase DBMS but without using the OPM toolkit. For GSDB 2.0, an OPM view (see `http://gizmo.lbl.gov/DM_TOOLS/OPM/opm_4.html`) has been constructed using the OPM Retrofitting tool; this view allows GSDB to be accessed using the OPM query tools.[7] GSDB 2.0 is structured around one main class of objects, `Entry`, whose objects represent DNA sequences identified by accession numbers; the actual sequences (strings) are represented by objects of another class, `Sequence`. GSDB 2.0 also contains objects representing various entities, including genes, products, sources, and references.

Both GDB and GSDB have a `Gene` class. In GSDB 2.0, genes are considered to be a kind of `Feature`, and are characterized by gene names and references to external MBDs, such as GDB, that contain additional information on genes. In GDB, genes are represented by objects of class `Gene` and are characterized by information that includes the reason a genomic region is considered a gene, links to gene families the gene belongs to, mapping information, and references to derived sequences.

Sequences are represented in GSDB by objects of class `Sequence`. Sequence data include the actual sequence, sequence length, and information on the source of the sequence. Sequence information in GDB is represented by objects of class `SequenceLink`. These objects contain annotations linking primary GDB objects to external sequence MBDs such as GSDB, as well as information regarding the beginning and end points of sequences.

Both GDB and GSDB contain classes representing products. In GDB, products are limited to gene products, while in GSDB a product can be associated with any feature. In both GDB and GSDB, these classes seem primarily to serve as a way of referencing external MBDs, such as protein MBDs.

Both GSDB and GDB contain data representing references and/or citations. In GSDB, a `Reference` object is considered as a kind of (i.e., a specialization of) `Feature` object. References in GSDB are characterized by titles, publication status, lists of authors and editors, and external references to the Medline bibliographic database. In GDB, citations are represented by objects of class `Citation` and are further classified in subclasses of `Citation` representing books, journals, articles and so on.

4.2. Examples of Queries Expressed over GDB and GSDB

The following OPM multidatabase queries are examples of typical queries expressed over GDB 6.0 and GSDB 2.0. These queries were suggested by Chris Fields of the National Center for Genome Resources, Santa Fe, and were specified with help provided by Ken Fasman and Stan Letovsky of the Johns Hopkins School of Medicine, Baltimore and Carol Harger of the National Center for Genome Resources.

Query 1: Find the protein kinase genes on chromosome 4. To identify protein kinase genes in GSDB it is necessary to first find protein kinase products in the GSDB `Product` class, and then find `Genes` associated with the same `Feature` as the `Product`. The corresponding `Gene` in GDB can then be accessed by following the `gdb_xref` attribute from the GSDB `Gene`, if present, and equating it with the GDB `accessionID` attribute. Some string reformatting was needed in this query in order to resolve discrepancies between the representations of accession numbers in GDB and GSDB; this reformatting was implemented using functions built into the OPM multidatabase query language, but are ignored in the queries shown below. In order to test whether a `Gene` occurs on chromosome 4, one can then follow the path in GDB from `Gene` to `MapElement` to `Map` to `Chromosome`.

```
SELECT GDB:Gene.displayName, GDB:Gene.accessionID, Feature.products.name
FROM GSDB:Feature, GDB:Gene
WHERE Feature.products.name MATCH "%protein kinase%"
   AND Feature.genes.gdb_xref = GDB:Gene.accessionID
   AND GDB:Gene.mapElements.map.chromosome.displayName = "4";
```

Query 2: Find sequenced regions on chromosome 17 with length greater than 100,000. Map elements on chromosome 17 are selected from the GDB class `MapElement` using the path from class `MapElement` to class `Map` to class `Chromosome`. Links from `MapElement` objects to GSDB `Entry` objects are found using the GDB `Se/quenceLink` class. From the GSDB `Entry` the corresponding sequence can be found and tested to see if its length is greater than 100,000.

```
SELECT Entry.accession_number, Entry.sequence.length
FROM GDB:MapElement, GDB:SequenceLink, GSDB:Entry
WHERE MapElement.map.chromosome = "17"
   AND SequenceLink.dBObject = MapElement.segment
   AND SequenceLink.externalDB.displayName = "GSDB"
   AND SequenceLink.accessionID = Entry.accession_number
   AND Entry.sequences.length > 100000;
```

Query 3: Find the sequences of ESTs mapped between 4q21.1 - 21.2. Currently this query requires two sub-queries: the first sub-query finds the coordinate range and the second sub-query finds ESTs with coordinates in that range and their sequences. Planned extensions to the multidatabase query system will allow this query to be expressed as a single OPM query.

The first part of the query finds the coordinates of the points q21.1 and q21.2 in the `Cytogenetic` Map of chromosome 4:

```
SELECT MapElement.coordinate, MapElement.point, MapElement.segment.displayName
FROM GDB:MapElement
WHERE MapElement.map.objectClass = "CytogeneticMap"
   AND MapElement.map.chromosome.displayName = "4"
   AND MapElement.segment.displayName IN {"q21.1", "q21.2"};
```

Next, one can retrieve the expressed `Amplimers` occurring between these coordinates and lookup the corresponding sequence in GSDB.

```
SELECT Amplimer.displayName, Entry.accession_number,
       Entry.sequences.length, Entry.sequences.sequence
FROM GDB:Amplimer, GDB:SequenceLink, GSDB:Entry
WHERE Amplimer.isExpressed = "Yes"
   AND Amplimer.mapElements.map.chromosome.displayName = "4"
   AND Amplimer.mapElements.sortCoord >= START_COORD
   AND Amplimer.mapElements.sortCoord <= END_COORD
   AND SequenceLink.dbObject = Amplimer
   AND SequenceLink.externalDB.displayName = "GSDB"
   AND SequenceLink.accessionID = Entry.accession_number;
```

where `START_COORD` and `END_COORD` are the values from the previous query.

5. PURSUING THE OPM TOOL BASED STRATEGY

We have developed some of the tools required for implementing our strategy for exploring heterogeneous MBDs in the context of OPM. In this section we will describe the current state of our tools and discuss our plans for pursuing the implementation of this strategy.

5.1. Constructing OPM Views

The current version of the OPM Retrofitting tool can be applied to MBDs developed with Sybase and can be adapted straightforwardly to MBDs developed with other commercial relational DBMSs, such as Oracle and Informix. We are in the process of extending the OPM Retrofitting tool to MBDs developed using non-relational DBMSs, such as ACeDB, and/or defined using non-relational DDLs, such as ASN.1. These extensions will broaden the range of MBDs for which an OPM view can be constructed, and that can thus be included in an OPM multidatabase system.

5.2. The Multidatabase Directory

As mentioned in section 3, central to our strategy of assembling MBDs into a multidatabase system is a Multidatabase Directory that includes general information on MBDs, MBD links, and MBD schemas. In our current implementation, the MBD Schema Library consists only of the OPM schema, associated schema documentation, and mapping information for each MBD. OPM supports extensive schema documentation capabilities: each class or attribute in an OPM schema can be associated with description and user-specified properties; for a controlled value class, each controlled value can also be associated with its description. Therefore, detailed schema descriptions can be embedded in an OPM schema definition. We are not aware of any other data models that support such documentation capabilities. Nevertheless, this is still not adequate for assisting users in examining and understanding the semantics of MBDs, nor in specifying and interpreting multidatabase queries.

We plan to develop an extended Multidatabase Directory as an independent resource that will provide support for examining and understanding MBDs as well as help scientists in specifying queries across multiple MBDs. The MBD Schema Library part of this Directory will contain schemas for MBDs expressed not only in OPM but in a variety of different DDLs as well, including each MBD's native DDL and several DDLs (e.g., ASN.1 and ACeDB), which are widely used within the molecular biology community. Consequently, scientists interested in a particular MBD will be able to view the MBD schema in a DDL with which they are familiar. The versions of an MBD schema represented in different DDLs will be generated using *schema conversion* tools that will follow the iterative schema conversion methodology underlying the OPM Retrofitting tool. The MBD Schema Library will also contain abstract *overview* schemas, in which related schema components will be grouped together into higher-level components in order to provide a more concise and comprehensible high-level view of the MBDs.

MBD schema documentation will contain *sample data* that will help to reveal schema nuances that are not evident in the schema definition. Further observing how the same or similar data are represented in different MBDs will help to give insights into how to exchange data between MBDs. Sample data will be annotated in order to explain the significance of its various components.

As a development and maintenance resource, the Multidatabase Directory will provide facilities for constructing, extending, and maintaining (revising, updating) information on MBDs. These facilities will include tools for constructing abstract overview schemas and schema and data converters for transforming schemas expressed in an MBD's native DDL into schemas expressed in alternative DDLs. Since MBD schemas evolve over time, the MBD Schema Library will support *schema versioning* and will include tools for keeping track of MBD schema changes. *Schema annotation* facilities will allow scientists to share their understanding and/or view of MBD schemas and thus contribute to enhancing the comprehensibility and value of MBD schema documentation. Search engines will be provided for identifying MBDs relevant to a particular topic, and for quickly determining the relevant parts of a particular MBD.

Certain MBDs provide additional tools such as sequence analysis programs for analyzing a DNA sequence. Such data analysis tools can also be employed in a multidatabase system, so the Multidatabase Directory needs to be extended in order to include information regarding software support.

5.3. Supporting Multidatabase Queries

The current (first) version of the OPM multidatabase query translator[8] has been developed between October 1995 and January 1996. This version of the translator supports the expression of queries that combine (join) and manipulate data from multiple MBDs, and relies on information on the OPM schemas and remote access facilities of these MBDs contained in the Multidatabase Directory.

The multidatabase query processing strategy currently pursued, involves two stages:

1. OPM multidatabase queries are decomposed into component OPM queries for each component database involved in the query, where single-database OPM queries are evaluated using the existing OPM query translator.[7]

2. Data retrieved from each single-database OPM query are assembled locally into the result of the multidatabase query, where the local query processor is capable of performing joins and evaluating conditions over complex nested data-structures.

Although this query processing strategy is very simple and general, it can be inefficient for certain types of queries. For example, for evaluating a query that selects a small number of genes from the GDB class `Gene` and then finds the related genes in the GSDB `Gene` class, it would be inefficient to retrieve all the GDB and GSDB genes separately and then compare their accession numbers, rather than just looking up the GSDB genes using the accession numbers of the genes retrieved from GDB. A more efficient query strategy could find an order for the subqueries and evaluate them in sequence, so that the results of each subquery would be used to restrict the next subquery in the sequence. However such a strategy would be considerably more difficult to implement, since it would require statistics on sizes of individual classes and the selectivity of constraints in order to determine an optimal evaluation order. Although we consider pursuing such strategies in the future, in the short term we plan to increase the efficiency of multidatabase query processing by using *inter-database links*.

Inter-database links are known connections between heterogeneous databases that are recorded in the Multidatabase Directory together with the metadata on component databases. An example of an interdatabase link is the link between the `Gene` class in GSDB and the `Gene` class in GDB, represented by attribute `gdb_xref` of class `Gene` in GSDB; this attribute contains GDB accession numbers and thus indirectly points to GDB `Gene` objects. Following such a link allows retrieving from a component database only the objects that are involved in specific links, instead of retrieving all the objects in a class, where following inter-database links predetermines a query evaluation order.

From the perspective of a user constructing OPM multidatabase queries, inter-database links look like regular OPM abstract attributes (which represent intra-database links), except that the result of following such a link will be an object in another database rather than an object in a different class of the same database. Thus the Multidatabase Directory will associate an attribute name with each inter-database link, thus augmenting the list of attribute names that are already associated with an OPM class. These attributes can then be used for including the inter-database links in attribute paths in a query.

It should be noted that inter-database links do not subsume the general multidatabase joins already implemented, but rather complement them: multiple MBDs can be queried using multi-database joins (as done in our current implementation), inter-database links, or a combination of the two. This means that users are not confined to using the links already determined and included in the MBD Link Library, but can determine their own correspondences between databases as well. Using a combination of multidatabase joins, inter-database links, and other locally performed data manipulations, it should be possible to express very general and efficient multidatabase queries.

In order to assist users in understanding the semantics of multidatabase queries, the OPM multidatabase query processor will also provide support for interpreting queries in terms of the semantics of both the target MBDs and the query processing operations.

5.4. Technological Alternatives

There are commercial distributed-join software tools, such as the Sybase *Enterprise CONNECT* family of products, that allow querying multiple relational databases. It should be noted that such tools do not help constructing multidatabase systems: one still needs to understand the component databases in their relational representation, their semantics and links. The OPM tools support higher level representations of databases, using abstract constructs that are better suited for representing biological data. In addition, the Multidatabase Directory simplifies substantially the task of exploring and understanding multiple databases.

A distributed-join tool could underlie the processing of OPM multidatabase queries, where an OPM multidatabase query would first be translated into multidatabase SQL queries. The multidatabase SQL queries could be then processed by the distributed-join tool and the query results could be then converted into OPM data format. This query processing strategy is different from our current strategy of translating an OPM multidatabase query into OPM queries over individual databases.

Although we plan to examine this alternative query processing strategy in terms of cost and performance, we are aware of several problems inherent to this alternative. First, a distributed-join tool can be used only for a set of data sources supported by the tool: usually major commercial DBMSs or widely used standards, but not the more specialized data-sources frequently used for molecular biology databases (e.g., ASN.1, ACeDB). For data sources that are not supported by such a tool, additional programming would be still required. Moreover, such a tool is restricted to the constructs of a standard relational query language (e.g., SQL). Such query languages have been found to be overly restrictive and difficult to use for querying complex MBDs: for example, an SQL query over the relational schema for GSDB 2.0 will in general involve substantially more tables, and will be considerably more complex, than an equivalent OPM query expressed over the OPM view of GSDB. Furthermore, the OPM multidatabase query translator is based on a more powerful nested relational algebra which supports directly operations on nested sets and complex data structures. Finally, using a distributed-join tool for processing OPM multidatabase queries will make the performance of the OPM multidatabase query translator dependent on this tool. With our current query processing approach, we have the flexibility of experimenting with any query optimization strategy and hopefully achieve better query performance.

ACKNOWLEDGMENTS

We would like to thank Otto Ritter for sharing with us his insights into the problems and techniques involved in developing a warehouse MBD.

REFERENCES

1. Blake, J., et al. Inter-Connection of Biological Databases: Exploring Different Levels of Molecular Biology Database Federation. In Ref. 20.
2. Bright, M. W., Hurson, A. R., and Pakzad, H. A Taxonomy and Current Issues in Multidatabase Systems. *IEEE Computer*, 25(3), pp. 50–59, 1992.
3. Buneman, P., Davidson, S., Hart, K., Overton, C., and Wong, L. A Data Transformation System for Biological Data Sources. In *Proc. of the 21st Int. Conference on Very Large Data Bases*, pp. 158–169, 1995.
4. Chen, I. A., and Markowitz, V. M. An Overview of the Object-Protocol Model (OPM) and OPM Data Management Tools. *Information Systems*, 20(5), pp. 393–418, 1995.
5. Chen, I. A., and Markowitz, V. M., OPM Schema Translator 4.0, Reference Manual, Technical Report LBL-35582 (revised), 1995.
6. Chen, I. A., and Markowitz, V. M., Constructing and Maintaining Scientific Database Views, Technical Report LBL-38359, 1996.
7. Chen, I. A., Markowitz, V. M., and Szeto, E., The OPM Query Translator, Technical Report LBL-33706, 1995. Available at `http://gizmo.lbl.gov/opm.html`.
8. Chen, I. A., Kosky, A., Markowitz, V. M., and Szeto, E., OPM*QS: The Object-Protocol Model Multidatabase Query System, Technical Report LBL-38181, 1995.
9. Cozza, S., Reed, E. C., Salit, J., Chang, W., Marr, T. Genome Topographer: A Next Generation Genome Database System (Abstract), presented at the meeting on Genome Mapping and Sequencing, Cold Spring Harbor Laboratory, Cold Spring Harbor, 1994.
10. Etzold, T., and Argos, P. SRS, An Indexing and Retrieval Tools for Flat File Data Libraries. *Computer Applications of Biosciences*, **9**, 1, pp. 49–57, 1993. See also `http://www.embl-heidelberg.de/srs/srsc`.
11. Etzold, T., and Argos, P. Transforming a Set of Biological Flat File Libraries to a Fast Access Network. *Computer Applications of Biosciences*, **9**, 1, pp. 58–64, 1993.
12. Fasman, K. H., Letovsky, S. I., Cottingham, R. W., and Kingsbury, D. T. Improvements to the GDB Human Genome Data Base. Nucleic Acids Research, Vol. 24, No. 1, pp. 57–63, 1996. See also `http://wwwtest.gdb.org/gdb/about.html`.
13. Goto, S., Akiyama, Y., and Kanehisa, M. LinkDB: A Database of Cross Links Between Molecular Biology Databases. In Ref. 20.
14. Heimbigner, D., and McLeod, D. A Federated Architecture for Information Management, *ACM Transactions on Office Information Systems* 3(3), pp. 253–278, 1983.

15. Karp, P., Report of the 1st Meeting on Interconnection of Molecular Biology Databases, Stanford, California, 1994; http://www.sri.ai.com/people/pkarp/mimbd/mimbd-94.html.
16. Karp, P., A Strategy for Database Interoperation, Journal of Computational Biology, Vol 2, No 4, 1995.
17. Kosky, A., Davidson, S., and Buneman, P. Semantics of Database Transformations. Technical Report MS-CIS-95-25, University of Pennsylvania, 1995.
18. Miller, R. J., Ioannidis, Y. E., and Ramakrishnan, R., The Use of Information Capacity in Schema Integration and Translation, *Proc. of the 19th International Conference on Very Large Databases*, 1993, pp. 120–133.
19. Ritter, O. The Integrated Genomic Database. In *Computational Methods in Genome Research* (S. Suhai, ed.), pp. 57–73, Plenum, 1994. See also http://genome.dkfz-heidelberg.de:80/igd/start_igd_doc.html.
20. Second Meeting on Interconnection of Molecular Biology Databases, Cambridge, United Kingdom, 1995, http://www-genome.wi.mit.edu/informatics/abstracts.html.
21. Sheth, A. P., and Larson, J. A. Federated Database Systems for Managing Distributed, Heterogeneous, and Autonomous Databases. *ACM Computing Surveys*, 22(3), pp. 183–236, 1990.
22. Shuler, G. D., Epstein, J. A., Ohkawa, H., Kans, J. A. Entrez. In *Methods in Enzymology*, (R. Doolittle, ed.). Academic Press, Inc. In press. See also http://www3.ncbi.nlm.nih.gov/Entrez/.

14

COMPREHENSIVE GENOME INFORMATION SYSTEMS

Otto Ritter

Department of Molecular Biophysics
German Cancer Research Center (DKFZ)
Im Neuenheimer Feld 280
D-69120 Heidelberg, Germany
Phone (+49-6221) 42-2372; fax (+49-6221) 42-2333
e-mail o.ritter@dkfz-heidelberg.de

1. INTRODUCTION

The science, technology, and art of both genome research (also known as genomics) and practical computer science and information technology (also known as informatics) dramatically reshape our modern life. These two disciplines also quite interestingly influence and inspire each other. Of the many exciting issues within that interaction, we will focus on how informatics contributes to the capturing, organization, maintenance, and evolution of domain knowledge in genomics, and whether it can help us in building a truly comprehensive global reference system.

Comprehensive information systems should enable a logically integrated (i.e., inclusive, consistent, and conceptualy homogeneous) access to a collection (static or dynamic) of primary information resources in the domain of interest.

Specifically, comprehensive genome information systems (CGIS) should unify, at the syntactic, semantic and pragmatic levels, a consistent subset of the union of structural and functional contents of genome-related informational and analytical resources on nucleotide and protein structures and sequences, genomic loci and regions, genetic and physical maps of different types, comparative maps, pedigrees and phenotypes, experimental reagents, experimental protocols, bibliography, a.o.

Databases, knowledge systems, and other information systems employ languages (of varying degrees of formality) for the representation of their contents and behavior, and for the communication with their environments (users, administrators, peer software). Usually, the data modeling, manipulation, and query languages are formal, while the semantics of represented concepts may be captured implicitly in natural language documentation or explicitly in natural or semi-formal languages.

Theoretical and Computational Methods in Genome Research, edited by Suhai
Plenum Press, New York, 1997

Representation languages may have more than one kind of semantics. We can distinguish:

- **Reference semantics** to identify designation, or how language symbols and expressions correspond to known concepts in the subject domain.
- **Truth and proof semantics** to identify the logical truth or falsity of symbols, or their syntactic validity.
- **Denotational semantics** to specify the correspondence between the language control expressions and the intended computation in all implementations.
- **Interface semantics** to characterize system's modular decomposition and to support the abstraction of protocols and function prototypes.
- **Action semantics** to characterize how language elements are used to cause action.

More complete account on semantics in knowledge systems can be found in [1].

Thus, the representation and communication languages of CGIS need to be expressive and consistent at all syntactic and semantic levels. In the following, we shall see that there are major obstacles to representation consistency. In my opinion, the hard problems are due to inherent terminological ambiguity and the lack of accessible and explicitly represented semantics in most of the existing genomic databases and related software applications.

Genome informatics is a young and highly interdisciplinary field. It borrows terminology from many branches of biology, mathematics, computer science, and other more recent disciplines such as artificial intelligence. One consequence of that is a relatively high degree of terminological ambiguity and confusion.

A number of basic concepts are highly semantically overloaded (e.g., gene, locus, object, object- oriented, mapping, function, data, information, knowledge, etc.), their intended meaning is a function of the originator's internal state (the author may be human or software), and of the immediate context (sentence, paragraph, data object). This ambiguity is not necessarily counterproductive in the case of self-contained publications: here a term's intended meaning is often stated explicitly (by reference or by definition) or it can be intuited from its consistent usage throughout the surrounding text. However, if we try (manually or by software) to collate several such texts of different origins and contexts, we suddenly face a dramatic shift towards ambiguity and uncertainity.

2. THE SUBJECT DOMAIN OF GENOMICS

We understand (i.e., can reduce to the laws of chemistry and physics) some of the life of complete organic systems such as viruses, monocellular organisms, or individual cells and multicellular structures in higher organisms, and we employ this new understanding in medicine and biotechnology with undeniable effects. Genetics and molecular biology, or genomics, constitute most of the intellectual framework for such an understanding. Unlike chemistry or physics, however, genomics doesn't so much study collective properties of anonymous entities, whose individual identity is either irrelevant or beyond practical reach (e.g., particles, atoms, or small molecules), it rather works with a large number of identified physical entities and abstract concepts (e.g., chromosomes, DNA clones, genes, pedigrees). One could argue that a given DNA clone is indeed a collection of anonymous DNA molecules. However, the clone has a lot of identity beyond just its name: it has unique properties which cannot be compressed into quantitative analytical models. Computer scientists would say that the type system of genomic knowledge is by orders of magnitude larger than the type systems of knowledge in chemistry or physics.

Genetics studies observable traits and the measure of their variation (preservation or change) in related individuals. Most of these traits are not inherited directly as such (i.e., in their extensional form)— it is the molecular instructions for these traits that are inherited (i.e, the intensional form). These instructions constitute genetic information, their logical units (corresponding to algorithms or programs in a simplistic computing metaphore) are genes, and the physical media for information storage are large polymers of nucleic acids (DNA or RNA) called chromosomes. The sum of all chromosomes in one cell of an individual constitues a genome. We have relatively good knowledge, on the molecular level, about the basic principles and mechanisms which control the "housekeeping" of genetic information, such as its storage in the chromosomes, its replication and various transformations during cell division, and its final translation into other functional molecules - usually into proteins. We have, however, in most cases much less knowledge about the processes linking particular instances of genetic information (genotypes) with particular instances of directly observable or otherwise determinable traits (phenotypes). For a special class of so called monogenic phenotypes (traits controlled by a single gene) it is relatively easy to find the corresponding genotypes; in the general case of oligogenic or polygenic phenotypes this becomes a much more difficult if not currently intractable task. Furthermore, establishing a genotype-phenotype correspondence is usually only a small first step on the way of defining this correspondence operationally.

Currently the prominent aims of genome research, or genomics, are to establish sets of ideal genome maps (in different resolutions and distance measures) for human and several other species of some medical, biological, and/or industrial importance. Maps of the finest resolution are the consensus nucleotide sequences of the whole chromosomes or their significant parts. Maps of different sorts are combined into so called integrated maps, and genome maps of different species are combined into so called comparative maps. Finally, mapping information on genomic landmarks is combined with references to information on their variation (e.g., mutations), patterns of expression, and both normal and morbid function.

In the first approximation, genome differences between individuals within the studied species are being ignored, as are the differences between cells of individuals. In a few cases now, and much more in the future, local and global genome variability and related functionality will be studied, within a species, at the level of populations, individuals, individual developmental stages, tissues, etc. up to individual cells.

2.1. Genome Related Information Systems

Substantial portions of the genomic domain knowledge have been computerized and transformed into informational and analytical resources such as data libraries, databases, analytical compute servers, graphical visualization tools, etc.[2]. Major public resources are network- accessible via Internet. Many laboratories and commercial companies are keeping their resources behind firewalls for private access only, but they frequently put some data (public views) and some functionality on the Internet. The most popular enabling infrastructure for that kind of electronic publication has become the World Wide Web (WWW), where extensional information (data) is packaged into hyperlinked documents in misc. text and multi-media formats, usually accessible through the HTTP or FTP protocols, and intensional information (analytical functions, queries, and other services) is interfaced usually via ad-hoc semantic protocols over HTML-Forms, E-mail, or other types of more sophisticated synchronous invocation or asynchronous messaging.

The total volume of genome related information on the Internet is enormous and grows exponentionally. There are hundreds of relevant databases and hundreds of relevant functional resources. If we only look at the total sizes of data collections, my conservative estimate would by somewhere between 10^{10} and 10^{12} bytes, with a doubling period between 1 and 2 years. Of course, drawing the line between genome related and genome unrelated information is both subjective and fuzzy, but we can so far ignore this imprecision without loss of generality.

What makes the genome information space so difficult to work with, is not the relatively large volume of data alone, but this in combination with:

- large complexity of data and operations,
- fuzziness, incompleteness, and scatter (distributedness) of data and metadata,
- heterogeneity in representation,
- autonomy of implementation,
- large dynamicity.

The nice thing about genomic information on the Internet is that almost everything you want to know or compute, is somewhere there. The bad thing is that it usually takes a number of nontrivial steps to collate the information you need from a number of partial finds (retrieved documents, results of function calls). The information is there, but scattered over numerous places, involuted, and convoluted with what you don't need.

Borrowing from popular characterizations of relational database systems, I would say that there is an impedance mismatch between the subject-oriented genomic information services on the Internet, and the problem-oriented needs of the genome research community. It may be even worse: there is no standard query language (SQL) for all the "wild" WWW relations. Genome related information systems view and represent dramatically different scopes and subdomains of the genomic universe, they use different information models, representation techniques, naming conventions on both meta and proper levels, and they support different (often idiosyncratic) languages and interfaces for end-users to interact with the respective system's functionality and contents. End-users, on the other hand, would prefer a very limited number (ideally one) of integrated, consistent and complete interaction methods for all the relevant information. The emergence of WWW has solved only one (rather easy) aspect of interoperability.

Several bioinformatics projects try to layer consistent services on top of preexisting heterogeneous autonomous resources. We can group them into three classes:

- **Menu-based environments**. Many biocomputing centers and service providers construct environments on top of local and remote resources. These environments hardly provide more than consistent menu-based invocation methods with unified access to help and other documentation.
- **Link-based navigation**. These linking systems index contents and linkage information for a number of related resources, and usually support simple set-oriented queries, navigation, and retrieval.
- **Integrated environments**. Integrated environments provide for a homogeneous set of operations on a set of heterogeneous resources, e.g., one global query language. For a truly declarative query against a number of heterogeneous distributed resources, common conceptualizations have to exist (and it is irrelevant here whether these exist extensionally as global schemata/ontologies, or intensionally as implicit or computable concepts).

We can see the navigational systems as an extreme case of integration, where the shared schema is very simple: just documents and links for WWW, or more types of documents (objects, records) with attributes and references, but no semantic unification or conflict resolution.

2.2. Domain Knowledge of Genomics

The sum of basic domain knowledge of genomics breaks into 14 classes:
definitions of basic terms

1. set of known or assumed principles of the structural organization of basic types of entities and their assemblies or collections
2. set of known or assumed principles of the dynamic behavior of basic types of entities; both "isolated" and in their interactions
3. set of "laws", or propositions formulated in terms of 1.-3., assumed to be true
4. set of important hypotheses and goals
5. set of procedural descriptions of the basic types of experiments and their parts;
6. including algorithms for computational experiments and inferences
7. typology and nomenclature of experimental material (both biological and non-biological material and reagents)
8. set of primary experimental results and observations
9. complex web of secondary results, derivations, and inferences
10. relevant published literature
11. set of references to knowledge in external domains
12. information about the scientific community and its organizational structures (individuals, institutions, projects, etc).
13. knowledge about the community's informational and computational resources, including their metainformation and documentation
14. all the relevant contents and functionality of informational and computational resources

This is of course just one of many possible classifications. Classes 1 to 6, and partly also 7, together constitute an ontology, i.e., conceptualization of the domain terms, types, functions, relationships, and invariant properties. Part of this ontology would serve, in the classical database sense, as metadata (or type system) for the bulk body of factual data and derived concrete knowledge in classes 8 to 10.

Classes 11 and 12 are a straightforward and natural complement to the first ten ones. Class 13, with perhaps some parts of 1 to 7, comprises the ontologies for the knowledge in class 14. We can imagine that there exists a set of partial mappings from the union of classes 1 to 7 into 13, and a set of partial mappings from classes 8 through 12 into 14.

The ontologies of genomic information are models of the real world of genomics. Information systems are then models of their respective ontologies. The on-line encyclopaedia of the Principia Cybernetica Project [3] defines ontologies and models as follows.

Ontology is an explicit formal specification of how to represent the objects, concepts and other entities that are assumed to exist in some area of interest and the relationships that hold among them.

For AI systems, what "exists" is that which can be represented. When the knowledge about a domain is represented in a declarative language, the set of objects that can be represented is called the universe of discourse. We can describe the ontology of a program by defining a set of representational terms. Definitions associate the names of entities in the universe of discourse (e.g. classes, relations, functions or other objects) with human-readable text describing what the names mean, and formal axioms that constrain the interpretation and well-formed use of these terms. Formally, an ontology is the statement of a logical theory.

Model is a system $S = <W , M , E>$ with:

- A modeled system or world $W = <W, L>$ with states $W = \{w_i\}$ and actions or laws $L: W \rightarrow W$.
- A modeling system $M = <M, R>$ with internal model states, or representations $M = \{m_j\}$ and a set of rules, or a modeling function $R: M \rightarrow M$.
- And representation function $E: W \rightarrow M$.

When the functions L, R, and E commute (E is a homomorphic mapping), we say that S is a good model, and the modeling system M can predict the behavior of the world W.

3. CGIS INFORMAL DEFINITION

CGIS should mediate a high-level declarative access to all genomic information of interest (for a well defined profile of users), related analytical functions, and generic methods for visualization, communication, syntactic and semantic transformations, and other kinds of processing . High-level and declarative means that the network location of the underlying resources, their hardware, operating system and DBMS platforms, and their representation languages (ontology, data model, interface syntax and semantics) should be transparent to casual users, yet accessible to the experts on demand.

The information management component of CGIS should comprise efficient storage and retrieval, data access through query languages, textual and graphical browsing tools, and conventional programming language interfaces, access to a comprehensive set of data manipulation operations and tools, and support for local (i.e., client-side) data management.

One approximation to a CGIS is described next.

4. IGD AS A CASE STUDY OF CGIS

The Integrated Genomic Database (IGD) is an international collaboration [4] aiming to develop an open information management system for (primarily human) genome data and methods. In the period of 1994 to spring 1996, the first generation of IGD (IGD-1) integrated information from 15 public data collections into a single logical database accessible over the Internet, and provided a graphical front-end for managing and analyzing retrieved subsets of public data and/or sets of local experimental data.

4.1. IGD-1 Contents and Functionality

As a database, IGD-1 integrated and referenced genome related data from public sources, called 'Resource End Databases' (IGD-REDs). As of January 1996, IGD held over 2 million objects from the following sources: human DNA and protein sequences from EMBL, GenBank, SwissProt and PIR, cytogenetic maps, loci, probes, contacts and bibliography from GDB, all OMIM documents, YACs and STSs of the CEPH-GENETHON map, probes and contacts from the UK DNA Probe Bank, clones, probes, and contacts from RLDB, the EUROGEM subset of the CEPH database, EUROGEM and CHLC genetic maps, restriction enzymes from REBASE. Data from the IGD-REDs were periodically collected, reformatted, and exported to IGD servers, called 'Target End Databases' (IGD- TEDs).

As an analysis tool, IGD-1 provided uniform interface to many preexisting programs and program packages for structure and sequence analysis, genetic and physical map construction and analysis, etc. Users interacted with IGD through a set of locally installed 'Front End' tools (IGD-FRED). Most important parts of the IGD-FRED were the local database manager and interfaces to communication and analysis. Users could query the IGD-TED and download the resulting data into their local database. They could also put private data and analysis results into the local database.

The IGD-FRED front-end could work on arbitrary subsets of IGD-TED data, and it could manage local data as well. The schema supported storing detailed experimental data on sequencing and mapping projects, and could be extended locally. Data objects had links to external methods. These included URL's to their native sources and methods to call local or remote analytical programs, including the HUSAR/GCG package of 140+ sequence and structure analysis tools, a pedigree display tool, interface to genetic linkage analysis, and an interface to physical map assembly/analysis.

4.2. IGD-1 Architecture

IGD-1 implemented the so-called data warehouse model by translating all data into common format and loading them into a physical database. First, data were collected by either querying the IGD-REDs or getting their exported data sets. Second, export IGD-RED data were syntactically reformatted from the RED-native format (e.g., relational table dumps, ASN.1 stream, tagged-text, etc.) into the IGD-native format, .ace of ACEDB[5]; essentially a set of recursive tree structures. Third, export RED data already in the IGD format were semantically transformed, filtered, and enriched so as to be compatible with the IGD global schema. Fourth, syntactically and semantically correct data were uploaded into the IGD-TED database.

4.3. Future Work on IGD-GIS

During the next three years, the second generation IGD Genome Information System (IGD-GIS) will be developed. IGD-GIS will construct a virtual data and method space on top of genomic information resources. The virtual space will be accessible, at the user-interface level, as a single object database. One objective is to make the level of schema integration and the policy of object instantiation (dynamic on-demand vs. static warehousing) configurable. Major emphasis will be put on explicit ontological modeling of the subject domain and the IGD-GIS system itself.

5. SUMMARY

I believe that comprehensive genome information systems (in the sense of the above definition) can be constructed for a relatively large number of resources and correspondingly powerful sets of queries. The advantage of a CGIS over non-comprehensive systems would be the support for expensive operations which span across the boundaries of existing resources (e.g., a relational join across several databases). Thus, CGIS would facilitate work which is not feasible using non-comprehensive systems. Examples of such work include exploratory data analysis, data mining and knowledge discovery. The potential of better analysis and substantial new pattern (or knowledge) discoveries may justify the large costs and difficulties associated with the developments of comprehensive systems, but rigorous problem analysis and risk management strategies are strongly recommended for such endeavors.

ACKNOWLEDGMENTS

The work, referred to in section 4, is the result of the work of all the IGD partners and contributors.

IGD-1 was supported by the Commission of the European Union by grant CT93-0003 “Development of an integrated information management system for human genome data (IGD)”. Development of IGD-GIS is supported by the Commission of the European Union by grant CT96-0263 “The IGD Genome Information System”.

REFERENCES

1. M. Stefik. Introduction to Knowledge Systems. (1995) Morgan Kaufmann: San Francisco.
2. V. Markowitz and O. Ritter. Characterizing heterogeneous molecular biology database systems. J. Comp. Biol., Vol 2, Nr 4, 547–556 (1995)
3. The Principia Cybernetic Project on-line dictionary, http://pespmc1.vub.ac.be/.
4. O. Ritter, P. Kocab, M. Senger, D. Wolf, and S. Suhai: Prototype Implementation of the Integrated Genomic Database, Computers and Biomedical Research, 27, 97–115 (1994). See also http://genome.dkfz-heidelberg.de/igd/.
5. R. Durbin and J. Thierry-Mieg. The ACEDB Genome Database. In Computational Methods in Genome Research, edited by S.Suhai, Plenum, 57–73 (1994)

VISUALIZING THE GENOME

David B. Searls

Bioinformatics Group
SmithKline Beecham Pharmaceuticals

ABSTRACT

A number of visually oriented computational technologies show promise of aiding in the exploration of genome sequences. These encompass the fields of *graphical user interfaces, scientific visualization*, and *visual programming*. This review describes examples of each, as they relate to genomics, and argues that a synthesis of such techniques may be necessary to make understanding entire genomes a tractable human enterprise.

1. INTRODUCTION

The advent of comprehensive large-scale sequencing projects for the human genome, as well as ongoing efforts in the microbial and model organism arenas, are placing renewed pressure on computational tools for effectively dealing with the resulting flood of data. The sheer volume of sequence data, combined with the wide variety of analyses and annotations that must be performed on it, provides a number of challenges to the bioinformatics community. This review will deal with one such challenge in particular: that of presenting the information *visually* to the user, in such a way that it can be understood and manipulated effectively.

The visual presentation and manipulation of data is a central theme in biology. Biologists, after all, depend heavily on images. It is important to realize that published illustrations of gels and blots (for example) are not just data, but highly effective visualization tools, and perhaps even icons in the semiotic sense. Moreover, it is remarkable that the vast majority of biological knowledge and hypotheses are extant only in the form of highly schematized diagrams in textbooks and papers. This can be contrasted to fields such as physics, which has its share of icons (e.g., Feynman diagrams), but which more typically states its models and hypotheses by way of mathematical formulae.

There are at least three "visual" technologies that are likely to impinge on attempts to display genome information, each with its own unique characteristics but also overlapping

Theoretical and Computational Methods in Genome Research, edited by Suhai
Plenum Press, New York, 1997

concerns. Each has its separate research communities, and products, and typical constituencies. While they are often confused, there are clear distinctions to be made as to their motivations. They are as follows:

- *Graphical User Interfaces* (GUIs) are now a central facet of most computational systems, and have long been seen as key to the effective presentation and manipulation of genome data. While GUIs often display data, their primary motivation is to provide a convenient means of control to a user, generally through some consistent visual metaphor (e.g., the "desktop") with standardized control elements (pulldowns, popups, buttons, sliders, scrollbars, etc.).
- *Scientific Visualization* is usually thought of as the portrayal of vast quantities of numerical information by way of images, using form and color to help humans assimilate otherwise intractable amounts of data.[3] It is often associated with physics and with the use of volume and surface rendering to capture features of multidimensional data, mapping it to characteristics such as hue, reflectance, etc. Algorithmic visualization is a variant in which the workings of a program in operation are portrayed.
- *Visual Programming* allows a user to literally draw, point at, or demonstrate relationships among concepts or data as a means of *programming;* that is, a graphical representation is (generally) compiled into a more conventional program automatically.[5] Visual programming systems may be based on dataflow, spreadsheet, or object-oriented paradigms, and as well have been explored in functional, imperative, logic, and rule-based formats.[4]

In what follows, we will discuss each of these approaches in turn, giving examples from the author's own experience as well as other systems. The effective "visualization" of the genome is seen as essential, and this will likely involve all three of these technologies in harmonious combination.

2. GRAPHICAL USER INTERFACES

Comprehensive GUIs have been key components of a number of genome-related applications and databases, such as the *Encyclopedia of the Mouse Genome,*[10] the *System for Integrated Genome Map Assembly* (SIGMA),[6] and the *Genome Topographer* (GT),[11] to name just a few. It has been said, indeed, that the widely-used *ACEDB* database and genome analysis system[8] owes most of its success to its GUI. Both ACEDB and GT are large, object-oriented systems with a very comprehensive range of purposes. We first propose an alternative philosophy for GUI-supported design in this domain, then discuss several examples of applications developed and lessons learned.

2.1. Interface Componentry

In genome analysis software in general there is a basic tension between the virtues of a top-down, model-heavy, integrated software approach, with its relatively consistent and polished presentation, and a rapidly prototyped "lightweight" custom tool approach, with its sometimes underpowered representations and ragged interfaces. The latter is often necessarily the choice of developers, and the question arises as to whether there might not be an approach that supports such prototyping activities with some of the facilities of established systems. Ideally, this could be done without assuming the burden of a design philosophy that may eventually become unresponsive in the face of a changing domain. Recently there have been some notable efforts to address this perceived need through an approach that falls under the general rubric of *componentry*. Goodman and colleagues at the Whitehead (MIT) Genome Center have noted that "the choices that face the architect of a genome information system today are: (i) build it yourself so that it does exactly what you want, or (ii) adopt someone else's system and live with most of its quirks and limitations".[13] They also note that this situation acts as a barrier to innovative work in informatics in general. The solution proposed by these workers and others is to develop architectural frameworks into which software components can be reused in a "plug-and-play" fashion.

These observations and the author's experiences in providing informatics support to a number of large and small enterprises have led to an effort to develop genome-specific GUI componentry. This work was motivated by the observation that, even within a single group, there was a disturbing degree of reinvention in the development of a wide range of seemingly disparate GUIs. In large degree, it seemed that this could be attributed to the emergence of common themes in the graphical and operational requirements associated with certain aspects of genome data, for instance in notions such as *map* and *annotation*. To the extent that such domain-specific features could be abstracted and associated with generalized graphical objects, it seemed that a large component of the wasted effort could be obviated.

The general methodology adopted is to develop components of a typical bioinformatics GUI at the level of graphical *widgets*, conceived here as high-level objects created and managed by a controlling environment in such a way as to hide the details of placement, drawing, motion, interaction, etc. from the application programmer, to the extent possible. Typical widgets would be GUI objects such as buttons and scrollbars, but they can also be considerably more complex when capturing high-level yet generic functionality. For example, text windows are reused in similar ways in countless varieties of applications, so that it is possible to encapsulate recurring functions, such as editing capabilities, at the widget level. Where similar commonalities can be detected in a specific application domain, there exists the opportunity to develop a high-level, specialized widget set in that area.

The challenge, then, is to recognize the commonalities, e.g., between the dozen or so different styles of genome map, and collapse them into widgets that hide as much as possible from the application developer while leaving her the maximum latitude to use them in novel, even unexpected ways. Moreover, among different widgets there should be a highly parallel set of option parameters, mimicking the object orientation of typical widget sets. A widget set that did nothing more than encapsulate code for different types of maps, but still required those maps to be specified according to their many individual conventions, would be nothing more than a set of utilities. The hard work of capturing analogies, both obvious and subtle, among cytogenetic maps, genetic maps, somatic cell hybrid maps, pulsed-field gel maps, restriction maps, STS-content maps, cosmid contig maps, sequence fragment maps, and so on, must pay off in a command language that is concise, intelligently overloaded, and even, one hopes, illuminating to application developers. This will then serve to lower the barriers to entry and minimize the learning curve. This approach would appear to be an attractive alternative to monolithic, integrated software in many situations.

One particular advantage of this factorization of applications is that some wider issues related to reusability can be subsumed by widgets in useful and perhaps unanticipated ways.

For example, among the functions typically associated with graphical user interface widget sets are:

- *option management,* by which general characteristics of various classes of widgets, such as their background colors or default mouse behavior, are centrally managed in an "options database" or configuration file. This allows wholesale changes to be made all at once, so that applications can be dropped into a new environment and assume the look-and-feel of that environment. This functionality can, of course, be extended to apply to domain-specific characteristics.

- *selection management,* which, for example, allows highlighted text in one widget to be transported to another widget or even application. It is easy to imagine generalizing this notion to allow selections of genome intervals or loci to be moved from among map and sequence widgets, while being cast in the appropriate form depending on the widget. In particular, data exchange formats, as well as higher-level functions such as coordinate transformations between maps, can be neatly encapsulated.

- *geometry management,* perhaps the single most important aspect of any widget system, by which objects are moved, resized, or otherwise altered in appearance as a function of a changing environment, behind the scenes and without direct user intervention. In a genome-specific widget set, the user could be similarly relieved of responsibility for many critical aspects of the appearance of maps and map objects, some of which are currently considered to be tied to very specific application concerns (e.g., "rubberbanding" of map scales).

The trick, we believe, is to design such a widget set at the appropriate level of generality. Making the widgets *too* functional, in some sense, risks choking off opportunities for serendipitous re-use (often a hallmark of widget-based design). We deem it important to leave the semantics to the application, and deal as much as possible with just the characteristic appearance and behavior of the literal, visual objects. In fact, such a principle is probably only possible in a limited domain, which affords some peculiar visual "language" (what has been called a technical "sublanguage" in Natural Language Processing).

We have suggested that a toolkit based on a domain-specific widget set would promote a rapid-prototyping approach to building small applications, which however are mutually consistent and able to interact relatively easily. Such "hypertools" offer an alternative to tightly-integrated, model-heavy informatics environments (though it must be said that systems such as ACeDB and Genome Topographer have something of this flavor with respect to their algorithmic assists). It must also be added that current trends toward World Wide Web-based systems may in fact necessitate such an approach. It now seems inevitable that many services and databases will be made available via browsers that download not only static data but dynamic code fragments, or "applets," over the net. If tiny applications involving visual displays are to be disseminated via systems such as Sun's Java, it would seem that a standardized, lightweight, domain-specific widget set offers the best chance of achieving reasonable consistency among a legion of information vendors.

Our *bioTk* widget set, which attempts to embody these ideals, has been extensively described,[24] and is available via the WWW (see `http://www.cbil.upenn.edu/~dsearls/bioTk.html`). With recent developments surrounding Java, we have begun porting this system so as to create a system of classes called *bioWidgets*, corresponding to the widgets in bioTk. Figure 1 shows a sample application involving both the map and sequence classes.

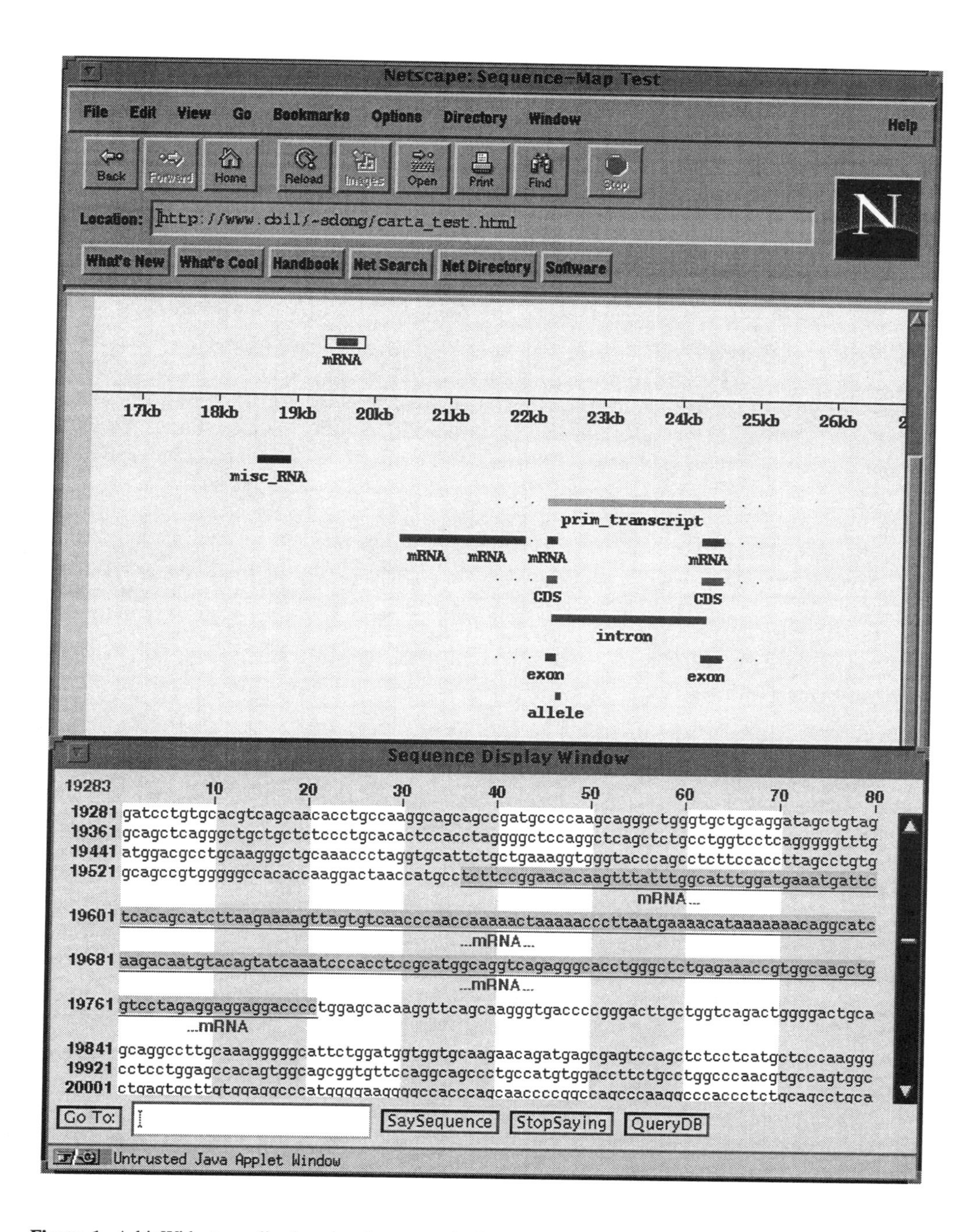

Figure 1. A bioWidgets application showing a map in the main browser window, and a separate window with a scrolling sequence, which interacts with the main map.

2.2. Examples

Figure 2 illustrates a facility developed for analyzing cosmid fingerprinting experiments. One interesting principle that emerged in this application is that the form of presentation of restriction fragment size data that was most easily accepted by biologists is one that most closely mimics the sizing gel itself. Although somewhat artificial, the display was immediately familiar and intuitively appealing to biologists, greatly shortening training time and promoting user acceptance.

After editing out of vector bands and other modifications, the band data were transferred to another tool, designed to display developing cosmid contigs in terms of individual restriction fragments. Because the laboratory approach permitted the analysis of tractable numbers of cosmids at any one time (by compartmentalizing them in bins established by overlapping YAC hybridizations), it was feasible to display each fragment and allow the user to manually move them among columns of similar-sized fragments, by drag-and-drop mouse operations. However, an algorithm was provided for automatically merging fragments into columns as well (a necessary first step). In addition, support for then ordering cosmids based on probability of overlap was added, using hill-climbing search. Once a putative cosmid ordering was arrived at, the user could reorder columns of fragments, again either manually or with an algorithmic assist, to try to account for all bands. This generally became an iterative process, as columns were reconstituted to best account for the cosmid ordering, and then the cosmids reordered. Because of errors in sizes and especially coincident bands, it was often necessary in this final process to use more sophisticated approaches to achieve a consistent ordering of fragment columns. This was done by reuse of the STS content mapping tool, with its simulated annealing algorithm, and heuristics which we had developed to deal with non-unique probes (which prove to be analogous to coincident bands). Thus, the original data are processed through a total of three separate applications to perform this task.

This case is presented to establish the themes of rapid prototyping, iterative development, data visualization, user feedback, interacting tools, and reuse, which we believe are important to address in this domain. However, as noted in the previous section, one frustration that results from this approach is duplication of effort at the level of basic GUI programming. In many of these applications we have dealt repeatedly with issues of displaying labels so as to avoid overlap, representing intervals and domain-specific operations on intervals such as inversions and translocations, implementing analogous drag-and-drop operations on clone or probe orders, redrawing intervals in a "stairstepped" fashion, and many other common operations. We have drawn three different styles of chromosome ideograms in different contexts, and displayed scrolling sequence data in several different ways, and in both cases reinvented methods for annotating the displays. We have created displays for half a dozen styles of maps, all of which had commonalities both obvious and subtle, yet in few cases have we had the luxury of abstracting and encapsulating GUI code for reuse. This was the original impetus for *bioTk/bioWidgets*, described in the preceding section.

In a collaboration with Dr. Maja Bućan of Penn's Department of Psychiatry, we have prototyped a tool to aid in exon trapping experiments. The GUI assists in assembling gene models from a set of trapped exons, as shown in Figure 2.2. Sequenced exons are displayed in a workspace and may be manually rearranged in two dimensions, perhaps reflecting partial ordering information available to the biologist. Six reading frames are displayed, and tagged at the putative splice junctions with color codes that aid the user in keeping

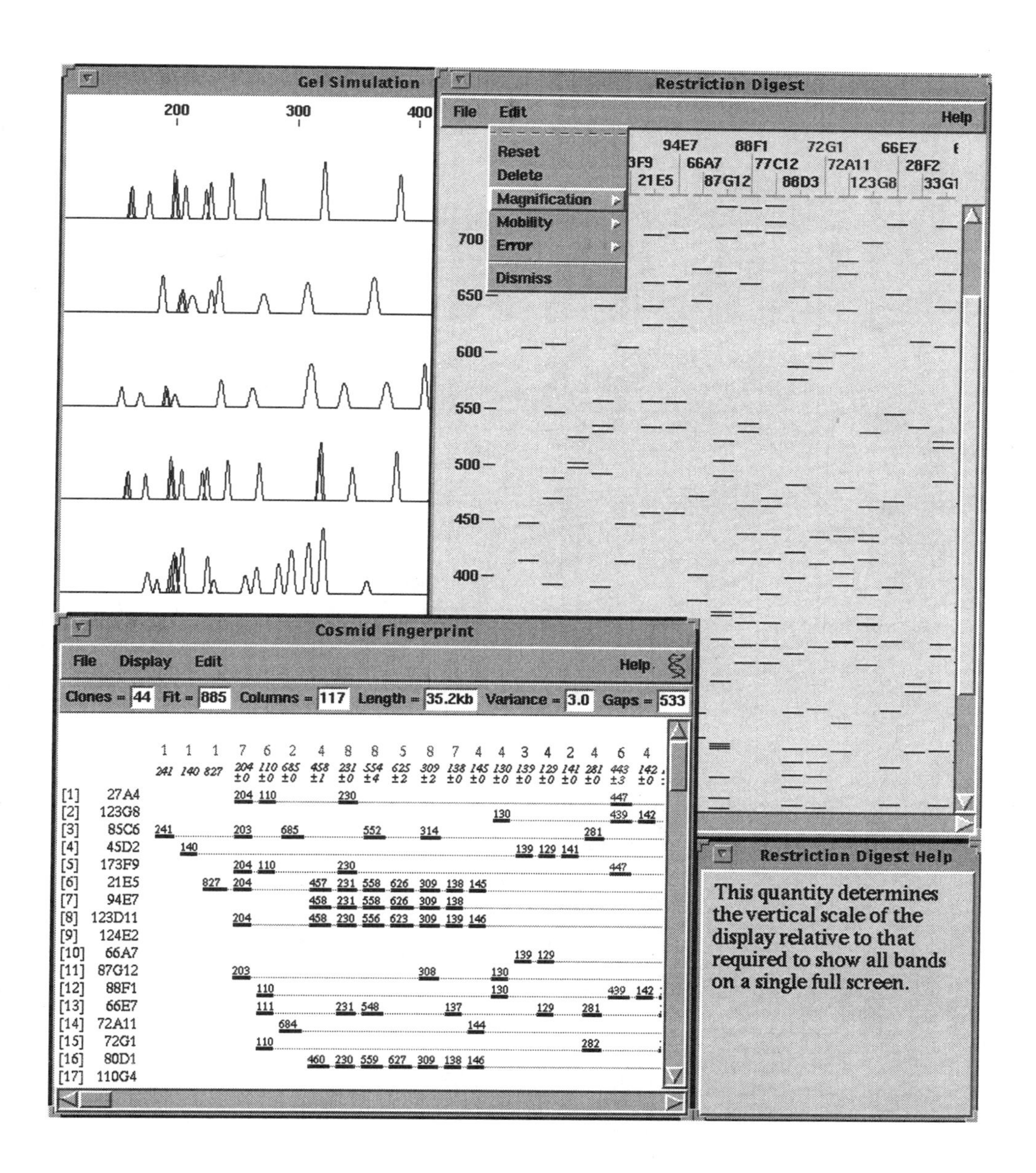

Figure 2. Cosmid fingerprinting tools, which take as input fragment sizes such as those shown in the gel simulation to the upper left, present them to the user for comparison and editing at upper right, and for ordering of common fragments at the lower left.

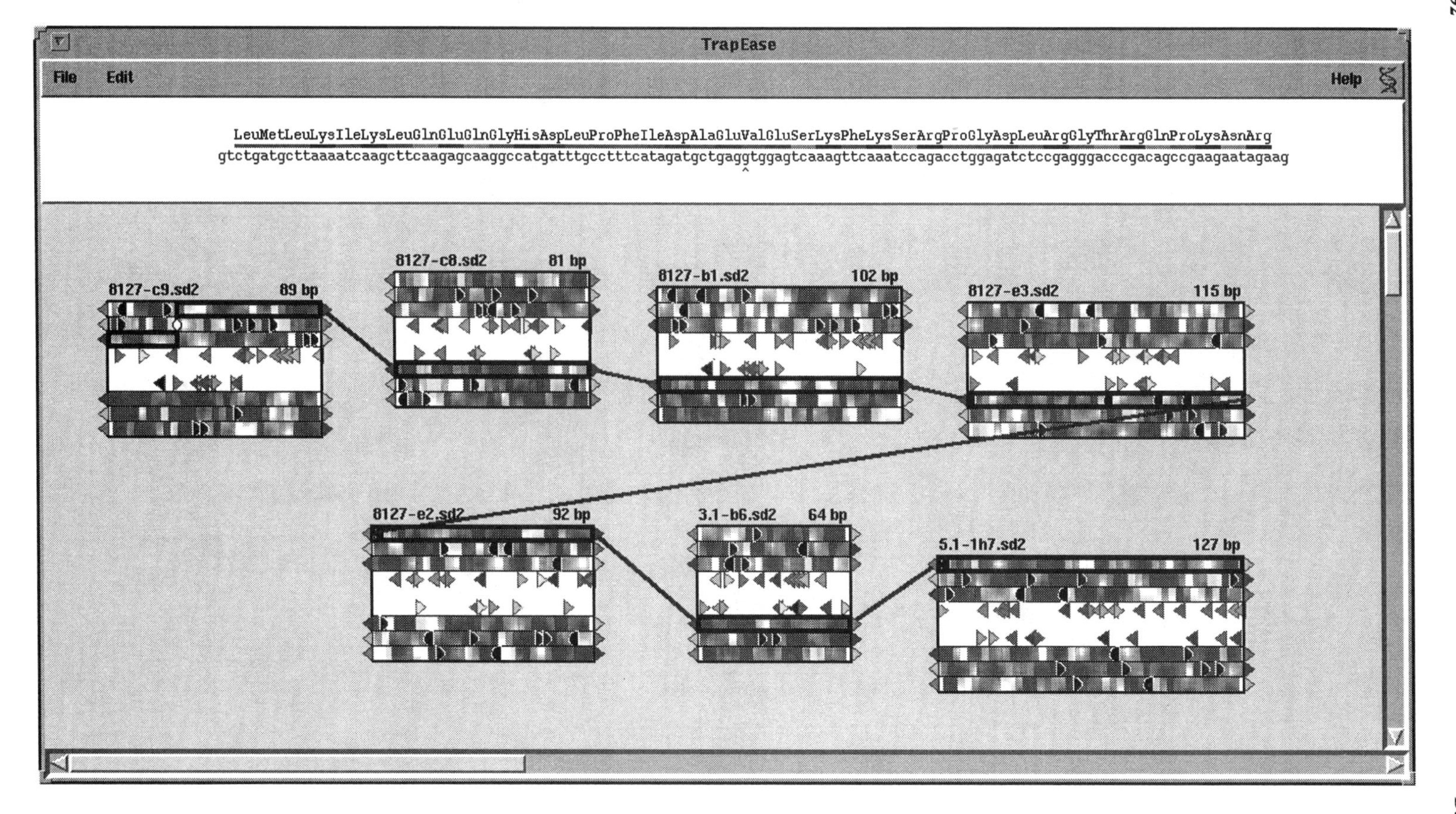

Figure 3. The TrapEase application for support of exon trapping experiments.

track of reading frame. Matching tags can be connected to develop hypotheses about gene structure. Start and stop codons are represented, as well as potential internal splice sites, according to their consensus match strength, in the event that the termini represent cryptic or alternative splices; allowance for sequencing frameshift errors is also made. The grayscale indicates exonic tendency in terms of hextuple frequencies. While the current interface is largely manual, it could be linked with algorithmic support for exploring the space of potential gene structures, and additional graphical features could allow for popping up sequence displays and annotation. Even so, it appears likely that definitive answers will be unlikely except in cases of clear homology, and that this system will be primarily an automated assist to the graphical user interface and the human intervention supported by it.

3. SCIENTIFIC VISUALIZATION

A continuing theme in bioinformatics has been the attempt to abstract and portray meaning at some higher level, from gray masses of sequence data. While this has often been attempted with the tools of signal processing, information theory, computational linguistics, and the like, the visual paradigm might be said to have most readily captured the imagination of biologists.

In large part these attempts at visualization can be seen as a struggle against the tyranny of the first dimension. Sequence and map data are intrinsically associated with linear molecules and chromosomes, and are thus most naturally displayed in an extended format that is easily understood at a surface level, but which obstructs the expression of associated properties, substructures, distant dependencies, periodicities, etc. Necessity has given rise to many imaginative and sometimes rather abstract techniques for scientific visualization of sequence data, e.g., Nussinov plots of secondary structure,[19] helical wheel plots of peptides to detect amphiphilic regions,[21] and topology/packing diagrams of secondary-structural motifs of proteins,[18] not to mention the more obvious uses of graphical plots of hydrophobicity,[17] etc.

In this section we describe several tools and techniques for visualization of both sequence and map data. One theme we will explore in particular is the extension of inherently linear data to two dimensions, in the true spirit of scientific visualization. Two-dimensional displays are very natural when pairwise comparisons of sequences are being depicted, as in the venerable dot-matrix plot, but otherwise making effective use of an additional dimension requires some imagination.

3.1. Sequence Visualization

The author (with tongue in cheek) can claim to have anticipated Java by several years. The RSVP (Rapid Sequence Visualization in PostScript) system[23] is a suite of sequence analysis routines implemented entirely in the page description language PostScript. RSVP can be used to perform limited analyses and visualization of DNA sequences according to a simple command language (see `http://www.cbil.upenn.edu/~dsearls/RSVP/RSVP_home.html`).

PostScript is a language not usually associated with general-purpose computing, and in fact compared to other sequence analysis software packages, RSVP is neither sophisticated nor particularly efficient. However, because it is implemented in PostScript, it has three features of interest:

- It produces attractive graphical output, by its very nature, for a surprising number of types of analyses; in fact, such visualization is its *only* form of output.
- It works with a wide variety of platforms and viewing software, and can even bypass your workstation altogether by executing directly on your PostScript print engine – an underutilized computational resource!
- When used via the World Wide Web, it can "execute" user programs by simply downloading the appropriate PostScript code, so that
  - execution cycles are supplied by the client machine rather than by a central server, avoiding overloads;
  - there are no security concerns or constraints as would normally surround an 'exec' of user-supplied code; and
  - the user can supply local file names as input, rather than having to type in or cut-and-paste extensive DNA sequences or scripts.

Thus, RSVP "applets" are PostScript code, to be executed in a safe fashion by a PostScript previewer, launched from a Web browsing tool such as NetScape. Because the analyses are reported exclusively by way of PostScript, a number of very specialized depictions of sequence features are supported. An example is shown in Figure 4, which was produced by the following RSVP code:

```
triple size
portrait mode
(gtcttctgaatccggcggt...atcttcggatcgtatcgg) setup
6 100 blue up direct repeats
numbering
tickmarks
[(CG) yellow ellipse] flags
no space
sequence
[(G) 1 (C) 1] 9 spectrum
5 100 red down inverted repeats
display
```

Recently Parsons[20] has developed a highly effective visualization tool for depicting repeated sequences, called *Miropeats*. Reminiscent of the loops used in RSVP and other systems, Miropeats goes further by showing repeats simultaneously both within and *between* extensive genomic sequences. The reader is referred also to similar techniques used to depict secondary structure in the GCG sequence analysis package.[7]

The notion of depicting not a single sequence, but a collection of sequences (or a consensus thereof), is the motivation for *sequence logos,* first put forth by Schneider and Stephens.[22] In this technique, which also makes good use of PostScript, both the relative distribution of residue types and the total amount of information in each position of a consensus, are depicted by "stacking" the corresponding letters with appropriate scaling. While not the most esthetically-pleasing of images, they do serve as an attractive alternative to visually scanning extensive aligned blocks of sequence data, and they have gained a definite following.

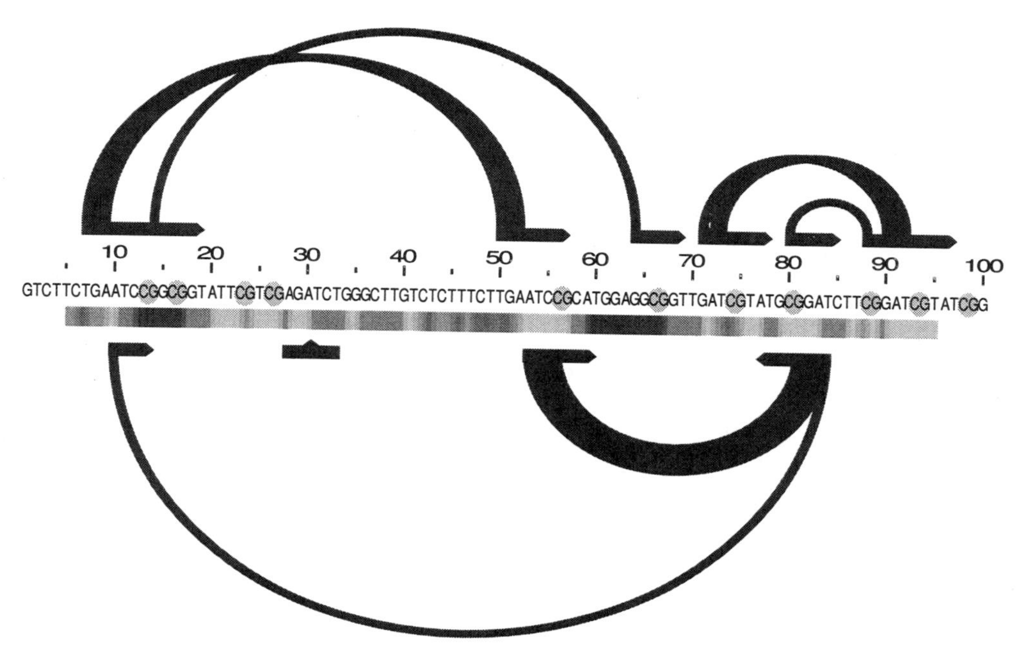

Figure 4. RSVP display showing direct repeats (top) and inverted repeats (bottom) using the `repeats` command, as well as G+C content using the `spectrum` command.

One attempt to depict DNA sequences in two dimensions, which has not achieved widespread use, is based on a *fractal* representation. This was originally suggested[16] as an instance of the "chaos game" algorithm in which successive points are plotted on a plane according to transformations chosen at random from a set – a so-called *iterated function system.* In this technique (which is available in RSVP) the sequence is depicted by setting up a unit square with each nucleotide base labelling one corner. Beginning at the center, a sequence is read a base at a time and at each step a dot is placed halfway between the current position and the corner indicated by that base. The resulting cloud of dots would converge on a uniform distribution given random sequence, and presumably would take on readily apparent shapes given significant nonuniformities of real sequence.

Regions of such fractal plots actually represent "words" of sequence. For example, any point within the upper right quadrant of the plot can only have arrived there as the result of having moved halfway towards the upper right vertex, from anywhere in the overall plot. In other words, any point in that quadrant represents some prefix of the overall sequence ending in the base labelling that vertex (G, for instance). Similarly, any point in the upper right quadrant of *that* quadrant represents a sequence ending in two successive G's. In fact any region obtained by dividing the plot in both dimensions a total of k times will contain points resulting from prefixes ending in a particular unique word of length k. Thus the fractal plot is actually a graphic depiction of all the words in a sequence, up to a word length determined by the resolution of the points in the plot; with sufficient resolution, in fact, any coordinate of the plot could represent the entire history of a sequence up to that point.

Solovyev, Lim, and others[28] have investigated many classes of DNA and protein sequences using such fractal analyses, and created a number of tools and methods to aid in the interpretation of the resulting plots. For example, Figure 5 shows at the upper left a fractal plot for human introns, similar to plots published by them. Note the distinctive pattern of square regions that are more lightly shaded; the most prominent such sparse square – the lower right subquadrant of the upper right quadrant – corresponds to the word CG, known to be relatively rare in the genome of higher organisms. All of the successively smaller fractal "echoes" of this region repeated around the plot represent larger words containing the subword CG.

Note also in the figure the concentrations of points in the upper and lower left corners, and along lines, particularly horizontal lines and to a lesser extent verticals and diagonals. The points near the left corners represent runs of A's and T's, respectively, while the lines indicate words rich in two particular nucleotides. For example, the heavy horizontal along the base of the plot represents words dominated by interspersed C's and T's; this is echoed in the other strong horizontals, containing extensions of those (C+T)-rich words. In fact, introns of this type are known to have relatively long (C+T)-rich runs near their ends. These patterns, of course, could in fact be the result of nothing more significant than skewed overall base frequencies, and this can be seen as a shortcoming of fractal plots as a visualization tool. That is, lower-order effects such as uneven base frequencies can create strong patterns in the plot that could dominate and mask any higher-order effects due to consistent patterns in longer words.

With this in mind, the author and his colleague, Shan Dong, have created tools to produce *normalized* fractal plots, using iterated function systems that adjust to lower-order effects such as base frequency skews. Calling the standard fractal plot a *zero-order* plot, we define a *first-order* plot as one in which the quadrants are not created by evenly dividing the plane in halves each time, but rather by tiling the plane such that the area of each "quadrant" is proportional to the frequency of the corresponding base. Thus, if a data set is depleted in A's, for example, the upper left quadrant will shrink accordingly, and the adjusted iterated function system insures that this applies recursively to successively smaller subquadrants. It can be shown that a sequence chosen randomly but with different frequencies of each individual successive base, will still create a uniform distribution over such a plot.

The upper right panel of Figure 5 shows a first-order plot for the same intron data set as the upper left. While the horizontal lines are softened somewhat, roughly the same concentrations as before are seen, indicating that these effects are not due exclusively to differences in single base frequencies. However, what about dinucleotide frequencies, such as the known scarcity of CG pairs? The technique described above can be extended to create a *second-order* plot, shown at the lower left of Figure 5, which normalizes for dinucleotides. The most notable change is the disappearance of the open blocks due to rare CG's; the corresponding tile is again shown in outline with a cross-hair, and its greatly reduced area shows how the normalization adjusts to create the uniform distribution. The fact that there is no remnant of this pattern indicates that it is not part of any higher-order effect, such as a scarcity of specific longer words containing CG's. Note also that the horizontal lines due to (C+G)-rich words have all but disappeared, while the vertical lines are relatively more pronounced, in particular at the left-hand edge. This may indicate that concentrations of A's and T's in introns are actually a more significant effect once nucleotide and dinucleotide frequencies are accounted for. The lower right panel of Figure 5 shows a third-order plot, which is much smoother yet has suggestions of more subtle patterns.

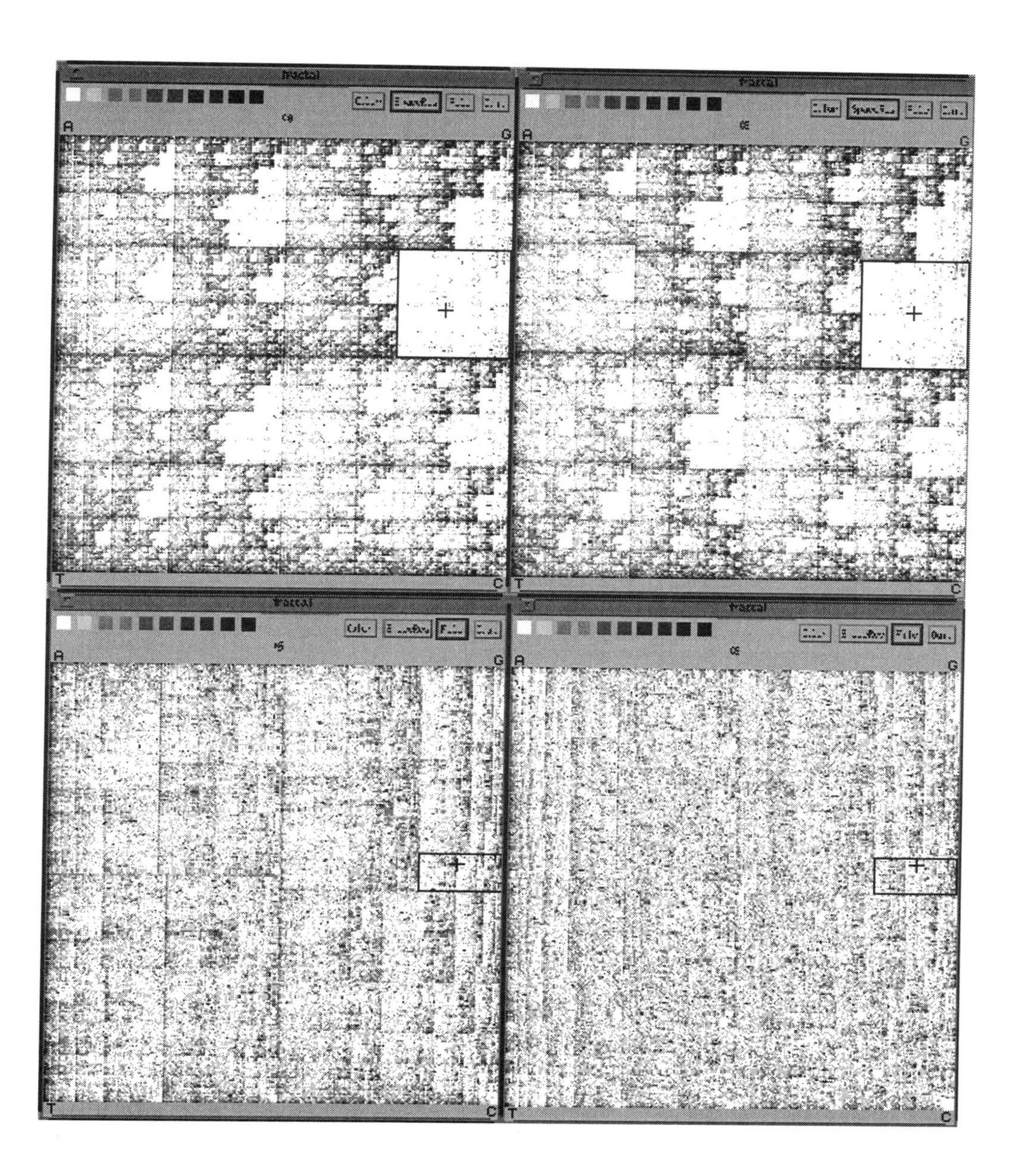

Figure 5. A fractal plot viewer showing higher-order skews in human intronic sequences.

3.2. Map Visualization

The notion of a map is essentially one dimensional, which again is natural insofar as DNA sequences and chromosomes are themselves inherently one dimensional in a physical sense. Use is often made of a second dimension, but this is almost invariably simply for the purpose of "stacking" additional map data on the dominant, single axis. Only recently have true two dimensional displays been explored, in which significant use is made of the second degree of freedom.

For example, Fain and co-workers[12] have promulgated an alternative form of genetic map representation which raises several interesting issues. They point out shortcomings of conventional human linkage maps, which show spans of likely locations of markers relative to some trusted framework map of a subset of markers. They believe that this rather artificial distinction can be avoided, and more comprehensive information displayed, by showing the markers as a two dimensional matrix, in which all pairwise order information is portrayed in the cells of the matrix. In conjunction with a different mapping method based on counting crossovers, this allows regions of greater or lesser confidence, and in particular the opportunity for alternative orderings, to be more easily visualized. Moreover, it is possible to combine data from several studies, and the authors argue that a major virtue of this approach is the opportunity to perform *map integration* using the extra degree of freedom.

We have described a somewhat analogous tool for physical maps, described in detail in Ref. 9. A serious issue in the visualization of map data arises when the ordering of markers is locally ambiguous. This can occur either because the data do not suffice to completely specify the order, or because the data is error-prone, or both. We have investigated visualization techniques that are helpful in both cases. The outcomes of stochastic methods like simulated annealing STS content mapping can be expected to vary for multiple runs, particularly for larger maps, reflecting both types of ambiguity. We do a large number of runs and consider the results to be a sampling of map orders whose notional energies lie near local minima in the space of all possible orderings. We then plot all these orderings, relative to some consensus, by listing the consensus order of probes on the vertical axis and then placing a point on the horizontal axis at the position in which each probe occurs in each sampled ordering, using grayscale intensity to indicate the number of superimposed points. In most cases this leads to a strong diagonal, with deviations that form characteristic patterns which are diagnostic of various forms of ambiguity. For example, if there is a subsequence of markers whose order is well-determined relative to each other, but whose orientation relative to the main diagonal is not well fixed by the data, it will tend to form an "X" pattern. Similarly, patterns emerge to indicate tendencies toward translocation or free ordering of probes within a region. This plot has proved to be very useful in terms of expressing the degree of confidence in "noisy" data, as well as indicating probes that could be rechecked to increase the quality of the map to the greatest extent.

We have since developed an "order editor" for this display, allowing the user to interactively examine individual orderings within the amalgamated display, to edit them, and to choose different ones as the central diagonal "anchor" order. This is shown in Figure 6. Also, an analogous display for fingerprint maps has been incorporated into a cosmid fingerprint tool, that essentially displays the overlap probabilities on a 2-D matrix, with algorithmic support for reordering. For the bioTk map widget, we intend to abstract these themes to create a general capability to visualize partial orders graphically, using order information supplied either as binary relations (e.g. pairwise crossover counts), as sets of

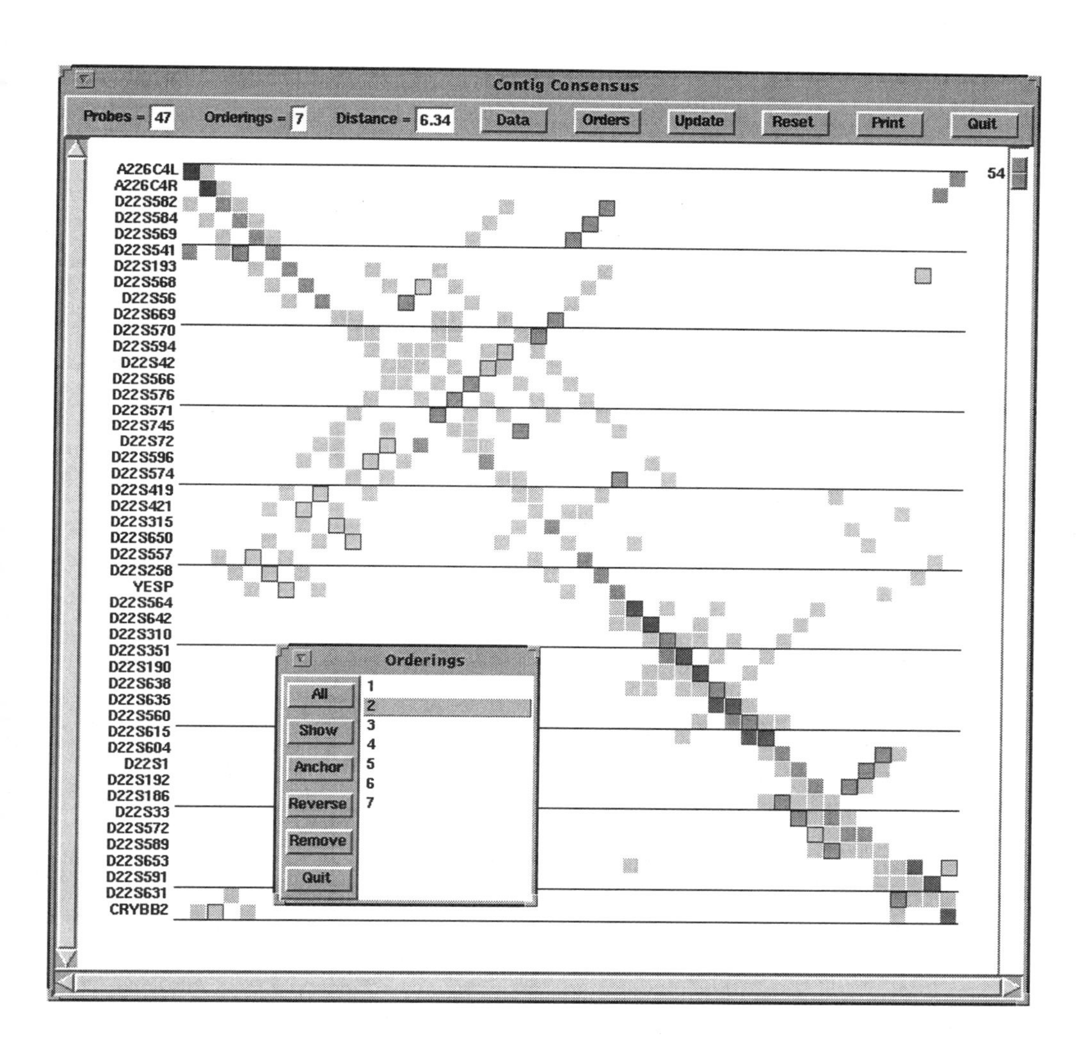

Figure 6. Consensus plot of markers ordered by a number of simulated annealing runs.

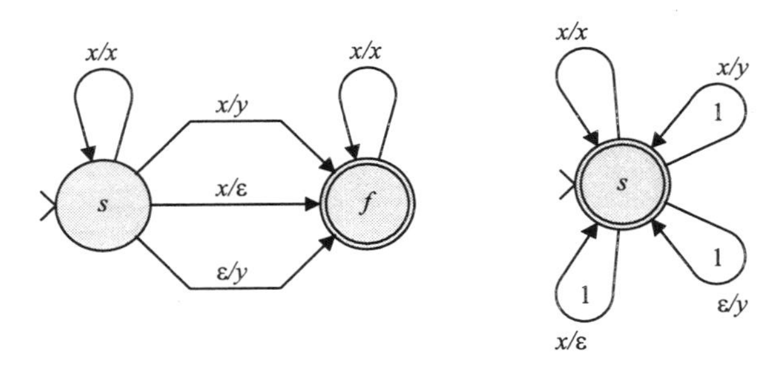

Figure 7. Finite transducers for single mutation (left) and edit (right).

alternative orders (e.g. from resampling strategies), or as more sophisticated data structures for representing alternative orders, such as PQ trees.[2] Such matrix displays, which can also have underlying algorithmic support, could be interchangeable with more traditional displays of the same data, providing a valuable generalization of many traditional one-dimensional map displays.

4. VISUAL PROGRAMMING

Visual programming systems are usually intended to support general-purpose programming, but but specialized tools hasve been developed as well, for instance to support certain forms of database query. We propose that *domain-specific* visual programming is a potential solution to the problem of how to involve biologists more fully in the process of algorithm design based on strong biological models.

The author has recently explored the relationship between certain forms of finite automata and dynamic programming alignment algorithms, with a view to creating such a visual programming system to ease the prototyping of new, model-based alignment algorithms.[25,26] We here briefly review the formal foundations of this approach.

Finite transducers are simply finite-state machines for which transitions have both input and output. Figure 7a shows how a finite transducer is used to model a single mutation occurring anywhere in a string, where the mutation could be a single-base substitution, deletion, or insertion (using transitions labelled by their input and output, separated by a foreslash, with x and y standing for any non-identical nucleotides). By simply merging the start and final states of the "mutation machine," we can produce an "edit machine" as shown in Figure 7b. This machine will make any number of non-overlapping mutations in a string, and in fact is capable of changing any string into any other. We have also introduced *weights* on the transitions, which are understood to be added to a running total with each move of the transducer. Then it can be seen that what is conventionally defined as the minimal edit distance between two strings is simply the minimal computation of this automaton.

We can consider each state of the transducer to be a matrix, and each term in the recurrence defining that matrix to be determined by outgoing transitions from that state. In fact it may be said that the automaton is simply an alternative form of the mathematical

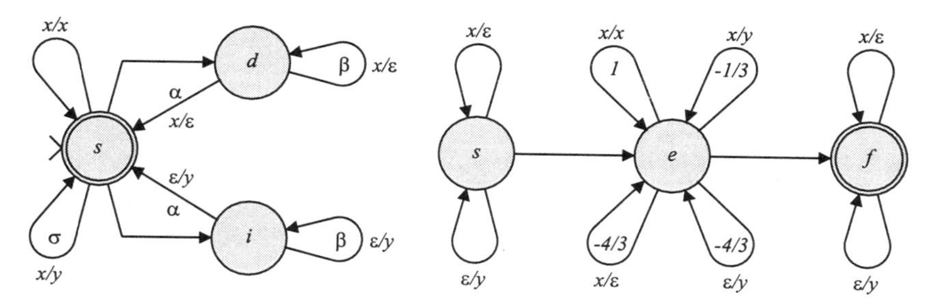

Figure 8. An affine gap (left) and a local alignment (right) transducer.

recurrence. In this case the edit transducer produces the classic edit distance recurrence, by a simple procedure:

$$s[i,j] = \min \begin{cases} s[i-1,j-1] & \text{if } \; x_i = y_j \\ s[i-1,j-1]+1 & \text{if } \; x_i \neq y_j \\ s[i-1,j]+1 & \\ s[i,j-1]+1 & s[0,0]=0 \end{cases} \tag{1}$$

Affine gaps are modelled as in Figure 8a, for gap initiation penalty α and gap extension penalty β, where typically $\beta < \alpha$, and substitution penalty σ. Again, the transducer is formally equivalent to a well-known recurrence, first derived algebraically by Gotoh.[14] More practical algorithms employ maximum similarity rather than minimum distance, and attempt to find local regions of similarity rather than global matches. Such algorithms can be modelled first by simply changing *min* to *max* and reinterpreting weights accordingly, and second by adding unweighted *scanning* transitions to find selected local regions. A transducer for local alignment is illustrated in Figure 8b. Here, only the transitions from the *e* state will be of relevance to the alignment. We can again produce a recurrence directly from the transducer.[26]

We have taken advantage of the relative ease with which novel alignment algorithms can be designed, by creating a visual programming system that makes use of a domain-specific drawing tool to specify an aligner, which is then translated automatically to code for the corresponding dynamic programming algorithm. It is hoped that this increased ease of design and experimentation will encourage the development of new algorithms that incorporate specific domain knowledge, in what we call model-based alignment. Such models have been created which are sensitive to reading frame, and which take account of phenomena such as simple tandem repeat polymorphisms.[26] Recently we have created an alignment machine that entails a model of gene structure, depicted in Figure 8. That is, implicit in the model is the notion of reading frame, start and stop codons, and splice junctions. In the figure, the states at the upper left align the putative start codons, and those at the lower left the stop codons. The *E* state captures the basic alignment, which may be interrupted by intronic states at the far right, entered and exited via splice junction dinucleotides. With this model we have been successful at comparing tubulin genes with disparate intron/exon structures, in such a way that gene structure is predicted with greater accuracy by means of the mutual information between two related genes. Similar work has been done by Hein and Stovlbaek[15]; we hope that the design of imaginative algorithms such as these will be facilitated by the tools we are creating.

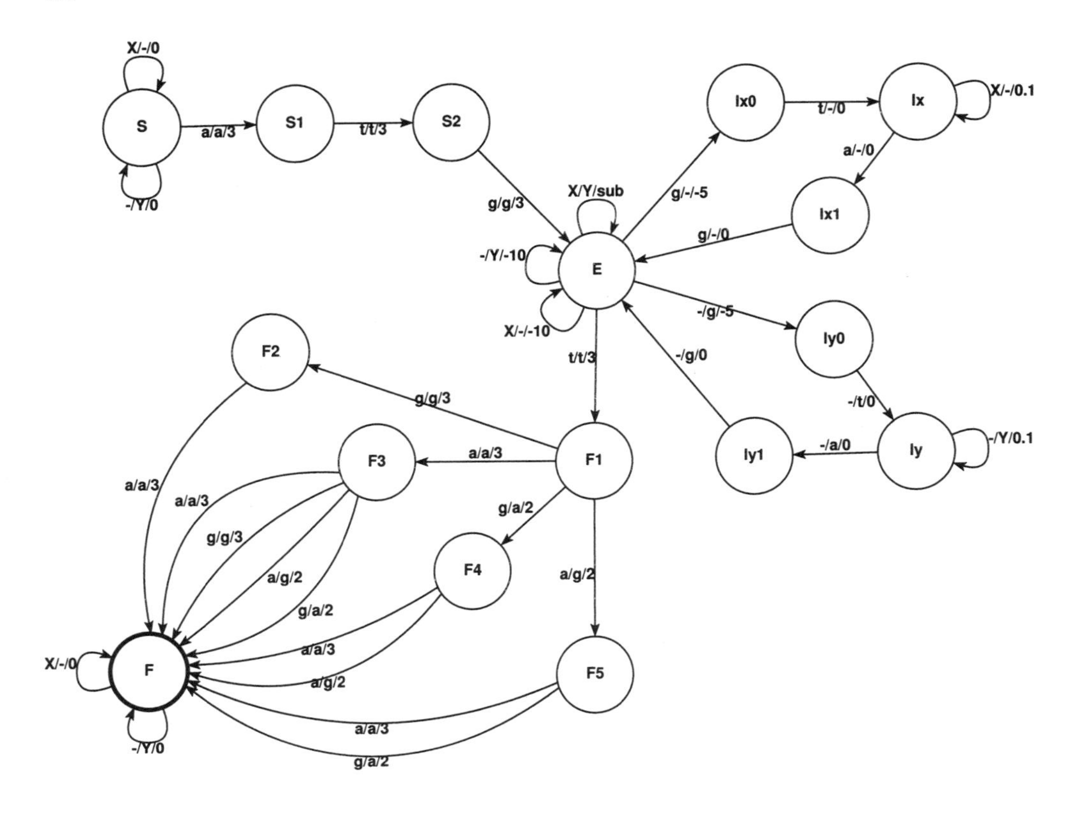

Figure 9. A gene alignment model.

5. CONCLUSION

The three technologies described above represent a useful categorization of the field, but ultimately they overlap in significant ways. For example, the question of how to provide a comprehensive view of extremely dense map information has often been addressed through pan-and-zoom viewers that make use of GUI techniques to address what is essentially a visualization problem. This is illustrated by recent research in the GUI field on *semantic zooming*, by which objects change qualitatively at different magnifications so as to reflect the users' likely needs at various levels.[1] Similarly, the notion of visual programming of alignment algorithms can be seen as a kind of algorithmic visualization; a visualization that helps biologists to understand the underlying model of an algorithm, or even to design it, will necessarily aid their comprehension of the results of their analyses. Naturally, the effectiveness of any visual programming system is critically affected by its GUI. By the same token, a highly specialized and sophisticated GUI controlling an algorithm can be considered to be a domain-specific visual programming system, precisely to the extent that it supports the kind of flexibility that is necessary for model-based programming. Stated another way, the ultimate control of a sequence analysis system is the ability to modify it to suit the state of biological knowledge of a sequence most closely – not just parameters, or substitution matrices, but actual mechanism.

We would argue that such flexibility will be a necessity in viewing and analyzing large-scale genomic data. Much is said about automated annotation of sequence data and warehousing of databases, but in the end much if not most of the useful results derived from sequence data are the result of interactive use of databases and sequence analysis algorithms, in the hands of skillful end-users. It is uncontroversial to assert that powerful GUIs will be necessary for such endeavors. Given the orders-of-magnitude increases expected in volumes of sequence data, it is just as clear that novel modalities for visualizing sequence and map data will be required. We believe that the ability to perform "dynamic annotation" with interfaces and algorithms that are easily adapted to rapidly-changing techniques and states of knowledge, will require all of these technologies together.

ACKNOWLEDGMENTS

This work was supported by the US Department of Energy under grant number 92ER61371, and by the National Institutes of Health under grant number P50HG00425. The author gratefully acknowledges the contributions of Shan Dong, G. Christian Overton, Dominic Bevilacqua, and Kevin Murphy.

REFERENCES

1. B. Bederson, L. Stead, and J. Hollan. Pad++: advances in multiscale interfaces. *Proceedings of ACM SIGCHI*, 1994.
2. K. S. Booth and G. S. Leuker. Testing for consecutive ones property, interval graphs, and planarity using PQ-tree algorithms. *J. Computer Systems Science*, 13:335–379, 1976.
3. K. W. Brodlie, L. A. Carpenter, R. A. Earnshaw, J. R. Gallop, R. J. Hubbold, A. M. Mumford, C. D. Osland, and P. Quarendon. *Scientific Visualization*. Springer-Verlag, Berlin/Heidelberg, 1992.
4. M. M. Burnett and M. J. Baker. A classification system for visual programming languages. *Journal of Visual Languages and Computing*, pages 287–300, 1994. `http://www.cs.orst.edu/ burnett/vpl.html`.
5. M. M. Burnett and D. McIntyre. Visual programming. *Computer*, 28(3):14–16, 1995.
6. M. Cinkosky and J. Fickett. System for Integrated Genome Map Assembly (SIGMA). `ftp://atlas.lanl.gov/pub/sigma`.
7. J. Devereaux. *GCG Program Manual*. Genetics Computer Group, Inc., Madison WI, 1991.
8. R. Durbin and J. Thierry-Mieg. A C. elegans Database (ACEDB). (unpublished).
9. C. Bell et al. Integration of physical, breakpoint, and genetic maps of chromosome 22: Localization of 587 yeast artificial chromosomes with 238 mapped markers. *Human Molecular Genetics*, 4(1):59–69, 1995.
10. J. Nadeau et al. The Encyclopedia of the Mouse Genome. `http://www.informatics.jax.org`.
11. T. Marr et al. Genome Topographer. `http://www.cb.cshl.org`.
12. P. R. Fain, E. N. Kort, P. F. Chance, K. Nguyen, D. F. Redd, M. J. Econs, and D. F. Barker. A 2D crossover-based map of the human X chromosome as a model for map integration. *Nature Genetics*, 9:261–266, 1995.
13. N. Goodman, S. Rozen, and L. Stein. The case for componentry in genome informatics systems. `http://www-genome.wi.mit.edu/informatics/componentry.html`.
14. O. Gotoh. An improved algorithm for matching biological sequences. *J. Mol. Biol.*, 162:705–708, 1982.
15. J. Hein and J. Stovlbaek. Genomic alignment. *J. Mol. Evol.*, 38:310–316, 1994.
16. H. J. Jeffrey. Chaos game representation of gene structure. *Nucleic Acids Research*, 18:2163–2170, 1990.
17. J. Kyte and R. F. Doolittle. A simple method for displaying the hydropathic character of a protein. *J. Mol. Biol.*, 157:105–132, 1982.
18. M. Levitt and C. Chothia. Structural patterns in globular proteins. *Nature*, 261:552–558, 1976.
19. R. Nussinov, G. Pieczenik, J. R. Griggs, and D. J. Kleitman. Algorithms for loop matchings. *SIAM J. Applied Math.*, 35:68–82, 1978.

20. J. D. Parsons. Miropeats – Graphical DNA-sequence comparisons. *CABIOS*, 11(6):615–619, 1995.
21. M. Schiffer and A. B. Edmundson. Use of helical wheels to represent the structures of proteins and to identify segments with helical potential. *Biophysics J.*, 7:121–135, 1967.
22. T. D. Schneider and R. M. Stephens. Sequence logos: A new way to display consensus sequences. *Nucleic Acids Research*, 18:6097–6100, 1990.
23. D. B. Searls. Doing sequence analysis with your printer. *CABIOS*, 9(4):421–426, 1993.
24. D. B. Searls. *bioTk:* componentry for genome informatics graphical user interfaces. *Gene*, 163(2):GC1–16, 1995. (appeared electronically in Gene-COMBIS).
25. D. B. Searls. Sequence alignment through pictures. *Trends in Genetics*, 12(1):35–37, 1996.
26. D. B. Searls and K. Murphy. Automata-theoretic models of mutation and alignment. *Proceedings of the Third International Conference on Intelligent Systems for Molecular Biology*, pages 341–349, 1995.
27. T. F. Smith and M. S. Waterman. Identification of common molecular sequences. *J. Mol. Biol.*, 147:195–197, 1981.
28. V. V. Solovyev, S. V. Korolev, and H. A. Lim. A new approach for the classification of functional regions of DNA sequences based on fractal representation. *Int'l. J. Genomic Res.*, 1:108–127, 1992.

16

DATA MANAGEMENT FOR LIGAND-BASED DRUG DESIGN

Karl Aberer, Klemens Hemm, and Manfred Hendlich*

GMD-IPSI
Dolivostr. 15, 64293 Darmstadt, Germany
e-mail: {hemm, aberer}@darmstadt.gmd.de

ABSTRACT

Developing a new drug is a cost- and time-consuming process. Rational drug design, i.e. the use of comptuational methods for finding or constructing new drugs, is expected to become an important factor in reducing these costs. We present database-oriented methods that support rational drug design by targetting on the processing of the explosively growing information that becomes available from biomolecular methods, in particular genome research. This information opens the way for new, structure-oriented methods in the search of drugs. The article addresses the biological expert who is interested in the application of database management approaches for domain specific information systems.

1. INTRODUCTION

Biological processes occur at a molecular level. These processes are based on the interaction of complex biological molecules that result from evolutionary processes. The effect of drugs is based on the way they affect natural biological processes. In order to gain a functional understanding for the manipulation of biological processes, biological molecules and their interaction with drugs have to be studied at a molecular level.

Without detailed understanding of biological processes new drugs have to be determined in an experimental way. Pharmaceutical industry has developed sophisticated methods for finding new drugs experimentally. Pools of existing substances are built, either by collecting natural and artificial substances or by modifying existing substances. Using experimental methods these pools are screened for candidates with potentially interesting effects. It is easy to imagine that this experimental approach is extremely expensive and

* Current address: E. Merck, Preclinical Research, 64271 Darmstadt, Germany. e-mail: hendlich@merck.de

Theoretical and Computational Methods in Genome Research, edited by Suhai
Plenum Press, New York, 1997

time-consuming. This is illustrated in Figure 1.1 which shows how cost as well as development time for new drugs is drastically increasing. Really new (types of) drugs are found in this way very rarely as compared to the effort spent. Thus the idea of designing drugs based on a detailed knowledge on how they work and interact in biological systems, so-called rational drug design, is of course very appealing. The hope is that this approach might lead to unprecedented progress in drug development.

Rational drug design is becoming a viable alternative in the search of new drugs as scientific results in various areas, like molecular biology or genome research, are recently becoming available on a large scale. At first isolated results, like structures of particularly interesting proteins, have formed the starting point for rational drug design. For example, visualization tools are used to support the intuition of the drug designer in analysing interesting protein_ligand interactions. With the explosion of the available data and knowledge however a more automatized support for rational drug design is becoming necessary. Today there exist huge databases of protein structures, the most prominent are being PDB (Brookhaven Protein Databank) [6], and small molecules, typically proprietary to pharmaceutical companies. Those databases are currently growing extremely fast in size. Thus for using this data for a systematic analysis in drug design, support by computational tools is required. In particular tools for managing and accessing huge data collections are a prerequisite to exploit the information that is hidden in large data collections.

Management of large data collections requires support at different levels. The first important observation is that different, heterogeneous data collections exist, that are relevant for drug design. This is a consequence of the fact that data collections are established according to the way the data is produced. For example, PDB systematically collects structural protein data that is produced since different methods for structure determination, like X_Ray or NMR, have been developed. To give another example, sequence databases are established as a consequence of efforts in analyzing genomes of several organisms. For an application using this data, like drug design, the different existing databases of interest thus need to be integrated. Therefore integration of heterogeneous databases is a primary requirement for an integrated system supporting drug designers in data analysis. The second endeavour is to find an appropriate representation for the existing data. The data collections are typically available in form of record_based files. These can be used to apply certain isolated analysis tools, but are an insufficient basis for systematic data processing. In the area of business processing data management systems have evolved for this purpose, like relational database systems. However, these systems are not equipped to rep-

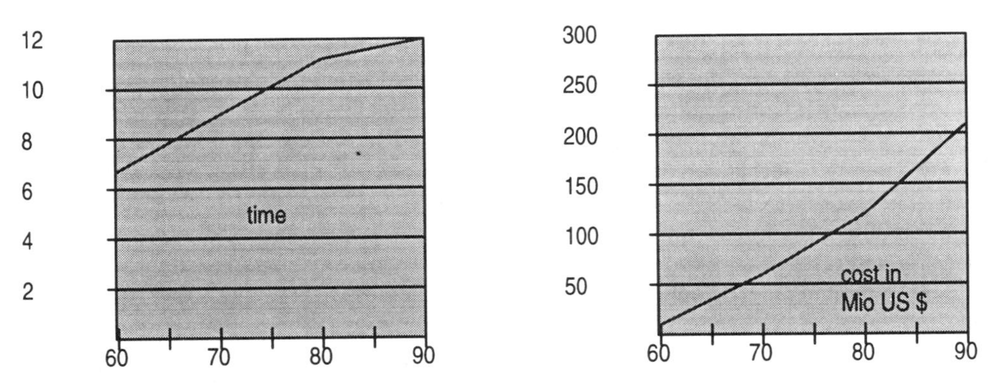

Figure 1.1. Time and costs for the development of a new drug.

resent the typical structures that occur with biological data, like sequences or spatial data. Therefore other types of systems or approaches for the efficient management of biological data are needed. The third important issue is the provision of flexible and efficient access mechanisms for data analysis and querying. Again business_oriented approaches, like SQL query processing, are too limited in scope. In particular ways have to be found to integrate specialized analysis tools, as they exist for biological data, for example fast sequence alignment or sequence search algorithms, into the system framework. On the other hand the retrieval mechanisms have to be flexible in order to quickly respond to the rapidly changing research environment and questions. Finally, as a fourth aspect, a system for supporting the management of data in the area of drug design needs to support the usage of visualization tools, as they are fundamental for the daily work of drug designers. This has to be supported in a flexible way, for example by dynamically visualizing interesting features of a molecule in a molecule viewer that have been generated by a preceding retrieval operation. Another aspect of visualization is the support of the user in finding his way in the complex and large data collections, for example by navigation and browsing tools. We have summarized the four steps of data processing that are the foundation for our approach in Figure 1.2.

With our approach we have built a system that covers all of the four aspects, starting from data integration, over data representation and retrieval, up to data visualization. As a core technology we have chosen object_oriented database technology. In particular the ability of object_oriented databases systems to support abstract data type definitions is advantageous for the representation of complex objects but also important for the management of complex operations that are needed for those objects. We have used this feature in particular for the implementation of database population operations, complex retrieval operations and visualization operations. In addition to using standard object-oriented database management technology it has turned out that some advanced features are required.

We have needed to develop a methodology to integrate the data. It was necessary to coordinate this step with domain experts who develop tools for analyzing data, and provide mechanisms to transform, correct, and interrelate data from the different source databases. This procedure consists of steps that allow different degrees of automatization. The problem we have solved is to enable the domain expert to describe the contents of the integrated database and the relations of data without overwhelming him with low level data processing tasks. This has been realized by providing him with specification mechanism for data transformation and integration at a relatively abstract level that is based on stand-

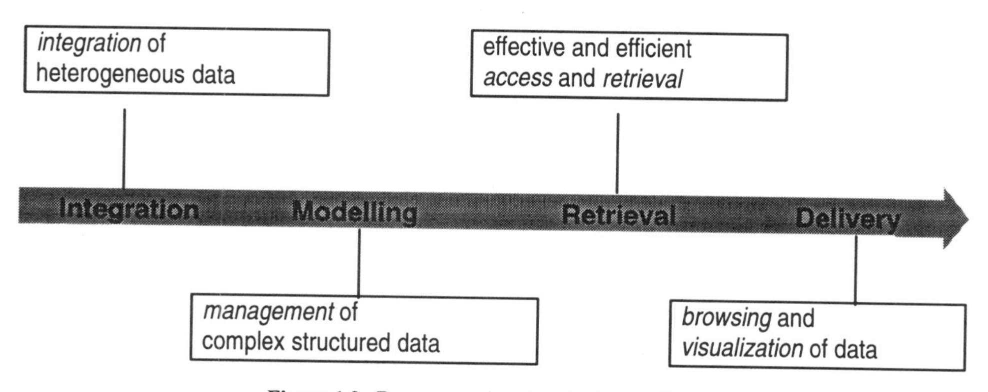

Figure 1.2. Data processing steps in the warehouse.

ard data transformations, like importing a complete protein structure in one step, that are provided as database methods.

For the representation of data it turned out that standard data type constructors, as provided by object-oriented database management systems, were not sufficient, or more precisely, not sufficiently efficient. Therefore we had to develop specialized data types, in addition to the existing ones, to support efficient management of, for example, sequence and structure data. To this end we had to separate the logical view and the physical representation of the data, a principle that is well known from relational database systems but not very frequently used in object-oriented database systems.

In order to support the use of methods in database queries we needed an extensible query language (in the style of SQL3 or OQL [7]). Using a declarative query language provides for sufficient flexibility in the formulation of requests, and combining this technology with the use of sophisticated domain-specific algorithms provides for the required efficient search mechanisms. For adequate processing of such queries application-specific cost models and optimization algorithms have been implemented. Here we rely particularly on extensible and rule-based query optimization technology.

Support for the generation of visualization information has been in particular WWW-oriented by generating HTML and VRML code. The visualizations comprise query entry forms and query result display, browsing through the data based with form-oriented object representations and visualizing molecules by different viewers. Also mechanisms for visually analyzing database contents for information discovery have been experimentally implemented.

Two aspects make the system unique:

1. the unique combination of biological data from different sources in combination with the algorithms to analyze and enrich the data during the integration process, and
2. the fully integrated system that supports the different steps in the data management process.

With regard to the first aspect the data treated by the system is unique, in particular due to the domain-specific, algorithmic methods used for data preparation [14] from PDB, the combination of this data with other resources and the combination of algorithms available on this data. The use of the different algorithms in the data preparation process is enabled by our integration methodology, while the use of retrieval and visualization algorithms is eased through the object-oriented approach. Extracting the ligand related data in PDB is a current research issue in various research groups. Our database is the first that provides the data prepared in this comprehensive form. We will not further go into this aspect of the system. More information can be found in different publications [15].

With regard to the second aspect we find various systems in the literature that cover the one or the other aspect we treat with our system. However to our knowledge none of them support the whole process of data manipulation from data integration up to visualization, in an integrated manner.

One of the systems that best matches ours with regard to comprehensibility is IGD [27] in the area of genome research. The system integrates data from various genome databases without hiding the origin of the data from the users. It is based on relational database technology, but uses an object-oriented front-end based on ACEDB [9] which also provides various visualization tools.

Different systems offer similar functionality at the level of data representation, retrieval and visualization, without using data from various resources. To mention in our context are in particular several systems for accessing protein data from PDB, like WHATIF/database [32], the Iditis product, P/FDM [17], or lately the developments of the Brookhaven Laboratory, offering an OPM (Object Protocol Model) based access mechanism. These systems vary widely in their functionality. Similar approaches can also be found in the area of genome research, e.g. GenBank [5].

With regard to data integration different approaches for establishing links between databases need to be mentioned, e.g. SRS [10]. However, due to their architecture these systems can not offer functionality that goes beyond pure navigation.

Clearly to be distinguished from our approach are those systems and tools that are centered around the major database resources or reference sites, which collect, manage and provide data from original research, like PDB, Swissprot [3], Genbank or PIR [4]. Those encounter slightly different problems as they have to treat overlapping data from one particular area, have autonomy over their data and are not focusing on data analysis issues. We rather use those resource databases to recombine and enhance this data from orthogonal areas - still having overlapping data as one problem among others - such that researchers can profit in the analysis of the existing data resources in an optimal way.

When working on the problem of integration and management of biological databases one has of course to consider developments in general computer science from different areas, like database integration, data warehousing, data mining, database systems, data visualization or user interfaces. We postpone the discussion of some actual deelopments regarding these issues when we come to the specific problems later.

In Section 2 we analyze the problem domain by identifying the data sources required and identifying some requirements from the user's viewpoint. In Section 3 we introduced the database integration approach, which is based on a data warehousing approach. In Section 4 the use of object-oriented data modelling and object-oriented database management systems is discussed. The different kinds of retrieval that are supported by the system are presented in Section 5 and the different visualization techniques in Section 6. In Section 7 we give some concluding remarks.

2. THE PROBLEM DOMAIN

2.1. Relevant Information and Algorithms for Drug Design

In order to identify the databases and tools required for rational drug design it is worthwhile to shortly discuss some of the basic principles that guide the search process for new drugs.

The straightforward approach to rational drug design is applying the so-called key-lock principle, i.e. to identify or construct receptors and ligands that have complementary physico_chemical and structural features at the binding site. For this purpose detailed structural data on the interaction of receptors and ligands is required. With some effort this information can be extracted from the PDB database and be made available for algorithmic * [26] and database-oriented analysis.

A more indirect approach to determine whether a substance is potentially interesting is to exploit similarity as a heuristic. Similarity can be applied either for small molecules

* Algorithmic methods to compute an approximation of the binding constant from the structural data are the focus of another part of the German national joint project RELIWE in which also this work has been performed.

assuming that molecules with similar structural features can have similar effects, or for receptor molecules (proteins), assuming that a substance that has an effect on a similar receptor will also have an effect on the target receptor. Detailed structural data can also help to analyse which aspects of a molecule (e.g. subgroup of ligands, residues in proteins) are relevant for an interaction. From this the notion of similarity can be refined by identifying the parts relevant for the interaction of a molecule, and thus only considering those parts for similarity measures.

Another indirect approach is to draw conclusions from mutation data. In many cases hints on the function of proteins are obtained by analyzing mutations (natural small changes) or modifications (artifical small changes) of a protein, respectively its genetic coding, that lead to substantial changes in function. In this way information on the binding site and its relevant functional information is derived. This approach is particularly promising in cases were detailed structural knowledge is missing, which is today the case for the (pharmaceutically) important class of GPCR proteins.

In order to support these different types of analyses different databases and algorithms are needed, of which we list now those that have been considered in our system.

- Protein structure data: Currently about 4000 protein structures are known in PDB. The information on the three-dimensional structure of proteins and complexes is taken from the PDB database. In the year 2000 it is expected to contain about 20.000 to 30.000 structures.
- Protein sequence data: More than 100.000 protein sequences are available in public databases. Our main sources are PDB and SWISSPROT. This information is additionally interlinked with the PIR and EMBL sequence databases.
- Protein secondary structure data: Solvent accessibility and secondary structure are taken from the DSSP database calculated by executing the DSSP-program for all PDB entries.
- Protein sequence alignment data and algorithms: Alignment data is taken from the HSSP database [28]. Alternatively alignments are computed by using the Needleman-Wunsch algorithm [22] or the FASTA algorithm [25].
- Protein families: This information is taken from EC and CATH codes.
- Protein mutations and modifications: This information is taken from SWISSPROT and specialized databases like PMD [23].
- Ligand structure data: A considerable fraction of the PDB entries also contains the coordinates of bound ligands. From the PDB database information on small molecules such as substrates, inhibitors, cofactors, prosthetic groups and metal ions is extracted.
- Ligand topology data: The topology for the small molecules extracted from PDB is derived using the Beilstein database [33], as well determined by special purpose algorithms and manual intervention.
- Ligand substructure algorithms: Substructure search for ligands is realized using subgraph matching algorithms as well as using fingerprints for a fast prefiltering. The notion of ligand similarity is based on fingerprints and maximal common substructures.
- Annotation data: Textual descriptions for biological effects, protein function or catalytic activities are currently taken from SWISSPROT and PDB, in-house information systems in pharmaceutical companies or literature.
- Visualization data: Different graphical formats, reprentations of binding sites or ligand topologies, are generated based on the structural data, e.g. using the Daylight software [8].

To combine these different types of data and algorithms for uniform access within one system is a desirable goal. In addition to the kind of data that is provided by the different databases, quality is a central issue. A large number of source databases is well known to contain errors or omissions from various sources, ranging from wrong interpretations of experimental results to simple typos. For many classes of such errors algorithms for correction are known. Also different kinds of information is contained implicitly in the data which should be made explicit for the information system. A typical example is the topology of ligands, which can be derived algorithmically in most cases uniquely from the atom coordinates in the PDB database. These types of improvements in the quality of data are an important aspect in the construction of the drug design information system.

In the following section we discuss what requirements from an information systems point of view need to be considered when attempting to achieve this goal.

2.2. User Requirements to an Integrated Drug Design Information System

As an illustration we first give an example of a typical question that a user might pose to an integrated drug design information system:

What is the homology of a thrombin H-chain and those proteins, that are contained in a protein-ligand-complex, whose ligand contains benzamidin as structural element?

Answering this query requires to access

- protein structure data from PDB data
- ligand topology data, that has to be derived from PDB using appropriate algorithms and ligand databases.
- homology data that is either taken from HSSP or derived by appropriate algorithms
- substructure algorithms for small molecules

Answering this query without information system support can easily grow out to a minor research project. To avoid this a uniform, problem-oriented view on the data, that hides technical details of data access as well as the origin of the data is required. Therefore a representation of the data has to be defined that directly reflects the domain model of the user. The query itself is declarative, such that a declarative, ad-hoc access needs to be supported. The query itself uses complex algorithms (sequence alignment) that must be smoothly integrated in the access mechanisms. Once a result is found, the user may be interested in browsing through the resulting objects, or navigating to neighboring objects along different types of references, for example inspecting the protein objects and their corresponding ligands.

All these requirements not only have to be satisfied functionally but also efficiently. Thus the physical design of the system must support the efficient processing of the different tasks required by the user. Which access mechanisms or algorithms are chosen is however not only determined by efficiency considerations but as well by quality considerations. In many cases the user might be satisfied with an approximate but fast answer. Therefore alternative mechanisms for computing the same functions are used, for example in the case of sequence alignment.

3. THE INTEGRATION APPROACH

3.1. Requirements and Basic Design Decisions

From the viewpoint of database integration the main issue for building our integrated database is the integration of databases at a data level. This is in contrast to the problem of integration at a schema level, which is the main focus of most computer science approaches to database integration [24][29]. In our case the schema integration problem is treated intellectually, i.e. database and domain specialists design the integrated view on the data as an application schema. This is a feasible approach as we are concentrating on a very narrow application domain. On the other hand migrating the data into the database is a more difficult problem, for which there exist no general tools or concepts. To approach the problem we make a few basic observations:

1. *All required data can be obtained in record-oriented files:* This first of all alleviates the problem of platform dependency, since we use ASCII as the least common denominator for interoperabillity of different platforms. On the other hand this immediately leads to the problem of expensive data access, e.g. by parsing, which makes some amount of preprocessing inevitable.
2. *Updates are not performed continuously:* Most of the resource databases perform periodic releases. Thus it appears sufficient to access this data only when a new release is provided.
3. *The resources are completely autonomous:* This is the most important source of problems for the database integration issue. Since the data providers are autonomous a high degree of heterogeneity at the data level is encountered. One finds semantically complex data in poorly structured representations. Database schemas are missing or implicit. Data is incomplete and erroneous. Data may only be available as the result of analysis programs or even only accessible in the literature. Thus to provide quality data for the scientific work an extensive analysis and enrichment process is required.

From these observations one can immediately derive a fundamental design decision. In database integration there are in general the two options to virtually integrate data from different databases by providing appropriate views on the source databases or to physically integrate the data by transforming the data and storing it redundantly. From a logical point of view the approaches do not differ. They are only different ways of physically realizing data integration. In the case of physical integration the integrated database often is called a data warehouse. This notion has developed for decision support systems in business applications where huge sets of data are aggregated for subsequent analysis [11]. In our application we are in a similar situation, since prior to data analysis extensive data preparation is required as described in observation 3. Also observation 1 supports our design decision, since for the efficient access to the file-based databases additional access mechanisms have to be provided. From observation 2 it follows that physical integration is feasible since no major problems with data consistency are to be expected.

To illustrate the complexity of the data integration process, and thus to clarify why we are following the data warehousing approach, we sketch a few steps that have been realized within our project for preparing the integrated database [14][15]. From PDB complexes of proteins and different types of ligands are algorithmically identified and are stored in the database together with the threedimensional models. The ligand topology needs to be reconstructed using algorithmic methods, small molecule databases and manual interaction. The topology is stored in the database and connected with the ligand objects identified in PDB. Additional information on the complex is derived then from sequence databases by accessing them using either keys or different sequence similarity searches. In this step also different sequence numbering schemes are aligned. Mutation data is derived either from the SWISSPROT database or the PMD database.

The example shows that data from a number of source databases has to be transformed into one target database. The source databases are organized in files, each file corresponding to one (or a few) objects, e.g. a protein, a sequence, or a ligand. The data within files is organized in record structures. Specific records may be optional. Values can be simply structured data or can be texts, sequences, or again record structures (e.g. the coordinates of an atom).

During the transformation between source and target databases complex restructuring of the data occurs including splitting and merging of objects, extracting keywords from text segments or restructuring of sequences. The structural relationships between source databases and the target database are sometimes relatively clear. For example, a given tuple representing the coordinates of an atom are straightforward to transform to a corresponding atom object in the target database, although the structural representation might be different. Other aspects of the data transformation cannot be described as data restructuring operations in an obvious way. So it may be the case that a PDB file does not contain data on proteins at all, but rather on nucleic acids. Such distinctions can only be performed by using heuristic algorithms that reflect substantial domain knowledge. These distinctions are however important in order to correctly identify those objects to which subsequent transformations can be applied correctly.

The example shows that the problem of integrating data is not just a problem of data model transformation and data restructuring, but requires steps that involve a deep analysis of data. From this it follows that different tasks in the integration process allow for different degrees of automation. Thus the following requirements can be identified to support a process that has been described in the previous example. It is necessary

1. to access heterogeneous file-based databases e.g. by parsing techniques.
2. to perform complex data restructuring in order to transform the data from the different heterogeneous data formats into the integrated warehouse.
3. to support correction of errors, data adjustments and data completion by using complex data analysis techniques and tools.
4. to be able to track any data back to its original resources for verification or visualization.
5. to allow the user to flexibly adapt the scope, structure and contents of the data warehouse to his needs without requiring support from the data management expert.
6. to relieve the user as far as possible from details on a technical level of data integration and transformation.
7. to minimize manual intervention as far as possible during the data integration process.
8. to consider updates in external formats and schema evolution.

Some of the requirements are contradictory such that a feasible compromise needs to be found. For example making the user able to adapt the integration process (5) is in conflict with requiring minimal technical understanding from his side (6). Fully automatizing the process (7) is opposed to the requirement that manual intervention might be required (3). So a feasible compromise needs to be found for the integration methodology.

3.2. Specification and Implementation of the Data Transformation and Integration Process

The basic approach is to identify those steps that can be fully automatized, we call these atomic transformations, and provide programs that can perform the atomic transformations. For example transforming one model of one PDB file into the database representation is one atomic transformation. We provide the domain specialist with an interface where he can compose a complex transformation specification from such atomic transformations to build up the integrated database. The specification of the complex transformation can be considered as metadata, i.e. data that describes the contents of the integrated database. This metadata is then used to automatically transform the source databases to the integrated database. The metadatabase defines a "parameter space", in which the user can control the integration process and define the contents of the data warehouse (see requirement (5)) while he is relieved from those parts of data integration and transformation that do never change. This approach allows to separate the more domain specific tasks of data analysis and preparation from the more data processing oriented tasks of physically building the data warehouse. Another benefit of using metadata is that it allows the user to define different databases for specialized purposes. Figure 2.1 illustrates the process.

It has turned out that different types of atomic transformation need to be supported. We discuss those different types at the example of an excerpt of a transformation specification given in Figure 2.2.

1. Creation of an Object in the Target Database. This is for example specified by

```
Protein pd11blc
```

which creates in the database a new protein object with name `pd11blc`. Sometimes this step is combined with the loading of associated data. For example

```
Chain: BLAC_STAAU BLAC_STAAU.sw
```

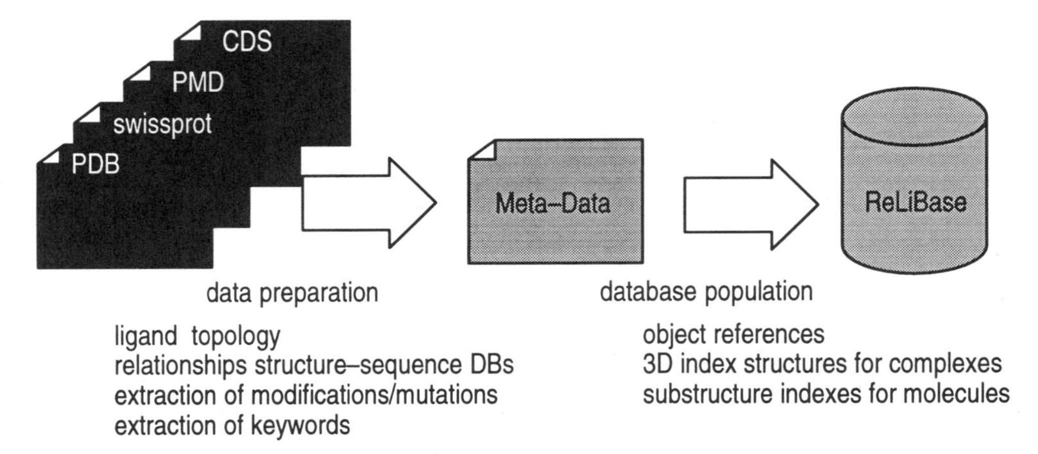

Figure 2.1. Integration approach for populating the integrated database.

creates a new chain object and sets the sequence information to that found in the

SWISSPROT file `BLAC_STAAU.sw`.

2. Update of Attributes of an Object in the Target Database. An example of such a specifications is

```
ProteinOrganism: pd11blc STAPHYLOCOCCUS AUREUS
```

setting the value of the organism attribute of the protein object with name `pd11blc` to `STAPHYLOCOCCUS AUREUS`. This type of specification can also trigger more complex methods in the database. For example

```
SSCorrection: pd11blc pdb1 - 31 57 -6
```

is a specification that adapts the sequence numbering found in a DSSP file to that of the corresponding sequence in the PDB file.

```
Protein                  pd11blc
ProteinModel:            pd11blc   pdb1  pd11blc.ent 1

ProteinOrganism:         pd11blc STAPHYLOCOCCUS AUREUS
ProteinKW:               pd11blc HYDROLASE; ANTIBIOTIC RESISTANCE; 3D-STRUC-
TURE;
                         SIGNAL; PLASMID; TRANSPOSABLE ELEMENT;

Chain:                   BLAC_STAAU BLAC_STAAU.sw
ProteinChain:            pd11blc - BLAC_STAAU
ChainSWISS:              BLAC_STAAU P00807;
ChainEMBL:               BLAC_STAAU X04121; SAPBLAZ;  X16471; SATNBLAZ;
M15526; PP5BLAZA;
ChainPIR:                BLAC_STAAU A01002; PNSAP;  A23600; A23600;
                                         S06757; S06757;  S11784;
ChainPROSITE:            BLAC_STAAU
ChainPMD:                BLAC_STAAU A911259
ChainMUTATION:           BLAC_STAAU 188     188   179   179    D->N
ModelCorrection:         pd11blc pdb1 1 215 -6
SSCorrection:            pd11blc pdb1 -    31    57  -6
HSSPCorrection:          pd11blc pdb1 -    31    57  -6
SecondaryStructure:      pd11blc pdb1 pd11blc.dssp

Ligand:                  CEM  HETERO  LABELLED CEM AND THE TRANS ENAMINE
                         IS LABELLED TEM(*INC* AND *INT*, RESPECTIVELY, IN
                         THE PAPER CITED ON*JRNL* RECORDS ABOVE)
LigandModel:             CEM  CEM-70-pd11blc  CEM-70-pd11blc.mol2
Complex:                 CEM_in_pd11blc
ComplexProtein:          CEM_in_pd11blc  pd11blc
ComplexLigand:           CEM_in_pd11blc  CEM

ComplexModel:            CEM_in_pd11blc  cm1
ComplexProteinModel:     CEM_in_pd11blc  cm1  pd11blc  pdb1
ComplexLigandModel:      CEM_in_pd11blc  cm1  CEM  CEM-70-pd11blc
```

Figure 2.2. An example transformation specification.

3. Creation of an Existence-Dependent Object in the Target Database. An example for this kind of specification is

```
LigandModel: CEM CEM-70-pd11blc CEM-70-pd11blc.mol2
```

which loads a threedimensional ligand structure from a file in MOL2 format. Each ligand can possess several models corresponding to different conformations. However, the model objects can only exist if the corresponding ligand object is contained in the database.

4. Establishing Relationships between Objects. This allows to establish relationships between existing objects in the database. For example after creating a new complex object it is necessary to determine which are the protein and ligand objects in the database that are part of the complex. This is specified in the example for a complex with name `CEM_in_pd11blc` by

```
ComplexProtein: CEM_in_pd11blc pd11blc
ComplexLigand:  CEM_in_pd11blc CEM
```

The transformation process is implemented by means of database methods following an object-oriented design. The data transformations corresponding to atomic transformation, that apply to the metadatabase entries, are realized as a methods of a class MetaDB. For each file type that can be integrated into the integrated database a corresponding class exists in the database schema of the database management system. For example there exists a class PDBfile that provides different parsing methods for PDB files. This establishes in the object-oriented database schema an integration layer in addition to the application layer, that contains the application schema. In Figure 2.3 the software architecture is depicted. We distinguish three layers: (1) the external file database layer, (2) the integration layer and (3) the application layer.

In addition to providing access methods for the different file types, the integration layer keeps a history of which information of external databases has been transformed in the database. This allows to track back how the information in the target database has been generated from the source databases (see requirement (4)). For this purpose corresponding links are supported from classes of the integration layer to the classes of the application layer in the target database and vice versa. This allows to use the original information in the source database files, when it is required. For example many visualization tools for proteins are able to interpret PDB files directly. In this case we can integrate such a visualization tool easily with our target database schema. Keeping these backward references in the application layer directly is problematic, since there is no one-to-one correspondence between objects in the source and target databases.

As a side effect of the integration process the application builds different index structures to support efficient searches later on.

- Fingerprints of ligands allow to quickly evaluate a necessary condition in the substructure test.
- The binding site atoms and residues of proteins are precalculated.
- Index structures on several kinds of attributes are provided, in particular the different names of objects. The latter is also useful to check the uniqueness of those names.

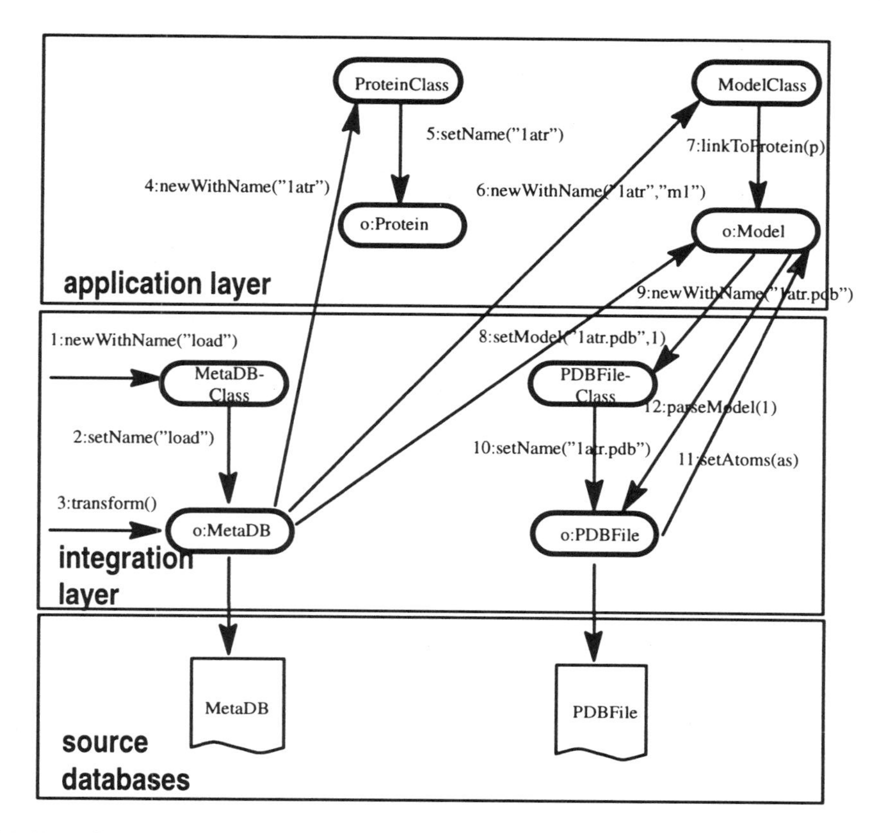

Figure 2.3. The software architecture for data integration. Boxes represent class and instance objects. The figure also indicates some methods calls executed for a sample integration step. Numbers indicate the calling order of the methods; the method names are self-explaining.

- A neighboring index for the spatial neighborhood relationships of the protein and ligand atoms can be precomputed.
- A neighborhood relationship for similar sequences based on the alignment algorithms provided by the database can be precomputed.

As the computation of some of these index structures, in particular the neighborhood relationships, can become extremely expensive their use has to be restricted for those cases where intensive support for querying will be required later.

4. DATA MANAGEMENT

4.1. Object-Oriented Database Management

The integrated database is implemented using the object_oriented database management system VODAK [30] developed at GMD_IPSI. The object_oriented data model is accepted to be appropriate for flexible modelling of the complex structures and operational semantics, that are needed for non_standard database application scenarios like molecular biology. VODAK provides full database management system functionality, including schema compiler, transaction management, object management, data dictionary, communication management and query processing. It runs on top of ObjectStore [20], which is used for low_level storage. Different features make VODAK particularly attractive for our purposes. The object_oriented data model VML [19] provides a metaclass concept for tailoring the datamodel to specific application domains. Thus semantic concepts that are characteristic for an application domain can be provided as new data modelling primitives, which extend the core data model. The declarative SQL_like query language allows the usage of method calls and path expressions within queries, and thus enables flexible access to the database. A powerful, extensible query optimization module, based on algebraic optimization techniques, allows to incorporate application_specific optimization strategies, which are particularly important when complex calculations or expensive access operations to external databases are involved in declarative queries.

4.2. Data Modelling

For data modelling we distinguish three different abstraction levels, namely the conceptual, the logical and the physical level, with a corresponding data model for each of the three levels. The conceptual level is used to design the application together with the domain experts, in our case drug designers, molecular biologists and biochemists. The data model uses a few very simple concepts such that it can be easily communicated with the domain experts without requiring experience in the design of database applications. It reduces the application model to the core concepts and leaves the detailed specifications of types and their corresponding constraints to the logical level. At the conceptual level the object classes and the basic relationships between these classes are determined. For the description of the conceptual model a graphical representation is used. At the logical level the object types and constraints are completely specified, including the object structure, relations to other objects and the implementations of the operations. For this purpose we use the object-oriented data modelling language VML, that is part of the VODAK database management system. Finally the logical model is mapped to the physical level, which uses the object model of C++ as data model. The implementation of the logical model within the physical data model determines the efficiency of the system. Although VODAK supports a fully automatic mapping from the logical to the physical level, there are possibilities to extend and adapt this mapping in order to optimize the system performance.

4.3. The Conceptual Level

The conceptual model of our database application is depicted in Figure 3.1. Object classes are represented as rectangular boxes e.g. Ligand, Protein, or Complex are different classes. Relationships between these classes are represented by arrows, where we distinguish 1:1, 1:n and n:m relationships. In addition to application-specific relationships between the classes, several general relationship types have been identified.

- Part-of relationship: The application model has several aggregation levels, where a complex object is composed of components. For example a residue is part of a chain, a chain is part of a protein, and a protein is part of a complex. This relationship can be refined by distinguishing whether an ordering relationship exists for the parts. For example a chain consists of sequentially ordered residues, while an ordering of the complex' components does not exist.

- Variant-relationship: For molecules we typically encounter the situation that for different reasons alternative threedimensional models exists. In the case of proteins these Either these correspond to different hypotheses or to the results of different modelling tools. In the case of ligands these correspond to different conformations of the molecule.
- Role-of relationship: An object can play different roles in different contexts. For example, a particular role of a chain is when it occurs within a protein. In this way the chain can occur multiply in the same protein by creating different role objects for each occurence without duplicating the chain data.

```
CLASS Chain
     INSTTYPE ChainInstType
     INIT                              Chain->initChain()
END;

CLASS ProteinChain METACLASS ROLE_SPECIALIZATION_CLASS
     INSTTYPE ProteinChainInstType
     INIT                              ProteinChain->defRoleClass(Chain)
END;

OBJECTTYPE ChainInstType;
     INTERFACE
         PROPERTIES
               name                 : STRING;
               residuechain         : ResidueChain;
               alignments           : {OneChainAlignment};
               mutations            : {ChainMutation};
               modifications        : {ChainMod};
               swissprotCode        : STRING;
               pirCode:             STRING
               numberOfResidues: INT;
         METHODS
               align(chain:OID)     :REAL;
END;

OBJECTTYPE ProteinChainInstType;
     INTERFACE
         PROPERTIES
               name                 : STRING;
               protein              : Protein;
               oneChainSSs          : {OneChainSS};
         METHODS
               linkToProteinByName(proteinName : STRING);
END;
```

Figure 3.1. Sample class and type definitions of an object-oriented database schema.

4.4. The Logical Level

We are using the object_oriented data model VML of VODAK for representation of data at the logical level. VML offers all standard object-oriented features like properties, methods, inheritance, object identity and complex data types. In addition VML provides some advanced modelling features that make the data model more flexible, like dynamic method dispatching, classes as first class objects, and mechanisms to introduce new semantic relationships. Thus, in addition to the standard semantic relationships, like specialization or aggregation, new semantic relationships between classes, like those discussed in the previous paragraph, can be defined. We illustrate the use of the data model by the example of a class definition in the application schema in Figure 3.2 with a concrete instance of a chain object given Figure 3.3.

In the example properties are used for three different purposes. The first is to represent those object properties that make up the internal state of the object. Examples are the property `name` with value `'LYCV_BPT4'` or the property `numberResidues` with value `'164'`. In contrast to relational modelling those properties can also contain complex values, like sets, lists or records. For example, ... Another use is to provide references into external databases, for example the properties `pirCode = 'A00875'` or `swissprotCode = 'P00720'`. Finally properties are used to relate the object with

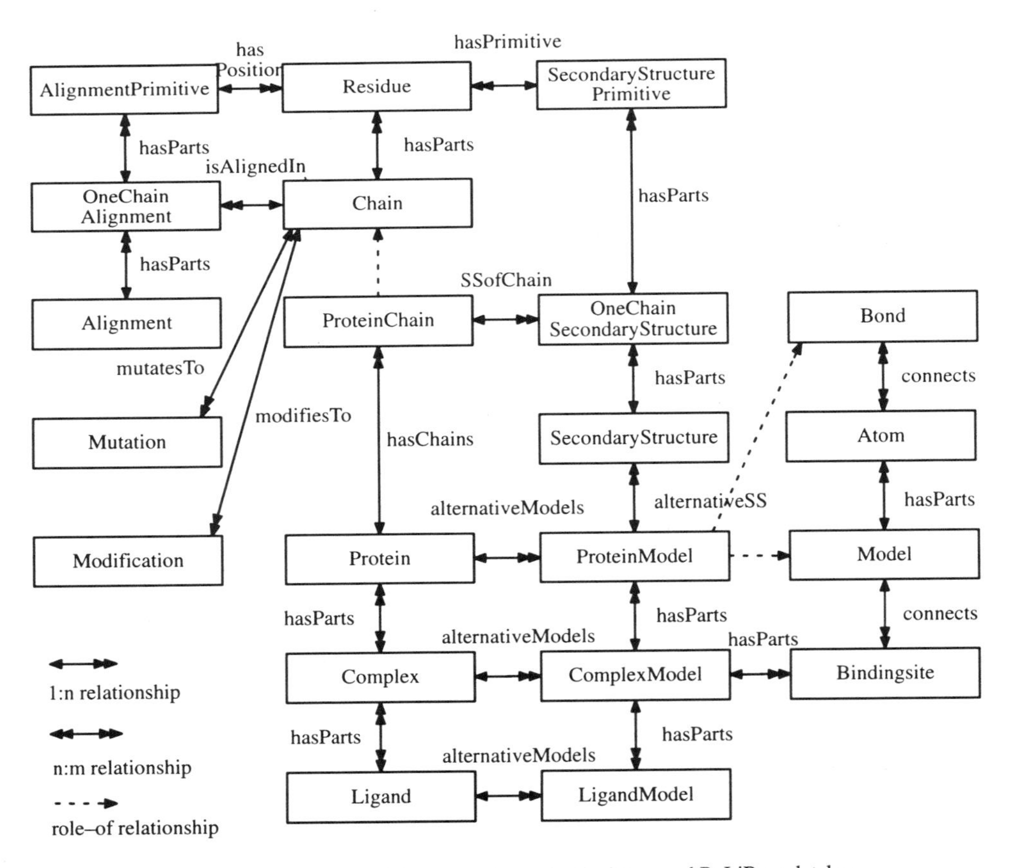

Figure 3.2. The conceptual database schema for the integrated ReLiBase database.

other objects in the database. In this case the type of the property is of the referenced class or a complex type composed of the referenced class. Examples of such properties are `residuechain` for a single valued reference or `{ChainMutation}` for a set-valued reference. These relationships can be used for navigation in the database and replace the joins that are typically used in relational database systems.

Methods are the object interface, that allows to read and modify the state of an object. The concept of methods allows to manage not only the state of an object but also the operations on the object. In this way the generic functionality of an application becomes part of the database. One use of methods is to encapsulate objects and to support consistent access. For example, the method `newWithName` not only creates a new chain object, but also checks for uniqueness of the chain name. Another use of methods is to provide complex algorithmic functions for the object, what can be used to avoid storing the materialized function redundantly. For example the implementation of the Needleman/Wunsch algorithm that is given by the method `align(chain)`.

At the conceptual level we have mentioned that certain class relationships, like part-of or role-of, are of a general nature. Due to the extensibility of the VML data model these relationships can be provided as data model extensions. An example for this is role specialization. At initialization of the class `ProteinChain,` which is a role specialization of the class `Chain` this is specified by the statement `ProteinChain->defRoleClass(Chain)`. The database system provides now the functionality required to consistently manage role objects, i.e. the specialization object inherits all methods and properties of the generalization object and a method `roleOf()` for determining the generalization object is provided. The implementation of this functionality is hidden from the application programmer.

4.5. Optimization at the Physical Level

The VML data model provides the application programmer with a high-level data model that abstracts from physical details of data management. These mechanisms at the physical level, like locking and buffering strategies, are either handled by the VML compiler or by the implementation of the database management system. Objects represent the unit of persistency, i.e. locking and object buffering are performed at the object granularity, and methods are the basic operations on persistent objects covered by transaction mechanisms. By using this data management default strategy the application programmer

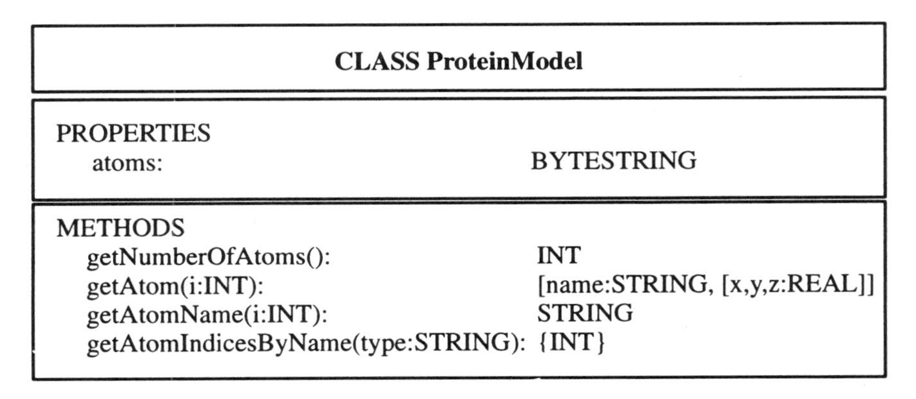

Figure 3.3. An object of class ProteinModel.

can obtain efficient implementations of object-oriented database applications without spending effort on physical implementation issues. The basic data type constructors, like set, list, array or dictionary allow to build fairly complex data structures with little effort that are efficiently managed. As an illustration we discuss the modelling of chain alignments. An alignment is a (monotone) mapping of the natural numbers 1, ..., n_{max} to the residue numbers of a residue sequence. A naive representation would be {[n: 1, res: r_1], [n: 2, res: r_2], ... , [n: n_{max}, res: r_{nmax}]}. We observe that this is also the only way to represent an alignment as a relation. Using the type constructor list we can instead model this more efficiently as a list <<r_1, r_2, ..., r_{nmax}>>.

However, like any default strategy, also this approach has its limitations. The mapping from the logical to the physical level has in some cases to be controlled by the application programmer in more detail. Although the concept of objects together with a few basic types and type constructors allows to represent almost any imaginable data structure in certain cases the representation might be too complex. Also the realization of data type operations by methods might be too expensive as in such cases different mechanisms like dynamic dispatching or transaction support are not required for accessing parts of the data structure, possibly in complex calculations. Therefore we exploit the principle of logical data independence. We provide the possibility to implement special purpose data types, which can be accessed by means of encapsulation at the logical level in the same way like an equivalent constructed type, but that are managed differently at the physical level.

A obvious example where such specialized data types are required is the representation of bitmap images. Although in principle a representation by using built-in array constructors is possible it is inefficient and does not provide adequate functionality (e.g. for image manipulation). A similar situation is encountered in biological applications for the threedimensional atom models of proteins. In those cases we need to hide the details of the data structure from the database management system. The approach is to use at a physical level binary large objects to represent the data within the database system and to implement special data types on top of these binary large objects. In this way efficient and consistent management of the data in multi-user mode is still ensured at the granularity of the binary large object. Such a mapping of the logical level to the physical level is always reasonable when a natural clustering of data is given (as it is the case with images or protein models).

In the example given in Figure 3.4 an object of the class ProteinModel represents the atom coordinates of a threedimensional model of a protein. The data of one atom consists of the atom name and the coordinate triple [x,y,z] (in practice more properties of an atom are stored in this data structure). A typical protein model consists of a few thousand atoms. All the atom data is encoded in one BYTESTRING. Access methods with specialized implementations allow to efficiently access the model data. For example, the method `getAtom` extracts the name and the coordinates `x`, `y` and `z` of atom `i` and converts it into a VML data structure. In this way the data stored in a ProteinModel object is fully available to VML applications and can be used for example in queries. Besides basic access methods it might be advantageous to additionally provide complex access methods for frequent access patterns and to move in this way algorithmic processing from the database system into the implementation of the specialized data type. For example, we encapsulate the search for all atoms with a given name into a special search method `getAtomIndicesByName`, that extracts all atoms with a given name from the encoded bytestring representation. This can be implemented considerably more efficiently than by repeated method calls on the ProteinModel object using the method `getAtom`.

CLASS ProteinModel

PROPERTIES
atoms: BYTESTRING

METHODS
getNumberOfAtoms(): INT
getAtom(i:INT): [name:STRING, [x,y,z:REAL]]
getAtomName(i:INT): STRING
getAtomIndicesByName(type:STRING): {INT}

Figure 3.4. An object of class ProteinModel.

5. RETRIEVAL

5.1. Retrieval Approaches

Retrieval capabilities of existing database systems, as they are used in molecular biology, belong in general to one of the following categories.

Browsing and Navigation. The user can inspect the database content by browsing through the content of classes, applying simple keyword-driven search and navigating along predefined links. Such functionality is typically offered through hypertext-like interfaces, like WWW. Systems like Entrez or AceDB [9] combine these retrieval facilities typically with comfortable user interfaces.

Search Algorithms. . Many systems offer highly efficient algorithms to search in extremely large databases or over many different databases for answering fixed or parameterized queries. The search is often supported by application-specific index structures. Examples for this approach are some of the search algorithms incorporated in WHATIF [32] or the BLAST [2] tool. The search interfaces of these tools are either provided within the interfaces of whole systems or as standalone scripts.

Declarative Querying. The notion of declarative query language stems from the area of database management systems, in particular relational database systems [21], where SQL is the most prominent example. Query languages are designed to access databases in an application-independent way. They allow a declarative specification of the information need, and are backed by powerful query optimization and evaluation components. With declarative query languages the user specifies his information need ("what"), while the system decides on the optimal operational strategy to obtain this information ("how"). Query languages are typically provided through an interpreter or are embedded in programming languages.

Data Mining. Only retrieving the data, that has been previously stored in the database is not sufficient to cover all information needs of the end user. New information needs to be derived from the stored data [11]. This new information can be statistical information (e.g.percentage of complexes that have two histidins in the binding site). Another task is object clustering and determination of non explicitly represented object relationships (e.g. the group of complexes with ligands containing the benzamidine substructure mostly share some bindingsite property). This can be performed by different algorithmic methods or by using visualization tools, like Lyberworld [13], which is a tool that allows visual clustering and is discussed in more detail in Section 6.

Each of the four approaches has its own advantages and disadvantages. Browsing and navigation allows the user to interactively explore a database, but cannot support the evaluation of complex requests. Search algorithms answer special queries with extreme efficiency, but are obviously also extremely inflexible. Declarative query languages offer a very flexible and expressive retrieval capability, but their availability today is mainly restricted to relational database systems, which are not adequate to model the complex biomolecular data in many cases. The most striking feature of data mining systems is their ability to extract new information from a database, but they are certainly not suited to support simple and fast access.

5.2. Retrieval with Object-Oriented Database Systems

The primary access to object-oriented databases is through their programming language interface. Within the programming language it is possible to access properties of objects, to follow references to other objects, perform browsing and navigation, to perform simple selections, and to invoke methods on objects. These are for example the capabilities that are typically offered by C++-based object-oriented database systems [20]. Thus, leaving aside the user interface layer, object-oriented database systems support both browsing and navigation and the usage of search algorithms, which can be encapsulated in methods.

A major drawback of many object-oriented database systems is that they do not support declarative querying capabilities [18]. Although substantial research on declarative object-oriented query languages has been performed, only few systems really offer advanced query language and query processing features. Efficiently supporting declarative querying capabilities in object-oriented database systems is an important step towards integrating all of the three database access mechanisms, namely browsing, algorithms, and declarative querying. To a certain degree also data mining techniques can be supported as many data mining systems access databases by an SQL-based interface.

We illustrate the integration of the different retrieval capabilities by the following example query that gives the formal representation of the example question that has been discussed in Section 2.1.

```
ACCESS    c.name, pc.residuechain->align(pc2.residuechain)
FROM      pc IN db1::ProteinChain,,
          c IN db1::Complex,,
          pc2 IN c.protein.proteinchains,,
          l IN c.ligands,,
          lm IN l.models
WHERE     pc.name == 'H' AND pc.protein.name == 'pdb1dwd' AND
          lm.bonds->hasSubStructureS('NC(=N)c1ccccc1')
```

This is a declarative query in the VODAK query language VQL against the RELIWE receptor-ligand database (ReLiBase). Navigation is necessary (e.g. c.protein.proteinchains), because sequence and structure information is encapsulated in different classes (i.e. ProteinChain, Complex) related to the protein class. An externally implemented alignment algorithm (Needleman/Wunsch [22]) has been incorporated as a database method align and a substructure search algorithm by a method hasSubStructureS which takes a Smiles code as parameter. The query is declarative as it leaves for example open, in which order the classes are traversed, the objects within classes are visited, or the conditions are evaluated.

5.3. Query Optimization

As we have stated declarative query languages to a large degree satisfy the requirements, that are needed to support retrieval. As data independence is used as a concept for separating logical and physical aspects of data representation, we have also to consider a separation of logical and physical aspects for the operational execution of database queries. The efficiency mainly depends on how the declarative specification is mapped to a concrete execution plan. Stressing the analogy to data modelling we again must be able to control and tune this mapping. Since queries are more dynamic than a database schema, we use a (runtime) query optimizer, that can be extended with application-specific information on the physical model to perform the optimal mapping of declarative queries to an operational execution of the query. In this mapping we need to consider the following aspects:

Methods. One aspect of query optimization is to avoid use of expensive methods. For some queries the method execution costs are the dominating factor, and optimization strategies have to avoid too frequent method calls. Thus knowledge on the execution costs of methods is made explicit in our system.

Data Types. For relational systems it is practicable to use the same efficient storage structures in all applications, as the data model supports in principle only one datatype, namely relations. This holds no longer in scientific applications where many different scientific data types with appropriate storage structures must be supported. Object-oriented database systems allow to resolve this by providing efficiently implemented data types. In the context of declarative access however the logical access of a user and the optimal physical access to the scientific data may substantially differ.

Index Structures. Application specific index structures, e.g. for efficient access to spatial data, and special purpose physical data representations play a central role in the data warehouse. The correct usage of index structures and specialized access mechanisms can be a difficult task for the user. Knowledge on the semantics of these mechanisms is thus made available to the query optimizer, such that they can be hidden from the user in ad_hoc querying.

External Systems. Sometimes access mechanisms for external databases are provided. For example, in some cases we do not migrate data physically into the warehouse in the way previously described. Those databases often provide alternative access paths, that allow for more efficient and adequate processing of specialized requests. To exploit these alternative access paths, we make the query optimizer aware of the semantics of these.

Techniques for application-specific optimizations in scientific applications to support those aspects are gaining growing attention in research [17][34]. Within VODAK we have implemented an extensible query processing component [1]. We are using an algebraic and rule-based approach to query optimization. The query optimizer is generated by using the Volcano optimizer generator [12] and we can generate application-specific query optimization modules. We are currently experimenting with application-specific optimization rules that exploit the particular semantics of database methods of the integrated RELIWE protein-ligand database ReLiBase. Our goal is to arrive at a general methodology for the definition of application-specific optimization rules exploiting method semantics by the schema designer and their integration into query optimization.

As an example consider the following query for retrieving close atom pairs, that occurs often as a subquery of more complex requests.

```
ACCESS   atom1, atom2
FROM     atoms IN db1::Atoms,,
         atom1 IN atoms->getAtomIndices()
         atom2 IN atoms->getAtomIndices()
WHERE    atoms->distance(atom1,atom2) <= 3.5 AND
         atoms.model.protein.name == '1atr'
```

The ACCESS-clause specifies as the result of the query the indexes of the selected atom pairs. Within the FROM-clause query variables are bound to their domains, either class extensions (db1::Atoms) or results of set-valued method calls (atoms->getAtomIndices()). The condition in the WHERE-clause expresses neighborhood by using the method call atoms->distance(atom1,atom2) <= 3.5.

Without optimization the cross product of all atoms of a protein is created. As a typical protein may contain up to 5000 atoms, this leads to a large intermediate result (up to 5000^2 = 25.000.000), for which the condition has to be evaluated. In the database an index structure for the atom neighborhood relationship can be precomputed, i.e., each atom has direct links to its immediate neighbors up to a certain threshold. A method belowThreshold makes optimal use of this index structure. A semantic optimization rule, that can be included by our techniques into the rule-based algebraic query optimizer, transforms the algebraic representation of this query *automatically* to one that corresponds to the following query

```
ACCESS   atoms->belowThreshold ( atoms->getAtomIndices(),,
                                 atoms->getAtomIndices(), 3.5)
FROM     atoms IN db1::Atoms,
WHERE    atoms.model.protein.name == '1atr'
```

6. VISUALIZATION

Visualization tools play a crucial role in the use of the integrated database. Biological data visualization tools are an important means to understand the structure and function of molecules in detail. Different visualizations of the same data can be useful in many situations. Thus we have integrated several visualization tools with our database. The advantage of tightly integrating visualization tools with the database system is that visualizations can be dynamically generated by the database management systems exploiting

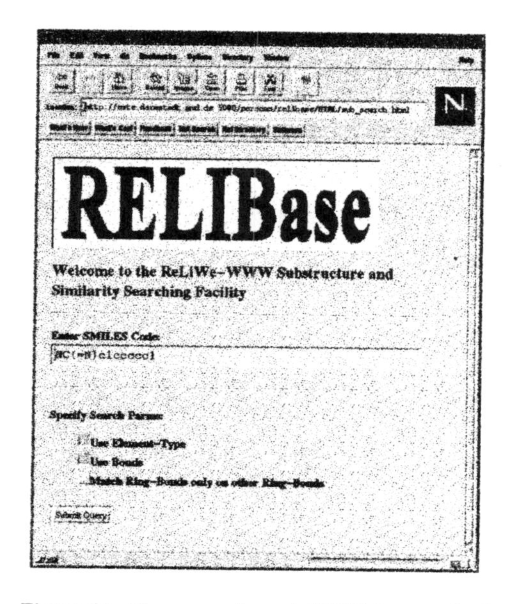

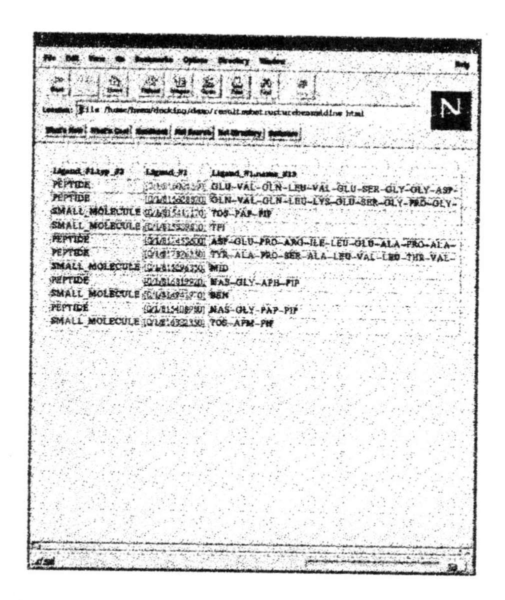

Figure 6.1. The preconfigured WWW-Inputmask for substructure searching and the dynamically generated WWW-result page.

database retrieval operations. With regard to visualization we are currently working in different directions to support different visualization paradigms.

To support data inspection, browsing and navigation we provide a generic HTML_based object viewer, that allows to navigate along object references, to display the objects attributes and to execute methods on objects. In this setting the object_oriented data model is a paradigm that is predestined to be accessed in a visual way. Queries can be posed via HTML, either as predefined (see Figure 6.1), parametrized queries or, by the more sophisticated user, by directly writing VQL queries. Query results are formatted in a way that the results become starting points for navigational access in HTML. It is planned to extend the system by tools for graphical composition of queries.

For viewing biomolecular data generic 3D viewers for VRML will be integrated, as well as specific biomolecular viewers like RASMOL or Whatif [31]. Specialized viewers operate only on a particular kind of data, for example protein structures. As we are integrating data from different sources viewers for the integrated data, e.g. for presenting complexes, may not be available. In such cases generic viewers like VRML_based ones are used. Another interesting aspect is to use visualized objects as starting points for consecutive navigation. For example, the attributes of an atom are displayed when clicking at the atom within a threedimensional visualization.

One major drawback of many biomolecular viewers is their limited flexibility. For example, RASMOL is specialized for displaying data provided in PDB format and visualizing protein information in certain predefined ways. On the other hand database queries can be used to dynamically generate spatial data that allows a more differentiated visualization of functional characteristics of molecules. We are using VRML as a generic specification language for spatial visualizations. The important aspect is, that we can generate VRML code dynamically in the database systems based on the results of other database operations and have generic visualization and interaction capabilities. For that purpose we

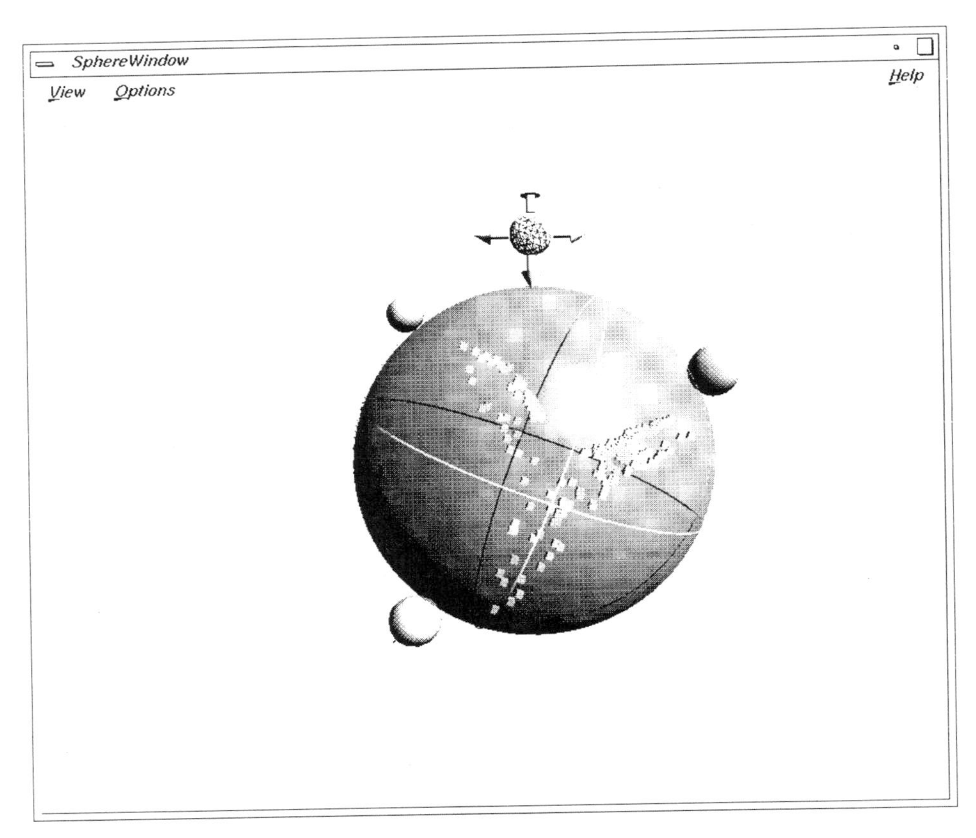

Figure 6.2. Data clustering with the Lyberworld coordinate sphere.

can provide database methods that generate visualization code. These methods can be directly invoked as part of a query, and perform the transformation of the pure query results into corresponding VRML-code.

Beyond pure visualization tools we have also integrated tools that support data mining in the integrated database. Lyberworld [13] is a system, that allows to perform a visual clustering of data, that has previously been extracted from the database. Arbitrary user defined queries can be used to return a ranked list of objects. Different queries over the same set of objects can deliver different rankings. For every object the ranking values form a coordinate tuple, that is graphically displayed in a coordinate system (see Figure 6.2). The coordinate axes can be modified in angle and stretching factor. In the spatial arrangement of the objects the user can identify clusters of objects or hidden relationships between different classes of objects.

7. CONCLUSION

The system described in this paper will be used for analysing the integrated databases under various aspects relevant in rational drug design. A typical question that is of interest concerns, for example, the analysis of the chemical diversity of ligands derived from the PDB. While in large small molecules databases due to the large number of entries a high chemical diversity is encountered, this is far from obvious for the small molecules identified in PDB. Other usages of the system are for statistical analysis of receptor-ligand interactions or the validation of receptor-ligand complexes.

The techniques for data integration, representation, retrieval and visualization described in this paper, that have been developed as result of dedicated application demands, are of a more general nature. They can be exploited for different other problem areas in scientific data management of which one of the most obvious examples are the data management issues in genome research.

8. REFERENCES

1. Aberer, K., Fischer, G.: Semantic Query Optimization for Methods in Object_Oriented Database Systems. International Conference on Data Engineering, March 1995, Taipei, Taiwan.
2. Altschul, S.F., Gish, W.Miller, W., Myers, E.W., Lipman, D.J.: Basic local alignment search tool. J. Mol. Biol., 215, p 403–410, 1990
3. Bairoch, A., Boeckmann, B.: The SWISS_PROT Protein Sequence Data Bank. Nucleic Acids Res., 19 (Sequences Suppl.), p 2247 - 2249, 1991.
4. Barker W.C., George D.G., Hunt L.T., Garavelli J.S.: The PIR protein sequence database. Nucleic Acids Res., 19 (Sequences Suppl.), p 2231 - 2236, 1991.
5. Benson: "GenBank", Nucleic Acids Res. 24, p 1–5, 1996.
6. Bernstein, F.C., Koetzle, T.F., Williams, G.J.B., Meyer, E.F. Jr., Brice, M.D., Rodgers, J.R., Kennard, O., Shimanouchi, T., Tasuni, T.: The Protein Data Bank: a computer based archival file for macromolecular structures. J. Mol. Biol. 112, p 535 - 542, 1977.
7. Cattell, R. G. G. (Ed.): Object Databases: The ODMG_93 Standard. Release 1.1, Morgan Kaufmann, San Francisco, 1994.
8. Daylight User Manual, Daylight Chemical Information System Inc., Irvine, CA 92715.
9. Durbin, R., Thierry-Mieg, J.: The ACEDB Genome Database, WWW page, http://probe.nalusda.gov:8000/acedocs/dkfz.html.
10. Etzold, T., Argos, P.: SRS an indexing and retrieval tool for flat file data libraries, Comput. Appl. Biosci. 9, p 49–57, 1993.
11. Fayyad, Usama M., Eds., Advances in knowledge discovery and data mining, Menlo Park, Calif., AAAI PressCambridge, Mass., London, MIT Press, 1996.
12. Graefe, G. : Volcano - An Extensible and Parallel Query Evaluation System, IEEE Transactions on Knowledge and Data Engineering. Vol. 6, No. 1, p 120–135, Feb 1994.
13. Hemmje, M.: LyberWorld - A 3D Graphical User Interface for Fulltext Retrieval. In: Proceedings of CHI '95, Video Summaries, May 1995.
14. Hendlich, M., Rippmann, F., Barnickel, G., to be published.
15. Hendlich, M., Rippmann, F., Barnickel, G., Automatic Assignment of Atom and Bond Types for Protein Ligands in the Brookhaven Protein Database, submitted.
16. Kabsch W., Sander C., Biopolymers 22, 2577–2637, 1983.
17. Kemp G.J.L., Jiao Z., Gray P.M.D., Fothergill J.E.: Combining Computation with Database Access in Biomolecular Computing, ADB 94, Vadstena, Sweden, p 317–335, 1994.
18. Kim, W.: Object-Oriented Database Systems: Promises, Reality and Future. In W. Kim (Ed.), Modern Database Systems, ACM Press, 1995.
19. Klas, W., Aberer, K., Neuhold, E.J. 1994. Object_Oriented Modeling for Hypermedia Systems using the VODAK Modelling Language (VML). In: Advances in Object_Oriented Database Systems, A. Dogac, T. Ozsu, A. Biliris, T. Sellis eds., NATO ASI Series F 130, p 389–434, Springer, Berlin, Heidelberg.
20. Lamb, C., Landis, G., et al.: The ObjectStore Database System. Comm. of the ACM, Vol. 34, No. 11, p 32–39, 1991.
21. Maier, D.: The Theory of Relational Databases, Computer Science Press, 1983.
22. Needleman, S.B., Wunsch, C. A.: General Method Applicable to the Search for Similarities in the Amino Acid Sequences of Two Proteins, Proc. of the National Academy of Science, Vol. 48, p 444–453, 1970.
23. Nishikawa,K., Ishino,S., Takenaka,H., Norioka,N., Hirai,T., Yao,T. and Seto,Y.: , Constructing a protein mutant database, Protein Engng, 7, 733, 1994.

24. Nukhres O., Elmagarmis A.K., Eds., Object_Oriented Multidatabase Systems, Prentice Hall, 1995.
25. Pearson W.R., Lipman, D.J., PNAS (1988) 85, p 2444–2448, 1988.
26. Rarey, M., Kramer, B., Lengauer, T., Klebe, G.: "Predicting Receptor-Ligand Interactions by an Incremental Construction Algorithm", accepted for Journal of Molecular Biology, 1996.
27. Ritter, O., Kocab, P., Senger, M., Wolf, D., Suhai, S.: "Prototype Implementation of the Integrated Genomic Database", Computers and Biomedical Research, 27, p 97–115, 1994.
28. Schneider: The HSSP database of protein structure-sequence alignments, Nucleic Acids Res. 24, p 201–205, 1996.
29. Sheth, A.P., Larson, J.A., Federated Database Systems for Managing Distributed, Hetereogenous, and Autonomous Databases, ACM Computing Surveys, vol. 22, no. 3, p 183 - 236, 1990.
30. VODAK V 4.0, User Manual, GMD technical report No. 910, 1995.
31. Vriend, G.: WHAT IF: A molecular modeling and drug design program., J. Mol. Graph. 8, 52–56, 1990.
32. Vriend, G., Sander, C., Stouten, P.F.W.: A novel search method for protein sequence-structure relations using property profiles, Protein Engineering, vol. 7, no.1, pp. 23–29, 1994.
33. Wollny, B., Process 3, p 64–66, 1995.
34. Wolniewcziz R., Graefe G.: Algebraic Optimization of Computations over Scientific Databases, Proc. of the 19th VLDB Conf., Dublin, Ireland, p 13–24, 1993.

17

PICTURING THE WORKING PROTEIN

Hans Frauenfelder[1] and Peter G. Wolynes

[1]Center for Nonlinear Studies
Los Alamos National Laboratory
Los Alamos, New Mexico 87545
[2]School of Chemical Sciences
University of Illinois
Urbana, Illinois 61801

1. INTRODUCTION

In discussions and in reading textbooks or papers, one often encounters two, implied or explicit, statements about proteins:

I. Folding leads from the unfolded state to the unique native structure.
II. This structure, as deduced from X-ray scattering, represents the "working" protein.

Both of these statements are misleading. Experiments and computations prove that a folded protein can assume a very large number of somewhat different structures or conformation substates, and protein reactions involve structures that are very different from the one deduced from X-ray scattering. Here we discuss both of these aspects. While the arguments, strictly speaking, refer to proteins they apply with minor changes also to DNA and RNA and are consequently also relevant to genome research.

2. THE ENERGY LANDSCAPE OF PROTEINS

The crucial concept needed to discuss the two problems raised above is that of an energy landscape. In simple systems, such as atoms or nuclei, the ground state has a unique structure and energy. In a complex system, however, the energy can depend on the arrangement of the atoms. Consider a protein that consists of N atoms. The structure of the protein can be characterized by giving the 3N coordinates of all atoms and this structure corresponds to a point in the conformation hyperspace of 3N dimensions. If the protein in its native state could only assume one particular structure, one point in the conformation space would uniquely describe the protein.

Theoretical and Computational Methods in Genome Research, edited by Suhai
Plenum Press, New York, 1997

A gedanken experiment shows that one point is not sufficient to identify a protein: A rotation of a side chain, for instance, will produce a new conformation with, most likely, only a small change in energy. If each amino acid in a protein can asume two positions, a protein with N=150 residues will have about $2^{150} \approx 10^{45}$ different conformations and hence will occupy not just one, but a very large number of points in the conformation space. We call each one of the conformations a conformation substate (CS). In going from one CS to another, the protein has to change its topology; atoms have to move. Such moves require overcoming an energy barrier. Each CS thus is a valley or crater in the 3Ndimensional conformation hyperspace.

Evidence for CS comes from many experiments[1 2 3] and from molecular dynamics and Monte Carlo computations.[4 5 6] The first unambiguous proof for CS came from flash photolysis experiments.[7] The most dramatic evidence is given by laser hole burning experiments.[8 9] Different substates have spectral lines with somewhat different center frequencies.[10 11] The line produced by a protein ensemble is consequently inhomogeneously broadened and is no longer a Lorentzian, but a Voigtian. If the homogeneous line width is much smaller than the homogeneous one, the existence of CS can be shown directly: Irradiating with a narrow laser line at a particular position may move the proteins with lines at that wavenumber to different substates. The result is a "hole" in the spectrum.

Experiments clearly show such holes.[12 13 14 15 16] The holes can be as narrow as 10^{-4} of the inhomogeneous width, implying that there are at least 10^4 CS present. Since many different conformations can lead to the same line position, the actual number of CS is much larger than 10^4.

The conclusion of these arguments is that proteins must be described by an energy landscape.[3 7 17 18] The details of the energy landscape are not yet fully known even for such a simple protein as myoglobin[19 20], but some general features are becoming clear and they are sketched in the one-dimensional cross section shown in Fig. 1. The figure shows a somewhat speculative plot of the free energy (Gibbs function) G as a function of a conformation coordinate. Three regions are distinguished. In the unfolded state, U, the polypeptide chain is essentially unstructured. In the native state, N, the protein has formed a compact globule which, on the average, can be assumed to be given by the X-ray or NMR structure. The exact nature of the molten globule or compact intermediate state is not unambiguously established.[21 22] For the present discussion we assume that molten globules consist of some compact, essentially correctly folded parts together with some partially or totally unfolded regions.

Additional information on the substates comes from thermodynamic data. Denote the number of substates in the unfolded (denatured) state by w_U, in the native (folded) state by w_N, the difference in free energy (Gibbs function) between the native and the unfolded state by ΔG_{UN}, between the native and the molten globule state by ΔG_{MN}. Consider now myoglobin (Mb) as a typical protein. Mb consists of about 150 residues. The thermodynamic parameters for Mb are[23] ΔG_{UN} = 50 kJ/mol and ΔH_{UN} = 150 kJ/mol, where ΔH_{UN} denotes the enthalpy difference between the unfolded and the native state. The difference in entropy, in dimensionless units, becomes

$$\Delta S_{UN}/R = (\Delta H_{UN} - \Delta G_{UN})/RT \approx 40 \tag{1}$$

Here R=8.31 J/K mol is the gas constant. The corresponding parameters for the molten globule are not as well known, but at the temperatures where proteins function, the molten globule is more stable than the unfolded structure.[21] With Figure 1 and these numbers, we can now discuss the two issues raised in the introduction.

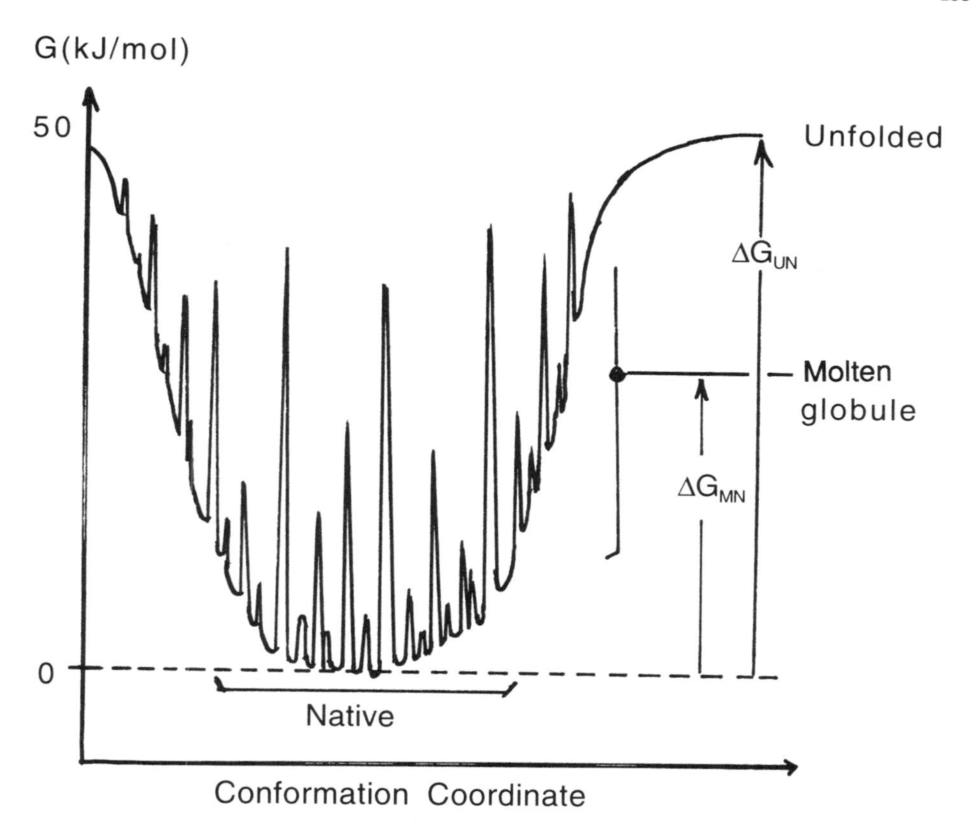

Figure 1. A one-dimensional cross section through the free-energy landscape of a globular protein with about 150 residues. N denotes the native state, MG the molten globular compact intermediate state, and U the unfolded (denatured) state. The number of states is given by w, the differences in free energy (Gibbs function) by ΔG.

3. FOLDING DOES NOT LEAD TO A UNIQUE STRUCTURE

In section 2 we discussed a broad range of experiments that suggest that the ground state of a protein is highly degenerate. The experiments are not sensitive enough to give a believable value for w_N, the number of substates. A lower limit can be obtained as follows: The experiments show that the CS are arranged in a hierarchy of at least five different tiers.[19] Hole burning experiments demonstrate that in one particular tier there must be far more than 10^4 CS.[12–16] If only three different dimensions are compressed onto the one-dimensional wave number scale explored in the hole-burning experiments, and if the CS not studied in the hole burning have about three CS, we get $\log w_N \geq 14$.

A speculative limit can be obtained by assuming that each side chain can assume two different positions. For Mb we then obtain $\log w_N > 40$. This lower limit can be improved by computer analysis.

A better limit could, in principle, be obtained by measuring the entropy of the protein and using the Boltzmann relation,

$$\ln w = S/R. \tag{2}$$

Usually the native state is taken as reference and the entropy of the unfolded state is determined with respect to this reference, as indicated in Eq.1. If log w_U were known, log w_N would follow. Assume, for instance, that each residue in Mb can take on ten different conformations so that log $w_U \approx 150$ log $10 \approx 150$. Eq. 1 then would yield log $w_N \approx 130$. There are, however, two problems with this estimate. First, we do not know the number of CS in the unfolded state reliably. Second, and more importantly, ΔS_{UN} in Eq. 1 is not given by the change in the conformation entropy alone. [24] In the unfolded state, hydrophobic and polar residues are exposed to the solvent, organize the water, and lower the entropy. Makhatadze and Privalov propose that this effect leads to a value of $\Delta S^{conf}/R \approx 6$ per residue or about 950 for Mb. This value implies log w_U>400 for unfolded Mb even if we assume that the standard state, native Mb, is in a single substate! Thus each individual residue would have, on the average, about 400 substates. Such complexity is difficult to visualize. Clearly much remains to be studied experimentally and understood theoretically and computationally.

The dominant lesson from these considerations is clear. The number of substates in the folded state is extremely large. Folding does not lead to a unique state, but to an ensemble of conformation substates. Most of these may have very similar structures, but there may also be substates that are very different, but cannot be seen with X-rays because they occur with less than about 10% probability. In Figure 1, the barriers between substates are drawn with very different heights, because experiments demonstrate that the substates are organized in a hierarchy.[19 20 25] The smallest barriers[20] have free energy heights of less than 100 J/mol, while large ones can reach 30 -50 kJ/mol.[26] The barrier heights are actually not determined by the bare protein alone; solvent damping plays a crucial role.[27 28 29 30] While this damping affects the rate of the protein motions, it should not influence the number of substates.

4. THE WORKING PROTEIN IS A DYNAMIC SYSTEM

We now turn to the second problem: What is the structure of the protein in action? Protein motions play a crucial role in protein function. The classical example is the entrance and exit of ligands such as dioxygen and carbon monoxide from myoglobin and hemoglobin. The X-ray structure of these proteins shows no channels through which the ligands could migrate. The protein must fluctuate in order to let the ligands pass. Many other examples of the importance of protein motions are known. The energy landscape in Figure 1 shows why such motions must occur and suggests how large they can be. The probability of a system to be in a substate with free energy G at the temperature T is given by $\exp\{-G/RT\}dG/RT$, where R is the gas constant. The probability of finding the proteins in substates with free energies above a values G_0 is given by

$$P(G > G_0) = \exp\{-G_0/RT\}. \tag{3}$$

The free energy of the unfolded state in Mb is about 50 kJ/mol. The probability of being in the unfolded state is therefore

$$P(U) \cong 10^{-9} \tag{4}$$

Thus there will be a finite probability that a protein in the native state is fully unfolded. This probability may be too small to affect the function dramatically. The situation is different for the molten globules. The free energy of a molten globule substate is smaller than ΔG_{UN}, and the probability of a protein being in such substate can be large. Typical protein reactions occur with characteristic times between ms and ps. It is consequently possible that many of these reactions occur when the protein is in a molten globule substate, with a structure far different from the one shown in X-ray structures.

Evidence for very large fluctuations comes from different experiments. The simplest one may be the entrance of molecules into myoglobin as studied by fluorescence quenchinq.[31] Dioxygen, anthraquinonesulfonate (AQS), and methylviologen enter with nearly the same rate, they have the same activation enthalpy for quenching, and the same dependence on solvent viscosity.[32 33] A straightforward explanation for this observation is through the gate model[27]: The ligand enters when a protein fluctuation is large enough to let the ligand pass. The observed activation enthalpy is that for the fluctuations of the protein, not for steric hindrance during the entrance of the ligand. AQS can be represented by an ellipsoid with short axis of about 5 Å. Mb must consequently open channels of at least this diameter. Moreover, the value of the rate coefficient implies that many channels must open. Molten globule substates may satisfy all these conditions.[34]

The conclusion that fluctuations open large channels is reinforced by the rate coefficients for entrance of isocyanides (isonitriles) after photodissociation.[35 36] These large molecules enter myoglobin with rate coefficients that are only one to three orders of magnitude smaller than that for CO or O_2. Further support for the existence of partially or fully unfolded substates in the native state comes from hydrogen exchange experiments and from simulations.[37]

5. CONCLUSIONS

The results discussed here can be summarized as follows:

1. The native state is not unique; it consists of a very large number of conformation substates. Most of these possess structures that are closely related to the average (X-ray) structure, but the existence of very different structures ("misfolded") cannot be excluded.
2. Proteins have appreciable probability of being in a molten globule state. The well organized structure shown in texts and papers is consequently misleading. Gregorio Weber's description of the protein as a screaming and kicking entity is more apt. These substates may be crucial for the function of many proteins.

This work was performed under the auspices of the U. S. Department of Energy.

REFERENCES

1. H. Frauenfelder, F. Parak, and R.D. Young, Ann. Rev. Biophys. Biophys. Chem. 17, 451–479 (1988).
2. H. Frauenfelder, G. U. Nienhaus, and R.D. Young, in Disorder Effects on Relaxational Processes. R. Richert and A. Blumen, Eds. Springer-Verlag, Berlin, 1994.
3. H. Frauenfelder, S. G. Sligar, and P. G. Wolynes, Science 254, 1598–1603 (1991).
4. R. Elber and M. Karplus, Science 235, 318–321 (1985).
5. T. Noguti and N. Go, Proteins 5, 97–138 (1989).
6. A. Garcia, Phys. Rev. Lett. 68, 2696–2699 (1992).
7. R. H. Austin, K. W. Beeson, L. Eisenstein, H. Frauenfelder, and I. C. Gunsalus, Biochemistry 14, 5355–5373 (1975).

8. R. Jankowiak, J. M. Hayes, and G. J. Small, Chem. Reviews 93, 1471–1502 (1993).
9. J. Friedrich, Methods Enzym. 246, 226–259 (1995).
10. N. Aqmon and J.J. Hopfield, J. Chem. Phys. 79, 2042–2053 (1983).
11. N. Agmon, Biochemistry 27, 3507–3511 (1988).
12. J. Friedrich, H. Scheer, B. Zickendraht-Wendelstadt, and D. Haarer, J. Chem. Phys.74, 2260–2266 (1981).
13. G. Boxer, D. S. Gottfried, D. J. Lockhart, and T. R. Middendorf, J. Chem. Phys. 86, 2439–2441 (1987).
14. J. Zollfrank, J. Friedrich, and F. Parak, Biophys. J., 61, 716–724 (1992).
15. N. R. S. Reddy, P. A. Lyle, and G. J. Small, Photosynthesis Research 31, 167–194 (1992).
16. J. Friedrich, J. Gafert, J. Zollfrank, J. Vanderkooi, and J. Fidy, Proc. Natl. Acad. Sci. USA 91, 1029–1033 (1994).
17. P. G. Wolynes, J. N. Onucic, and D. Thirumalai, Science 267, 1619–1620 (1995).
18. J. N. Onuchic, P. G. Wolynes, Z. Luthey-Schulten, and N. D. Socci, Proc. Natl. Acad. Sci. USA 92, 3626–3630 (1995).
19. H. Frauenfelder, Nature structural biology, 2, 821–823 (1995).
20. D. Thorn-Leeson and D. A. Wiersma, Nature structural biology, 2, 848–851 (1995).
21. E. Freire, Annu. Rev. Biophys. Biomol. Struct. 24, 141–165 (1995).
22. A. L. Fink, Annu. Rev. Biophys. Biomol. Struct. 24, 495–522 (1995).
23. P. L. Privalov, Annu. Rev. Biophys. Biophys. Chem. 18, 47–69 (1989).
24. G. I. Makhatadze and P. L. Privalov, Protein Science 5, 507–510 (1996).
25. A. Ansari, J. Berendzen, S. F. Bowne, H. Frauenfelder, I. E. T. Iben, T. B. Sauke, E. Shyamsunder, and R. D. Young, Proc. Natl. Acad. Sci. USA 82, 5000–5004(1985).
26. H. Frauenfelder, N. A. Alberding, A. Ansari, D. Braunstein, B. R. Cowen, M. K. Hong, I. E. T. Iben, J. B. Johnson, S. Luck, M. C. Marden, J. R. Mourant, P. Ormos, L. Reinisch, R. Scholl, A. Schulte, E. Shyamsunder, L. B. Sorensen, P. J. Steinbach, A. Xie, R. D. Young, and K. T. Yue, J. Phys. Chem. 94, 1024–1037 (1990).
27. D. Beece, L. Eisenstein, H. Frauenfelder, D. Good, M. C. Marden, L. Reinisch, A. H. Reynolds, L. B. Sorensen, and K. T. Yue, Biochemistry 19, 5147–5157 (1980).
28. M. Settles, F. Post, D. Müller, A. Schulte, and W. Doster, Biophys. Chem. 43, 107–116 (1992).
29. A. Ansari, C. M. Jones, E. R. Henry, J. Hofrichter, and W. A. Eaton, Science 256, 1796–1798 (1992).
30. S. Yedgar, C. Tetreau, B. Gavish, and D. Lavalette, Biophys. J. 68, 665–670 (1995).
31. J. R. Lakowicz and G. Weber, Biochemistry 12, 4171–4179 (1973).
32. N. Barboy and J. Feitelson, Biochemistry 26, 3240–3244 (1987).
33. N. Barboy and J. Feitelson, Biochemistry 28, 5450–5456 (1989).
34. D. Xie and E. Freire, J. Mol. Biol. 242, 62–80 (1994).
35. M. P. Mims, A. G. Porras, J. S. Olson, R. W. Noble, and J. A. Peterson, J. biol. Chem. 258, 14219–14232 (1983).
36. E. E. Di Iorio, K. Winterhalter, and G. M. Giacometti. Biophys. J. 51, 357–362 (1987).
37. D. W. Miller and K. A. Dill, Protein Science 4, 1860–1873 (1995).

18

HIV-1 PROTEASE AND ITS INHIBITORS

Maciej Geller,[1,2] Joanna Trylska,[2] and Jan Antosiewicz[2]

[1]Interdisciplinary Centre for Mathemathical and Computational Modelling
University of Warsaw
Banacha 2, 02-097 Warsaw, Poland
[2]Department of Biophysics
Institute of Experimental Physics
University of Warsaw
Zwirki & Wigury 93, 02-089 Warsaw, Poland

ABSTRACT

HIV-1 protease (HIV-1 PR), one of the three enzymes encoded by the viral genome and vital for its replication, is natural target for chemotherapy. The three-dimensional X-ray structure of the native form, and of its complexes with various inhibitors, provide a basis for understanding the physicochemical properties of the protease. Formation of the catalytically active homodimeric form of the protease, and of its complexes with inhibitors, may be determined by two physically different types of interactions: electrostatic and hydrophobic. Selected problems related to rational drug design will be discussed; rigidity and flexibility of the enzyme; analysis of electrostatic and hydrophobic interactions in the interface region; dissociative inhibition in the intertwining region; electrostatic field of the binding site; protonation state of the catalytic aspartates; location of internal water molecules; presentation of selected inhibitors. From a methodological point of view, HIV-1 PR is an exellent object for testing different theoretical methods applied to computer-aided drug design.

1. INTRODUCTION

To postulate that the sequence of amino acid residues determines the three-dimensional structure of a protein, which constitutes a physical basis for its biological function, is one of the central dogmas of molecular biology. Prediction of the 3D-structure of a protein, based on its sequence, is one of the major challenges in biophysics and biochemistry, known as the protein folding problem. For a long time, determination of the structure denoted, however, only the static location of the heavy nuclei of a molecule by means of X-ray crystallography or, in the case of smaller molecules, by NMR methods. This is of course a fundamental step, but only the first (i) in the process of understanding how protein structure relates to its function. The main next steps appear to be: (ii) understanding its static and *dynamic* conformational properties, e.g., determination of biologically significant local minima of potential energy (and/or free enthalpy) not only in a vacuum or in the crystal, but also in the natural environment, as well as estimation of the possible pathways for their changes; (iii) understanding the mechanism of its intermolecular interactions with different types of ligands; (iv) understanding the mechanism of rearrangement of the molecule during chemical reactions.

Studies of proteolytic enzymes, especially digestive proteases, have helped pave the way along steps (i–iv). It has been recognised that there are only four distinct mechanisms for cleveage of peptide bonds, resulting in the existence of four different groups of proteases. Moreover, apart from digestive enzymes, there are many more specific regulatory proteases containing catalytic domains related to a parent digestive enzyme[1]. Retroviral proteases belong to the regulatory group of aspartyl proteases and are related to such well-known cellular proteases as pepsin (digestive) and renin (regulatory)[2]. The postulated causative role of the HIV-1 virus, a member of the retroviral family, in AIDS disease has stimulated enormous efforts to resolve its structure and to understand the properties and functions of its constituent molecules.

HIV-1 protease (HIV-1 PR) together with reverse transcriptase (RT) and integrase (IN), form a triplet of enzymes encoded by the viral genome. The latter two enzymes are involved in copying the single-stranded RNA viral genome into the double-stranded DNA form (RT) and in its integration into the host genome (IN). The protease exhibits its main activity during the course of viral maturation.

The viral enzymes, and most of its structural proteins, are first translated as part of large polyprotein precursors of two types, Pr160(gag-pol) or Pr55(gag). Then, after budding of the premature virions, the HIV-1 PR cleaves both precursors at specific sites to yield (eight) mature polypeptides, including the protease itself[3]. Since protease acts as a homodimer, autoprocessing of the enzyme from the precursor seems to require proximity of the catalytic residues of the two precursor polyproteins[4]. This vital role in viral replication makes the protease a natural target for chemotherapy of AIDS[5,6]. Because it was the first HIV-1 enzyme with a three-dimensional structure resolved by X-ray diffraction, a large number of laboratories focused their attention on this molecule, leading to nearly eight hundred positions listed in a (MEDLINE) publication search using "HIV-1, protease" as keywords. The purpose of this paper is to present selected conformational features of the protease related to the design of its inhibitors.

2. STRUCTURE OF NATIVE ENZYME

HIV-1 PR, in contrast to the cellular aspartic acid proteinases, which are single-stranded proteins of about 350 residues, is active as a homodimer with 99 residues in each monomer[7]. Fig.1 shows that β-strands dominate the secondary structure of the enzyme. There is only one short α-helix in each monomer. For purposes of further analysis, the interface region formed by the nearly perfect C_{2v} symmetry of the dimer may be divided into seven subregions shown in Fig.2: (i) intertwining region, (ii) hydrophobic core, (iii) fireman's grip, (iv) catalytic site, (v) binding site, (vi) flaps, and (vii) 'side triplets' (see below).

The intertwining region (i), located at the bottom of the intact enzyme, consists of the five C and N terminal residues of both monomers. The termini of one monomer pene-

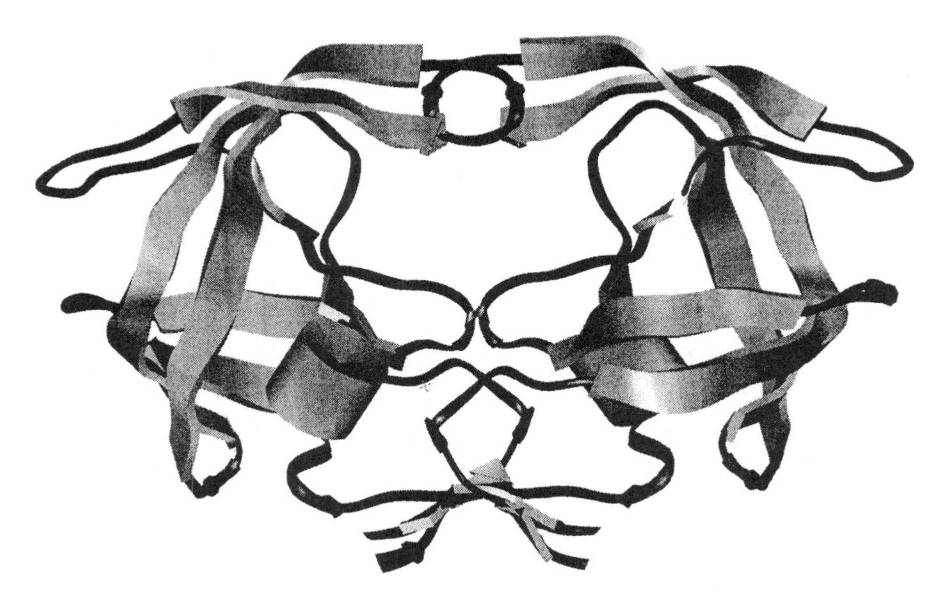

Figure 1. Secondary structure of native form of the HIV-1 protease

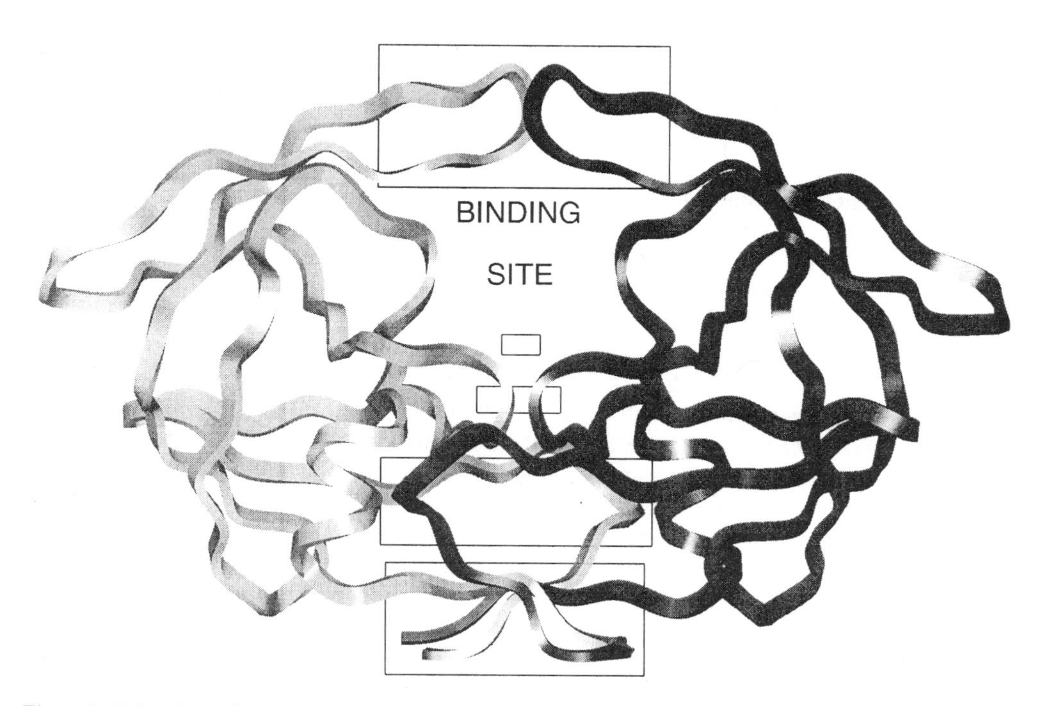

Figure 2. Subregions of the interface of the HIV-1 protease. In boxes,from the bottom: (i) intertwining region, (ii) hydrophobic core, (iii) fireman_s grip, (iv) catalytic site, (v) binding site, (vi) flaps.

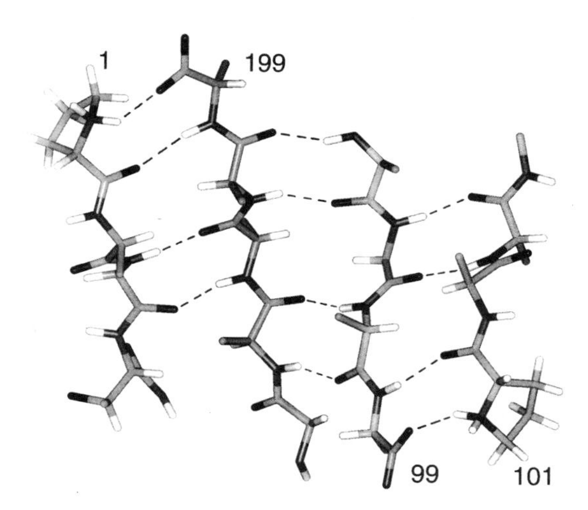

Figure 3. Network of hydrogen bonds in the intertwinig region of the HIV-1 protease. Only main chain : main chain atoms and hydrogen bonds are shown. Gray scale, on all Figures, from black up corresponds to: O, N, C, H atoms, respectively.

trate the regions of the corresponding termini of the other monomer, forming three hydrogen bond layers of the β-sheet type shown in Fig.3. Additionally, some hydrogen bonds may be formed between the side chains of these residues.

The second region of the interface (ii), located immediately above, is dominated by a different type of interactions, namely hydrophobic, which results in formation of the hydrophobic core, consisting of a large cluster of the hydrophobic side-chains of res.: Leu-5, Leu-23, Leu-24, Ile-93, Aba-95, Leu-97 and the corresponding residues from the second monomer, shown in Fig. 4. A smaller hydrophobic core may also be observed in the structure of the isolated monomer. It is formed by the side chains of residues: Leu-13, Leu-15, Ile-85, Leu-89, and Leu-90.

"Fireman's grip", the third part of the interface (iii), is formed by two symmetrical hydrogen bonds between the lone electron pairs of each the OG1 atom (Thr-26 or Thr-126) and the main chain N-H group of the Thr-126 or Thr-26, respectively. It is located between the hydrophobic core and the catalytic site (iv) formed by the equivalent acidic groups of Asp-25 and Asp-125, and is essential for maintaining close contact of these groups. Characteristic features of the catalytic acidic groups of Asp-25 and Asp-125 are their nearly perfect coplanarity and the close contact between the 'inner' OD1 oxygen atoms, (see Fig. 8). It equals 2.83A in the native dimer and 2.63A in the HIV-1 PR :MVT-101 complex. The 'outer' oxygen atoms which are exposed to the solvent are significantly farther, 4.5A. The protonation state of these aspartates plays an important role during catalysis. Moreover, as it will be discussed further, it may be important for binding of inhibitor.

The binding site (v) has the shape of an extended ravine formed by both monomers and perpendicular to the two-fold symmetry axis of the dimer. Entry to the ravine, from above, is controlled by two flexible flaps (vi) which form "a gate" for an approaching ligand. Each flap has a β-harpin form maintained by several hydrogen bonds. There are four hydrophobic pockets inside the side walls of the ravine. Symmetrically located 'side trip-

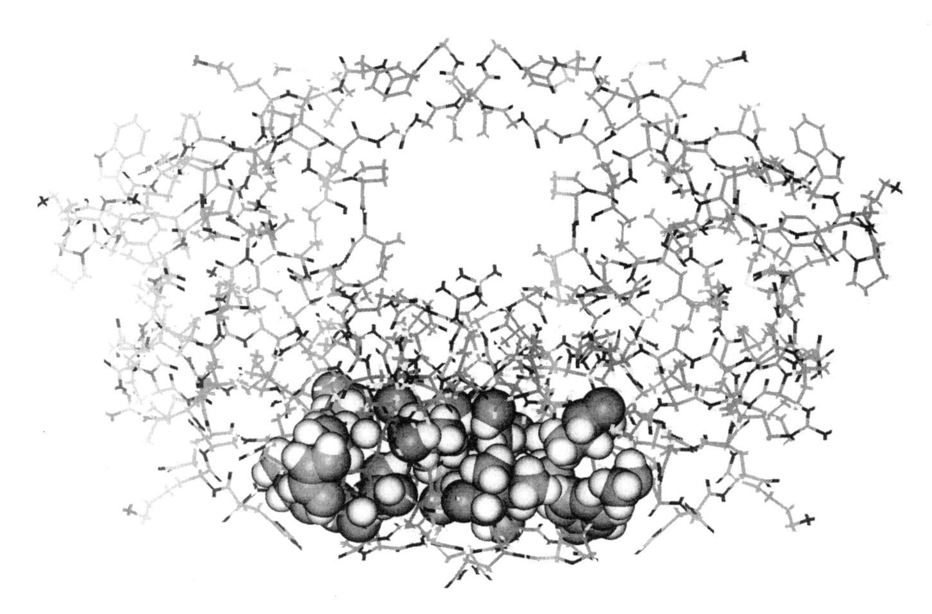

Figure 4. Hydrophobic core in the interface of the HIV-1 protease, shown as Van der Waals spheres (dark: C, light: H).

lets' (vii) are formed by side-chains of residues Arg-8, Asp-29, and Arg-187 on one side and the corresponding side chains on the other.

2.1 Dimer Formation

The area of the interface constitutes nearly one-third (28%) of the solvent accessible surface of each monomer (assuming that it maintains its dimer conformation). During the process the dimer formation, total solvent accesible area decreases significantly (by about 3860 A^2). Contributions to this change are listed in Table I.

Analysis of the interface shows that about 50% of its surface is hydrophobic and related mainly to the hydrophobic core (ii) area, the remaining part being largely responsible for the hydrogen bond network formation in the intertwining region (i). Assuming that 1 A^2 decrease of the hydrophobic surface, exposed to the water solvent, results in drop of

Table I. Solvent accesible surfaces of HIV-1 protease(a) [in A^2]

Object	Hydrophobic(b)	Other	Total
(1) Dimer	2660	7410	10070
(2) Monomers(c)	4640	9290	13930
(3) Interface(d)	1980	1880	3860

(a) Discover[37] Van der Waals radii for protein atoms and a probe radius of 1.4A for water;

(b) as hydrophobic: non-heteroatom groups of the side chains of Ala, Ile, Leu, Phe, Pro, Val, Thr, Trp, and Tyr;

(c) sum for both separated monomers in conformations of the dimer;

(d) 3 = 2 -1.

the free energy by 47 cal/mol (at room temperature)[8], one obtains the high value of about -90 kcal/mol as an estimated contribution to the free energy change during the dimer formation process. It should be borne in mind, however, that this estimate is based on two assumptions: first, the similarity of the conformation of the isolated monomer to that observed in the dimer, and second, of the dilute concentration of hydrophobic objects interacting in aqueous solvent for which the above-mentioned value of 27 cal/mol/A^2 was obtained.

To what extent these assumptions are valid, remains to be established. Because of the high binding affinity of monomeric subunits for each other (K_d in the nM range[9]), the solution structure of the native monomer is difficult to study experimentally. Some hints about the validity of the first assumption may be drawn from theoretical simulation methods. The molecular dynamics (MD) trajectory of the monomer in a water droplet showed that the general features of the tertiary structure were preserved and only the flaps and termini regions exhibited considerable flexibility[10]. As to the second assumption, one should note that formation of the dimer during or after budding takes place under condition of very high concentration of the polyproteins[4].

The conclusion that hydrophobic interactions determine proper dimer formation would, however, be incorrect. Experimental data show that mixing of a native protease with mutant protease subunits (missing 1–5 residues) results in inactivation of the former[11]. This points to the important role of interaction in the intertwining region for dimer formation. This feature forms the basis for a suggestion that inhibition of the enzyme activity may be achieved by interfering with dimerization[7], for instance by the C-terminal tetrapeptide, Thr-Leu-Asn-Phe[12], confirmed by experimental data (K_i = 45 μM)[13].

Designing of this type of inhibitor depends significantly on the termini conformation of the isolated monomer. To check it in detail, a 100 ps MD simulation was performed using the AMBER (4.0 version) package program[14] to obtain the trajectory of motion of both termini (res. 1–7 and 93–99, T=300 K) in the water droplet (radius 28 A) centered at the midpoint between C_α atoms of Ile-3 and Thr-96 according to protocol[15]. The only exception was that final equilibration of the whole movable system (termini + water) was not performed, being expected that the starting point for data collection would not be in a state of equilibrium. Although the general location of both backbone regions of the termini did not undergo significant change during simulation, the system was not in a state of equilibrium. Systematic motion inside the system was observed during the first 80 ps of the simulation. Side-chains of residues Ile-3, Leu-5, Leu-97, and Phe-99, separated by water molecules at the beginning of the simulation (Fig.5a), approached each other and formed a closely packed cluster (Fig.5b), with no water molecules between them after 80 ps. Because these residues have hydrophobic side chains, it seems that this movement is due to hydrophobic interactions. Additional evidence for such a conclusion is the stability of the total energy of the system (the mean value of the total energies in the first and last 5ps periods of simulation differed by only 1 kcal/mol), which shows that the observed motion must be mainly due to the entropy term. Total decrease of the water-accessible area of the termini, due to this movement, was 230 A^2. Decrease of the hydrophobic area of 112 A^2 corresponds to about -5 kcal/mol as a contribution to the change in free enthalpy. In attempts to design a dissociative inhibitor, one could consider a molecule which can interact with those regions of the termini not involved in this cluster formation, since separation of the cluster will result in an unfavorable increase of free energy because of exposure of the hydrophobic side-chains to water.

This is an example of the molecular structure of a protein which may contain many hidden features not obvious from the static X-ray structure. MD simulation of mutant en-

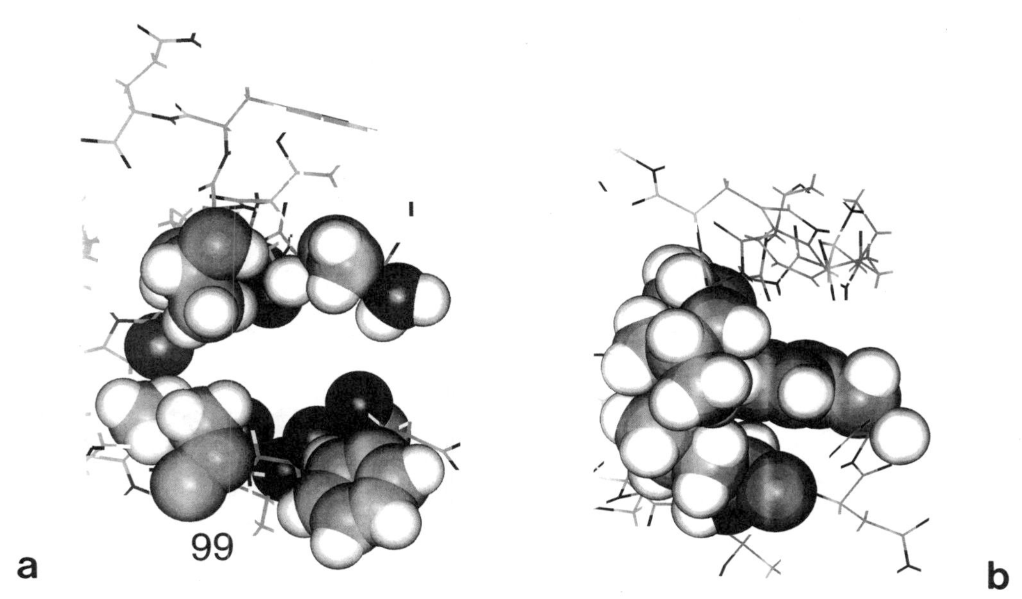

Figure 5. Formation of a hydrophobic cluster in the termini region of the HIV-1 monomer. Molecular Dynamic simulation (100ps). (a) starting (b) final conformation; Hydrophobic side chains are shown as Van der Waals spheres.

zyme, N88Q, showed that local net loss of one hydrogen bond, as a result of that point mutation, leads to significant disorder in the intertwining region which may be responsible for observed inactivation of the protease[16]. Usually, the intertwining parts of the protease are described as the main region responsible for stability of the dimer. Sometimes, however, trial studies may reveal some unexpected feature of a structure. During initial investigation of MD studies of the HIV-1 PR : MVT-101 complex[15], it was observed that, apart from the crucial water 301, four more water molecules located symmetrically, in pairs, in small cavities between the catalytic aspartates and the "side triplets" of charged residues (Arg-8, Asp-29, Arg-187 and Arg-108, Asp-129, Arg-87; respectively) on both entries to the binding side ravine, are important for enzyme conformation. Disruption of either of the triplets results in the entrance of several water molecules to the cavity and significant changes in the conformation of the catalytic site.

3. BINDING SITE

The unprecedented level of activity of crystallographers in investigations of complexes between the protease and designed inhibitors is obviously due to the hope of finding an efficient drug against AIDS disease. Starting from the first complex with the MVT-101 inhibitor[17], nearly 200 hundred such structures have been solved since 1989 in more than 20 laboratories. Despite the fact that most of them are unavailable because of commercial reason, some are sufficient to draw several general conclusions. Binding of a substrate/inhibitor introduces substantial conformational changes in the flaps region and

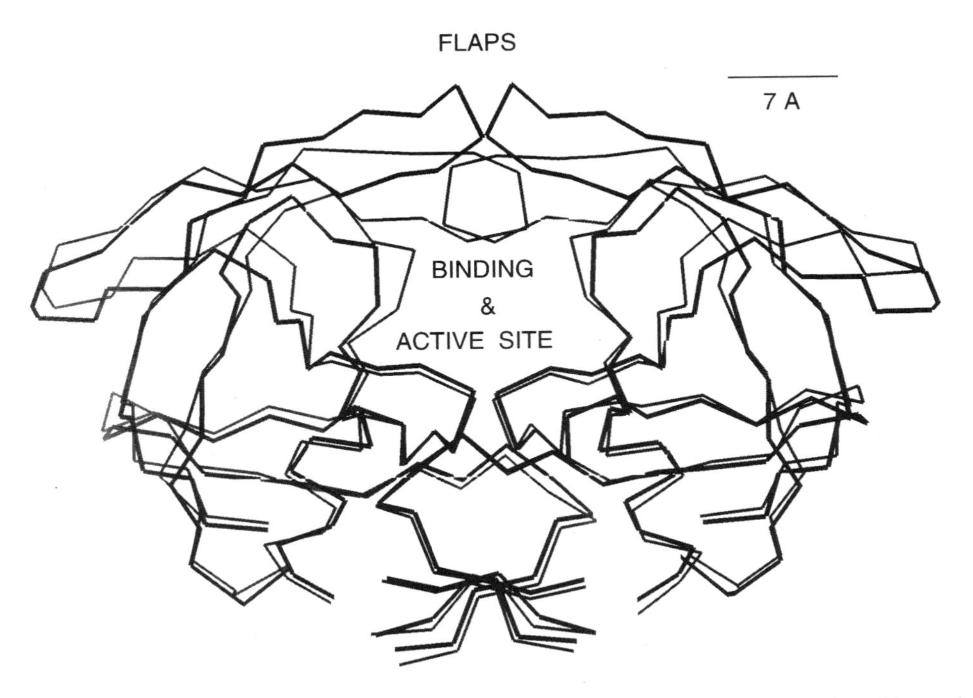

Figure 6. Changes of the CA traces of the HIV-1 protease upon binding of MVT-101 inhibitor; thin—native, thick—complexed forms.

some tightening of the binding site presented in Fig.6. The rms deviation for C_α atoms between both structures is 1.83A. Both flaps close the ravine of the binding site, forming a kind of partial roof, presented in Fig.7. The tips of the flaps move during this step as far as 7 A. MD simulations have revealed some domain communications related to this movement, resulting in "fulcrum" and "cantilever" identification.[18,19]

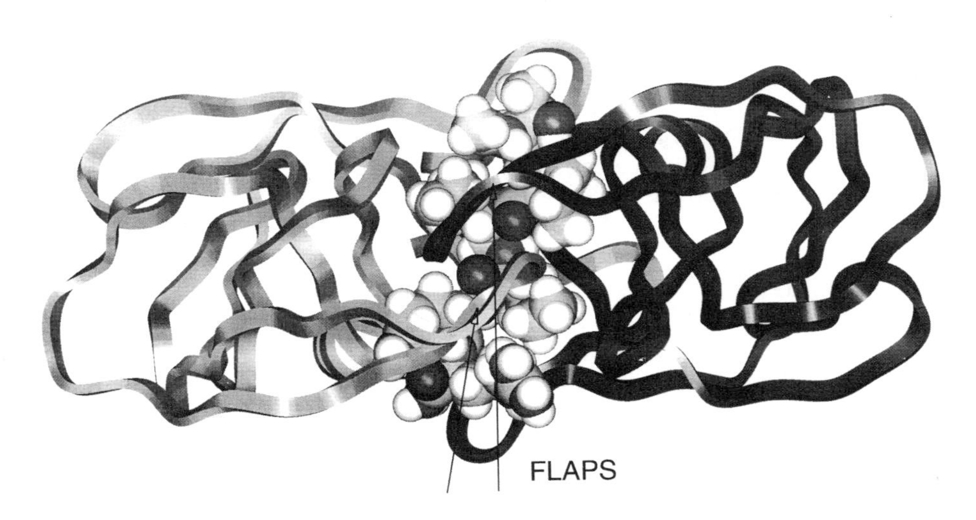

Figure 7. The HIV-1 protease complexed with the MVT-101 inhibitor. View from above.

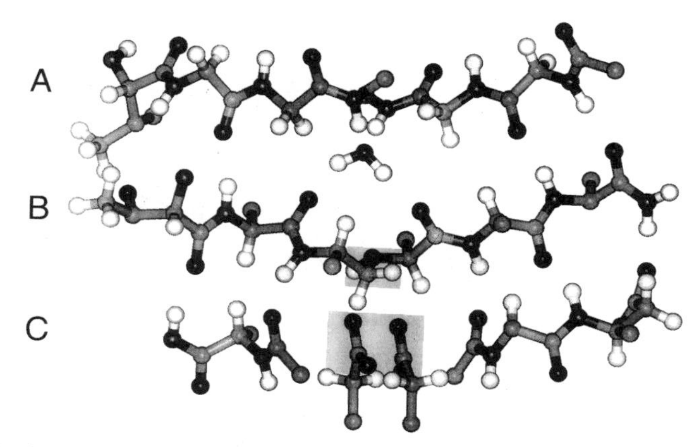

Figure 8. Network of the hydrogen bonds between inhibitor (MVT-101) and the HIV-1 protease. Central water molecule (301) is also shown. (A)—flaps region, (B)—inhibitor, (C)—catalytic site. Catalytic aspartates, (Asp - 25, Asp - 125) are marked by the large shadow box; reduced link is marked by the small box.

The general features of the binding site, using the ravine analogy, may be summarized as follows: bottom and roof are ready for hydrogen bond formation, while each side wall has two niches with mainly hydrophobic surfaces (see Figs.9,10). The physical properties of this environment result in peptide binding in an extended conformation, shown in Fig. 8, and forming a network of hydrogen bonds *via* main chain donor and acceptor groups, with lower and upper parts of the binding site. Its side chains may then displace themselves instead into the niches *via* hydrophobic interactions.

One additional characteristic feature is also worth noting. A molecule of water is observed, known as water 301, located like a hook hanging from the middle of the roof,

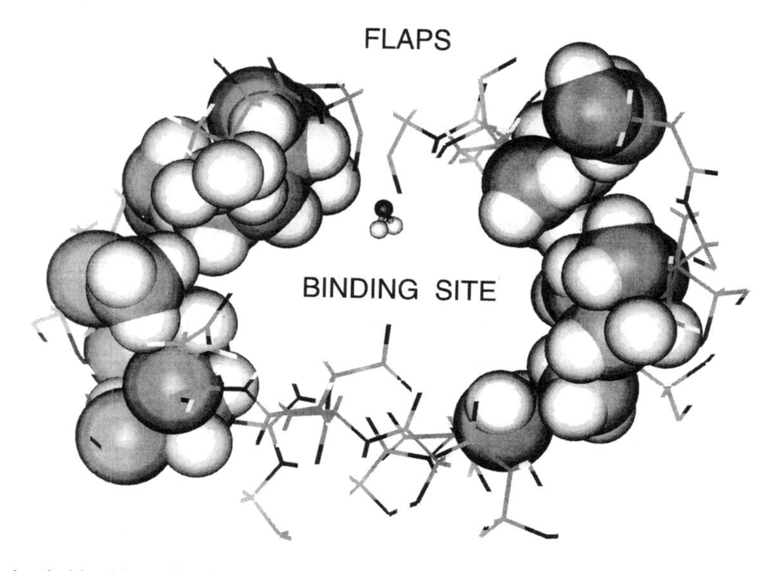

Figure 9. Hydrophobic side walls of the binding site of the HIV-1 PR, shown as Van der Waals spheres. (Colours: carbon—dark, hydrogen—white).

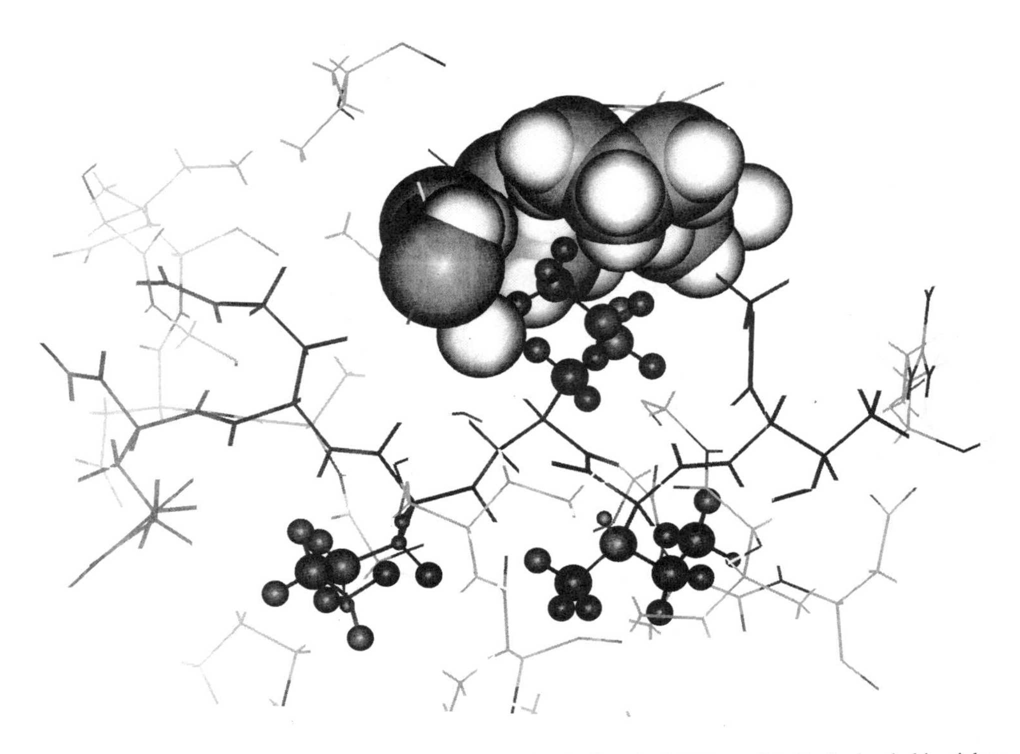

Figure 10. Hydrophobic side chains of the MVT-101 inhibitor (ball-and-stick) located in the hydrophobic niches of the HIV-1 protease. S1 site is shown as Van der Waals spheres.

forming two hydrogen bonds with both flaps, and the two other with the main chain carbonyl oxygen acceptors from two peptide bonds of the inhibitor. Mobility of the flaps, and certain flexibility of the side walls, may be usefull in accommodating the different aminoacid side-chains during processing of the polyprotein backbone. Analysis of about 20 complexes of HIV-1 PR with different inhibitors* showed that the conformation of the enzyme undergoes only small changes. The rms deviation for superposition of all these structures is only 0.6 A for C_α carbon atoms and 0.72A for main chain atoms. As shown in Fig.11, the main differences are observed for two external loops (res. 14–20 and 37–40) and, to smaller extent, for the side walls of the binding site. After exclusion of these two loops, the rms deviation drops to 0.38A for the C_α trace, and to 0.43A for the corresponding main chain atoms (and to 0.99A for all corresponding heavy atoms) of the enzyme. It is interesting that, despite the fact that only 48 residues in HIV-1 and HIV-2 proteases are

* 1aag, 1hih, 1hos, 1hps, 1hpv, 1htf, 1htg, 1hvb, 1hvi, 1hvj, 1hvk, 1hvr, 1sbg, hvp2, hvp3, ag-1002, pd-135390, pd-135392, ro-8959, ro-31–8558.

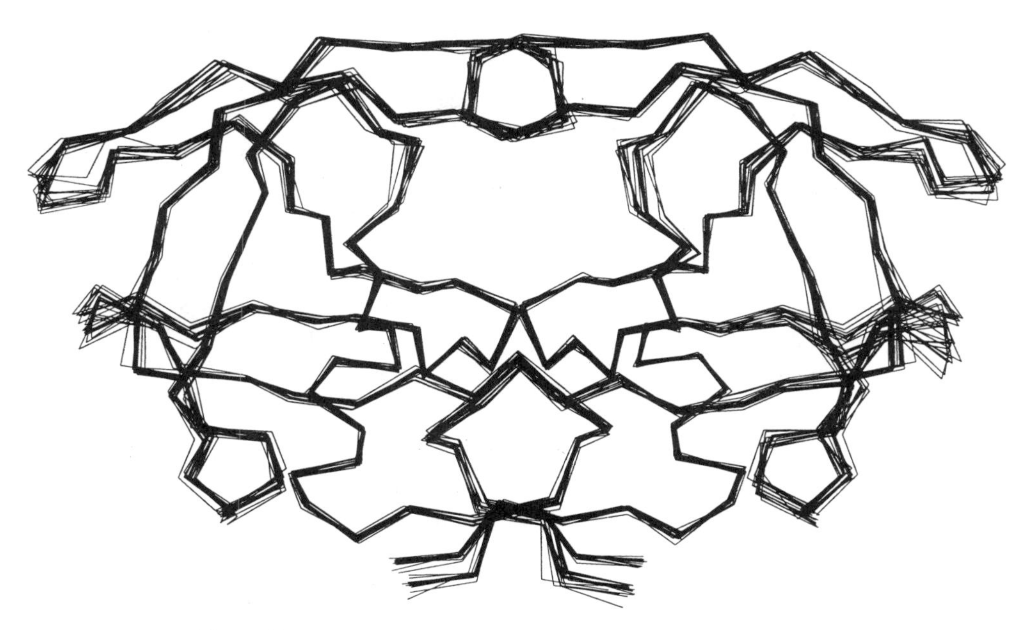

Figure 11. C_α-traces of the HIV-1 protease complexed with 20 inhibitors.

identical, the rms deviation for superposition of the C_α carbon atoms of both enzymes, complexed with the cgp-53820 inhibitor, is only 1.0A.

Accommodation of various types of inhibitors is partially related to the specificity of the protease. The ravine of the binding site of the protease is long enough to accomodate six residues, but it is known that the enzyme is active when the substrates are not shorter than seven residues. Sequences of the natural eight sites of cleavage of the polyprotein chain suggested initially bulky hydrophobic side chains at positions P1 and P1′, and hydrophobic or Asn/Gln residues in position P2/P2′. This may be rationalized by taking into account hydrophobic niches of the binding site, and the possibility of bending the side-chain of Asn or Gln in such a manner that their hydrophobic methylene groups interact with the hydrophobic surfaces of S2 or S2′ sites, while polar terminals form hydrogen bonds with the polar bottom of the ravine. However, in contrast to serine or cysteine proteases, specificity of the aspartyl proteases is ill-defined and, hence, also of HIV-1 PR[20]. Detailed analysis showed multi-site recognition of its substrates which spreads out from P4 to P4′[21]. Moreover, studies of nonviral proteins as substrates of the HIV-1 protease revealed[22] that no amino acid was required at any of the eight positions, P4-P4′, and many different residues may be located on both sides of the scissile bond. Perhaps the only rule is, that lysine in position from P2 to P2′ liquidates the substrate properties of the peptide.

4. DESIGNING HIV-1 PROTEASE INHIBITORS

Several review articles summarize results of investigations to date.[5,23–27] Design strategies may be divided into several, sometimes overlaping, groups according to the method for identifying a lead compound. In the first, known inhibitors of related enzymes are screened for activity against the protease. Initially, pepstatine A was known as a standard inhibitor of aspartyl proteases. Another approach (substrate-based methods) used known sequences of the cleavage sites of the polyprotein (or nonviral substrates) and replaced the scissile bond with some nonhydrolyzable analogs. Finally, two approaches (structure-based methods) are related to computer-aided drug design methods. Both may be applied if the three-dimensional structure of the target is known, and some criteria for "a good fit" are available. From a physical point of view, applied criteria vary from simple ones to the quite sophisticated[28]. Starting with some geometrical estimation of the goodness of fit, including hydrogen bond formation and the area of the hydrophobic contacts, and ending with calculation of the energy of interaction between the proposed ligand and the target. One of these theoretical approaches mentioned above starts with a search through existing data bases for a candidate for a new lead compound while the other starts from scratch. Indentification of a lead compound initiates the laborious process of its modifications to find the optimal inhibitor. This process may be facilitated by computer graphics and posible estimations of resulting changes of energy (or free energy) of interaction by molecular computer simulations, reviewed in[28,29,30].

The scissile bond was then replaced by several transition-state analogs that partially mimic the tetrahedral intermediate formed during the hydrolysis of a peptide. Such known reduced links of renin inhibitors are, for example, statine-like, hydroxyethylene, dihydroxyethylene, hydroxyethylamine, phosphinate, or reduced amide. The first crystallographically resolved inhibitor complexed with the HIV-1 PR was constructed on the basis of the X/p9 junction (Thr-Ile-Met-/-Met-Gln-Arg) of the polyprotein which, after modification and reduced amide type link, led to the MVT-101: N-acetyl-Thr-Ile-Nle-[CH2-NH]-Nle-Gln-Arg-NH2. These studies led to construction of several inhibitors in the nanomolar (or even subnanomolar) range of K_i[26]. The construction of the first, licenced in 1995, inhibitor of the HIV-1 PR (Ro-31–8959 from Hoffmann-La Roche), presented in Fig. 12, started from the *pol* Leu-Asn-Phe-/-Pro-Ile cleavage site and used a hydroxyethylamine junction instead of the scissile bond.

Initial efforts in these studies focused on modifications of the side-chains at positions P2-P2′ by changing the shape of their hydrophobic surfaces. This was possible because the hydrophobic side walls of the binding site are not rigid. For example, MD simulation showed that the tip of the side-chain of norleucine at the P1′ position, in contact with the S1′ surface, fluctuates with a standard deviation of 1.7A[15]. It was mentioned that the ravine of the binding site can accommodate up to 6 peptide residues. Experimental data indicate that effective binding of a substrate may depend even on the sequence of eight residues, with P4 and P4′ residues partially exposed to the solvent. This indicates that interaction of residues in these positions with the outer surface of the enzyme, as well as with the solvent, may be important for efficient catalysis or inhibition. On the other hand, the length of an inhibitor need not be so long, it may even be as short as three residues.

Among the structure-based investigations it is worth mentioning a variety of, so called, symmetry-based inhibitors[31]. Their design utilizes the C_2-symmetric structure of the HIV-1 protease homodimer, with the assumption that the inhibitor which reflects the symmetric active site of the enzyme may prove beneficial in terms of potency and selectivity. Fig. 13 presents one of these (DMP-323, K_i=0.27nM), which shows one additional interesting feature of the construction. The carbonyl oxygen of the urea group is located precisely in the position usually occupied by the oxygen from the central water molecule (301). It was argued that dispensability of this water molecule for binding simultaneously during the inhibitor:enzyme formation, and the highly rigid and proper for binding conformation, would be entropy favourable. It is also an example of a nonpeptide inhibitor, the

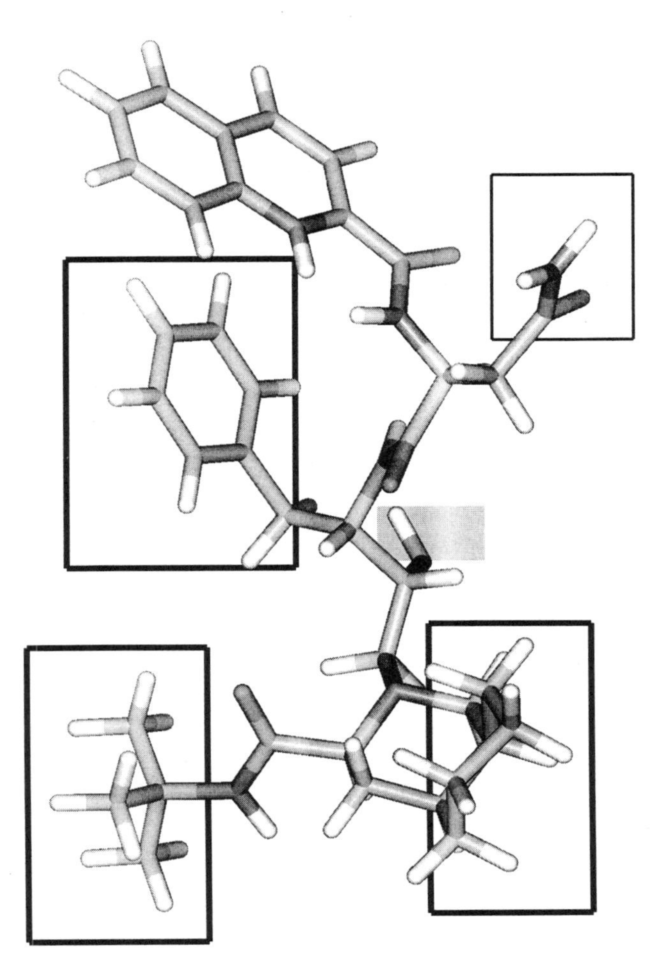

Figure 12. The first licenced inhibitor of the HIV-1 protease: Ro-31–8959 (invirase, saquinavir, Hoffman-La Roche). The three hydrophobic side-chains marked by heavy boxes; Gln residue, by thin box. The hydroxyl group of the reduced link is marked by the filled box.

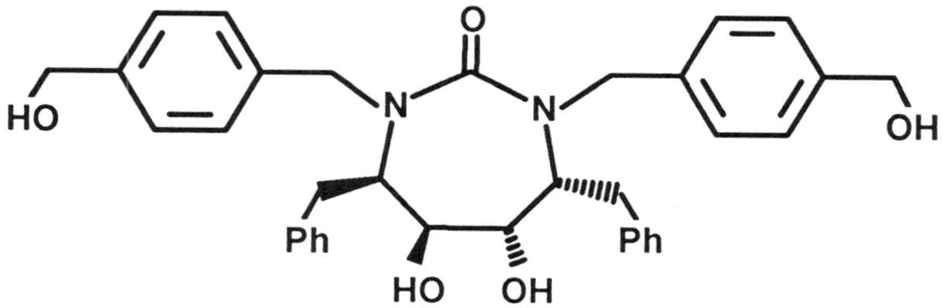

O
HO N N OH
Ph Ph
HO OH

Figure 13. Structure-based DMP-323 inhibitor. Carbonyl oxygen atom of the ureas grup is located exactly in the position usually occupied by the oxygen from the central water molecule (301) and forms two hydrogen bonds with the flaps.

construction of which started with a 3D database search. Two more examples of structure-based design are worth mentioning. The first is a haloperidol derivative, UCSF8, which binds to the binding site in a different way than peptidomimetic inhibitors, although somewhat different from the theoretically predicted location. The second is a hydrophobic fullerene molecule, C_{60}, with K_i=5.3 μM. While not spectacular, it shows that these methods may lead to some unexpected new compounds.

5. ELECTROSTATIC FIELDS AND PK$_a$'S OF CATALYTIC ASPARTATES

As mentioned above, both catalytic acidic groups (Asp-25, Asp-125) are nearly coplanar, with a short distance between "inner" OD1 atoms (for example, 2.63A for MVT-101). However, crystallography does not tell us precisely what is the protonation state of the ionizable groups. In the case of the protease it may influence both the mechanism of catalysis and the interaction between potential inhibitor and binding sites. For the native enzyme, there is a distinct electron density located symmetrically and slightly above "both outer" OD2 atoms. It is proposed[7] that a water molecule "is hidden" inside this density (but why not, for example, H_3O^+ ?). In the case of a peptidomimetic inhibitor, a nonhydrolyzable bond is located precisely in this site. The protonation state of the catalytic site has been subjected to both experimental and theoretical studies[15,32–36]. In all theoretical studies MD simulations were used for this purpose. The results indicate that the protonation states of both catalytic aspartates are not fixed and depend on the environment. Using the sequence convention (Asp-25, Asp-125), there are four possible states: (-1,-1) the dianionic,

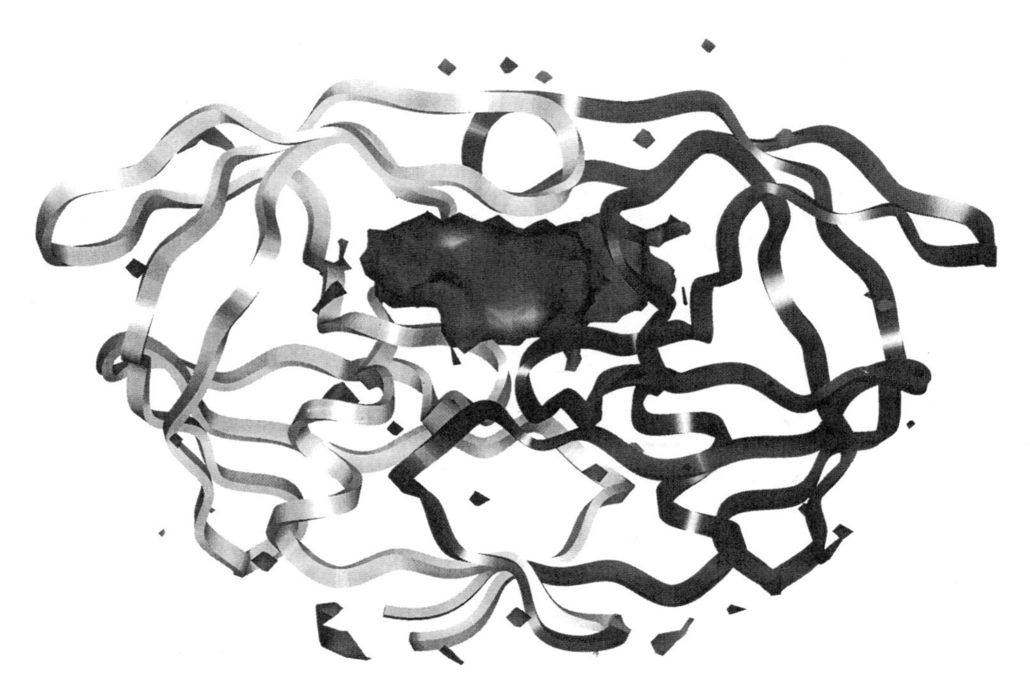

Figure 14. Contours of negative electrostatic potential (-5.0 kT/e) for the dianionic state of both catalytic aspartates of the native form of the HIV-1 protease.

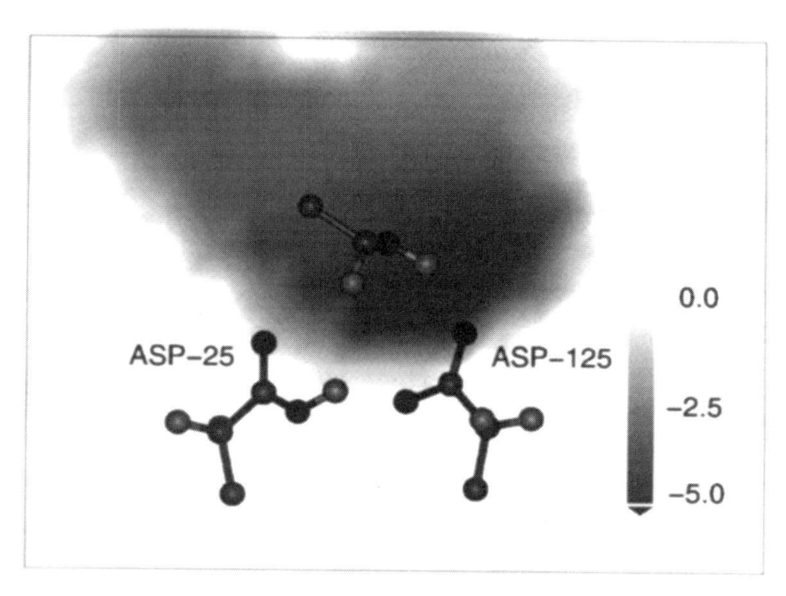

Figure 15. Slice of the electrostatic potential field for the (0,-1) protonation state in the catalytic region of the HIV-1 protease (native form). Location of the reduced link of the MVT-101 inhibitor [-NH-CH_2-] is also shown.

(0,-1) the monoanionic (protonation on Asp-25), (-1,0) (protonation on Asp-125) and (O,O) the neutral one.

Using the linearized Poisson-Boltzmann equation in the DelPhi program[37], changes in the electrostatic potential (ES) field for different protonation states of the hydrated HIV-1 protease (ionic strength of 0.145) were calculated, using AMBER force field 4.0 with non-standard net charges[15]. Fig.14 shows contour of negative ES value (- 5.0 kT/e) for (-1,-1) state. It is seen that the contour encompasses nearly the whole region of the binding site.

Fig.15 shows a slice of the ES potential for the (0,-1) state in the region of the catalytic aspartates. It may be seen that the protonation state of the catalytic aspartates strongly influences the ES field in which an inhibitor is located. Of course, distribution of

Table II. Calculated mean charges at pH 7.2 on ionizable groups in the catalytic region of the HIV-1 PR : MVT-101 complex and the corresponding pK_a's (in parenthesis), for different ionization models[a]

Model	Asp 25	Asp 125	INH
a	0.00	−1.00	0.99
	(22.6)	(−10.0)	(9.8)
b	0.00	−1.00	0.99
	(22.5)	(−15.0)	(9.8)
c	−0.04	−0.93	—
	(22.5)	(5.6)	
d	0.00	−1.00	0.53
	(22.2)	(−15.0)	(7.3)
e	−0.14	−0.86	—
	(24.0)	(−1.2)	

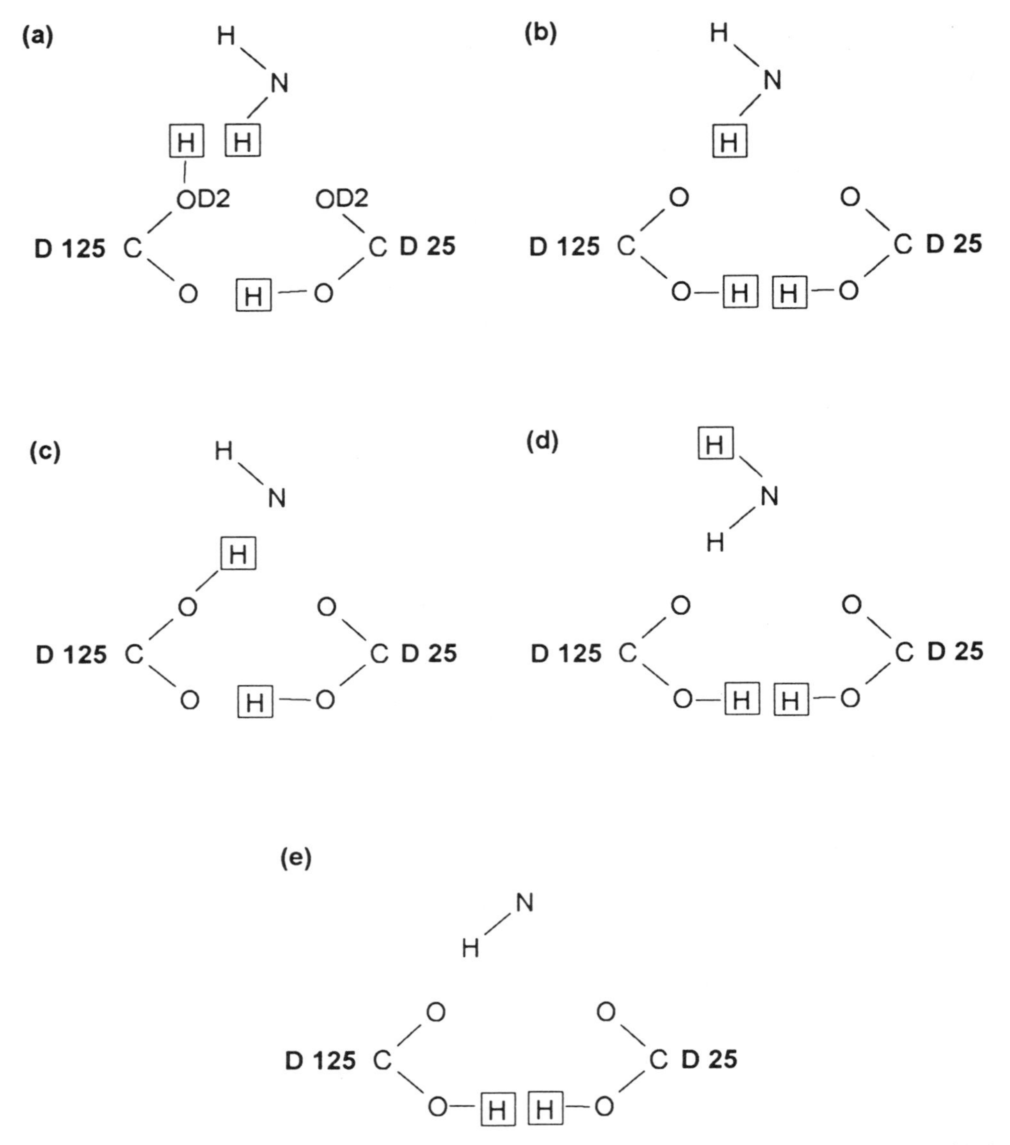

Figure 16. Models of the protonation states of the catalytic region of the HIV-1 protease complexed with the MVT-101 inhibitor. Boxes denote titrated hydrogen atoms.

electric charges of the inhibitor creates its own electrostatic field, and may also influence the pK_a's of the ionizable aspartates and *vice versa*.

A considerable progress in the prediction of ionization constants of titratable residues in proteins, has been made recently through application of the Poisson-Boltzmann model for electrostatic interactions[38–43]. One of these methods[42] has been applied in this study to calculate the pK_a's of both ionizable groups of the catalytic aspartates and the N-H group of the reduced link (-[-NH-CH_2-]-) in the complex of the HIV-1 PR and the MVT-101 inhibitor. Results for different models of the protonation states are summarized in Table II.

Only three protonation sites of the aspartates were studied because protonation of the OD2 oxygen of Asp-25 would result in a very short Van der Waals contact (1.2A) between the proton and the hydrogen of the CH_2 group from Nle-203 of the inhibitor. These results indicate that, in this complex, Asp-25 is protonated on the inner OD1 oxygen, while Asp-125 is deprotonated. The latter forms a hydrogen bond with the amide group of the inhibitor. Presented results indicate equal probability of charged and neutral forms of this group. More detailed analysis of energetic and entropic factors, involved in protonation equilibria, is under investigation. Formation of a hydrogen bond between Asp-25 and Asp-125 is in disagreement with some theoretical results, but in accord with others.[33] Crystallographic data indicate hydrogen bond formation between these groups because of the quite short distance between both inner oxygens (2.68A). Protonation of the inhibitor in the complex was suggested also by MD simulation.[15] There are no direct experimental data on this point.

6. CONCLUSIONS

Because of the AIDS pandemia, HIV-1 protease has become, in the past decade, probably the most studied object in the whole area of rational drug design. From the point of view of this branch of scientific activity the protease is a quite suitable molecule. It is a dimer, but not a large one, some parts of the enzyme are movable, its native structure and structure of its complexes depends on two main class of biological interactions: hydrogen bonds and hydrophobic interactions. Protonation states of some residues may not be standard. All the features pose a challenge for experimental, as well as computer-aided rational, drug design methods and influence significantly their progress.

The above discussion of inhibitors of the HIV-1 protease has not, of course, covered all aspects of the problem. Four points should be noted. Firstly, large attractive energy/free energy of interaction between inhibitor and enzyme does not denote high association constants for complex formation because it may not correspond to the change of free energy. Secondly, being a good inhibitor *in vitro* does not denote good antiviral activity because of the possible pharmacokinetic problems (transfer through the lipid membranes and cellular metabolism). Thirdly, possible toxic effects are not taken into account. Finally, as in the case of HIV-1 virus, lack of a proof-reading mechamism in the reverse transcriptase results in development of resistance to applied inhibitors[24] which may create serious clinical problems.

ACKNOWLEDGMENTS

Authors are indepted to Prof. D. Shugar for critical review of the manuscript. This study was partially supported by the Polish State Committee for Scientific Research (8 T11F 006 09). M.G. wishes to thank the Biosym/MSI Company for their grant. All drawings were printed out from Insight II molecular modeling system. The calculations were partially performed in the ICM at Warsaw University.

REFERENCES

1. Reich, E., Rifkin, D.B., and Shaw, E., eds. (1975), *'Proteases and Biological Control'*, Cold Spring Harbor Lab., Cold Spring Harbor, NY.

2. Dunn, B.M., ed. (1991), *'Structure and Function of the Aspartic Proteinases'*, Plenum, New York.
3. Kruslich, H.G., Oroszlan, S., and Wimmer, E. (1989), *'Viral Proteinases as Targets for Chemotherapy'*, Cold Spring Harbor Lab., Cold Spring Harbor, NY.
4. Gonda, M.A., Wong-Staal, F., Gallo, R.C., Clements, J.E., Narayan, O., and Gilden, R.V. (1985), *Science* 227, 173.
5. Tomasselli, A.G., Howe, W.J., Sawyer, T.K., Wlodawer, A., Heinrikson, R.L. (1991), *Chim. Oggi.*, 9: 6–27.
6. Wlodawer, A., and Erickson, J.W. (1993), *Annu. Rev. Biochem.*, 62: 543–585.
7. Wlodawer, A., Miller, M., Jaskolski, M., Sathyanarayana, B.K., Baldwin, E., Weber, I., Selk, L.M., Clawson, L., Schneider, J., and Kent, S.B.H. (1989), *Science*, 245: 616–621.
8. Sharp, K.A., Nicholls, A., Friedman, R., Honig, B. (1991), *Biochemistry*, 30, 9686–9697.
9. Darke, P.L. (1994), in *Methods in Enzymology*, vol.24, p.104–127, Academic Press.
10. Venable, R.M., Brooks, B.R., Carson, F.W. (1993), *Proteins: Struct., Funct., Genet.*, 15, 374–384.
11. Babe, L.M., Pichuantes, S., Craik, C.S. (1991), *Biochemistry* 30, 106.
12. Weber, I.T. (1990), *J.Biol.Chem.*, 265, 10492–10496.
13. Zhang, Z.Y., Poorman, R.A., Maggiora, L.L., Heinrikson, R.L., and Kezdy, F.J. (1991), *J.Biol.Chem.*, 266, 15591–15594.
14. Pearlman, D.A., Case, D.A., Caldwell, J., Singh, U.C., Weiner, P.K., and Kollman, P.A. (1992), 'Amber 4.0', University of California, San Francisco.
15. Geller, M., Miller, M., Swanson, S.M., Maizel J., *Proteins: Struct. Funct. Genet.*, (in press).
16. Hartre, W.E.Jr., Swaminathan, S., Beveridge, D.L., (1992), *Proteins: Struct. Funct. Genet.*, 13,175.
17. Miller, M., Schneider, J., Sathyanarayana, B.K., Toth, M.V., Marshall, G.R., Clawson, L., Selk, L., Kent, S.B.H., and Wlodawer, A. (1989), *Science*, 246, 1149–1152.
18. Harte, W.E.Jr., Swaminathan, S., Mansuri, M.M., Martin, J.C., Rosenberg, I.E., and Beveridge, D.L. (1990), *Proc. Natl. Acad. Sci. U.S.A.*,87, 8864.
19. Swiminathan, S., Harte, W.E., Beveridge, D.L. (1991), *A. Am. Chem. Soc.*, 113, 2717.
20. Tomasselli, A.G., and Heinrikson, R.L. (1994), in *'Methods in Enzymology'*, vol.241, p.279–301, Academic Press.
21. Dunn, B.M., Gustchina, A., Wlodawer, A., Kay, J. (1994), in *'Methods in Enzymology'*, vol.241, p.254–285, Academic Press.
22. Tomasselli, A.G., Hui J.O., Adams, L., Chosay, J., Lowery, D., Greenberg, B., Yem, A., Deibel, M.R., Zurcher-Neely, H., Heinrikson, R.L. (1991), *J. Biol. Chem.*, 266, 14548.
23. Wlodawer, A., Erickson, J.W. (1993), *Annu. Rev. Biochem.*, 62: 543–585.
24. De Clerck, E. (1995), *J. Med. Chem.*, 38, 2491–2517.
25. Ringe, D. (1994), in *'Methods in Enzymology'*, vol.241, p.157–177, Academic Press.
26. Vacca, J.P. (1994), in *'Methods in Enzymology'*, vol.241, p.311–334, Academic Press.
27. Dutta, A.S. (1993), in *'Pharmacological Library'*, vol.19, ed. Timmerman, p.482–523, Elsevier, Amsterdam-London-New York-Tokyo.
28. Miller M.D., Sheridan, R.P., Kearsley, S.K., and Underwood D.J. (1994), in *'Methods in Enzymology'*, vol.241, p.354, Academic Press.
29. Van Gunsteren, W.F., King, P.M., Mark, A.E. (1994), *Rev. Biophys.*, 27, 435
30. McCarrick, M.A., and Kollman, P. (1994), in *'Methods in Enzymology'*, vol.241, p.370, Academic Press.
31. Kempf, D.J. (1994) in *'Methods in Enzymology'*, vol.241, p.334–354, Academic Press.
32. Hyland, L.J., Tomaszek, T.A.Jr., Meek, T.D. (1991), *Biochemistry*, 30, 8454–8463.
33. Harte, W.E., Jr., Beveridge, D.L. (1994), in *'Methods in Enzymology'*, 241, 178–195, Academic Press.
34. Chatfield, D.C., Brooks, B.R. (1995), *J. Am. Chem. Soc.*, 117, 5561–5572.
35. Ferguson, D.M., Radmer, R.J., Kollman, P.A. (1991), *J. Med. Chem.*, 34, 2654.
36. Tropsha, A., Hermans, J. (1992), *Protein Eng.*, 5, 29.
37. InsightII (version 95.0), Biosym Technologies Inc.
38. Gilson, M. K., Honig, B. (1987), *Nature*, 330, 84–86.
39. Bashford, D., Karplus, M. (1990), *Biochemistry*, 29, 10219–10225.
40. Yang, A.S., Gunner, M.R., Sampogna, R., Sharp, K., Honig, B. (1993), *Proteins: Structure, Function and Genetics*, 15, 252–265.
41. Antosiewicz, J., McCammon, J.A., Gilson, M.K. (1994), *J. Mol. Biol.* 238, 415–436.
42. Antosiewicz, J., Briggs, J.M., Elcock, A.H., Gilson, M.K., McCammon, J.A. (1996), *J. Comp. Chem.*, (in press).
43. Gilson, M. K. (1993), *Proteins: Structure, Function and Genetics*, 15, 266–282.

19

DENSITY FUNCTIONAL AND NEURAL NETWORK ANALYSIS

Hydration Effects and Spectroscopic and Structural Correlations in Small Peptides and Amino Acids

K. J. Jalkanen,[1] S. Suhai,[2] and H. Bohr[3]

[1]Calvin College
Department of Chemistry
Grand Rapids, MI
[2]Department of Molecular Biophysics
German Cancer Research Center
Im Neuenheimer Feld 280, D-69121 Heidelberg, Germany
[3]Center for Biological Sequence Analysis
The Technical University of Denmark
DK-2800 Lyngby, Denmark

ABSTRACT

Density functional theory (DFT) calculations have been carried out for hydrated L-alanine, L-alanyl-L-alanine and N-acetyl L-alanine N'-methylamide and examined with respect to the effect of water on the structure, the vibrational frequencies, vibrational absorption (VA) and vibrational circular dichroism (VCD) intensities. The large changes due to hydration on the structures, relative stability of conformers, and in the VA and VCD spectra observed experimentally are reproduced by the DFT calculations. Furthermore a neural network was constructed for reproducing the inverse scattering data (infer the structural coordinates from spectroscopic data) that the DFT method could produce. Finally the neural network performances are used to monitor a sensitivity or dependence analysis of the importance of secondary structures.

1. PERSPECTIVE

Hydration is an important issue in genome research as exemplified by the structural change as one lowers the relative humidity of DNA below 75%, B-DNA converts to A-DNA. The phosphate groups in the A-helix bind fewer waters than do the phosphate groups

Theoretical and Computational Methods in Genome Research, edited by Suhai
Plenum Press, New York, 1997

in the B-helix, hence dehydration favors the B form of DNA. The effect of hydration on the binding of proteins to DNA and RNA is still not well understood and most modeling of the interaction of proteins with DNA and RNA does not treat the waters explicitly. In this work we have not tried to treat the binding of the proteins with DNA and RNA but to study the effect of hydration on the structural and spectroscopic changes in small biomolecules that function as model systems for DNA and proteins with hydration phenomena similar to the effect of hydration on the forms of DNA. Once the effect of hydration is understood at the molecular level for small peptides and later for proteins we can go on try to understand the effect of hydration on the binding and recognition process in protein-DNA/RNA complexes, and hence to understand at a molecular level the molecular biological processes and how they are mediated in aqueous solution and then in the cell. Many of the current models treat hydration macroscopically and do not include the structural and electronic effects due to the solvent microscopically or quantum mechanically. Our work here is an attempt to document the hydration effect in proteins at a microscopic level with the hope of pointing out some of the deficiencies in the current models and provide some directions and insights into possible improvements.

The effect of hydration on small peptides and amino acids is, in spite of their limited size, still an ubiquitous problem, hard to calculate, measure and to understand. Here we present some DFT calculations on hydrated L-alanine, L-alanyl-L-alanine and N-acetyl-L-alanine N'-methylamide (NALANMA) which will shed some light on the effect of water on the structures, vibrational frequencies, VA and VCD intensities. We also constructed an artificial neural network to solve the inverse scattering problem of retrieving structural information of the biomolecule from spectroscopic data, that is, vibrational frequencies, VA and VCD intensities of isolated NALANMA.

We take two routes to get from the spectroscopic data to predict the structure of our test molecule. One route is to use DFT to calculate all the possible structures and for all of the structures the corresponding frequencies, VA and VCD intensities and then compare to the experimental data. The other is to train the neural networks on a large combination of calculated correlations to infer or extrapolate new results. When going to large biomolecules one can determine whether there is a correlation between the best predicted structural details from spectroscopic data and the data connected to secondary structure stability. This is in order to see which spectroscopic data are the most important determiners of secondary structures, so that such information can be used to predict secondary structures.

The larger goal is to utilize neural networks for determining the structural minima. At these minima VA and VCD intensities are calculated by DFT in order to produce training data, that is, sets of spectroscopic data correlated with $\phi - \psi$ angles, for the network. Other methods such as X-ray crystallography and NMR have only been utilized to determine the native states of proteins. VA and VCD spectroscopy provide the possibility of determining the denatured states of proteins. The problem is to know the structures and VA and VCD spectra of the denatured states of proteins. Keiderling and coworkers have utilized neural network methodology to find correlations of VCD spectra with the native states of proteins by utilizing the known NMR and X-ray crystallographic structures. Our work compliments their work in providing correlations of VA and VCD spectra and the higher energy denatured states of peptides and proteins. These denatured states can be produced under various experimental conditions, that is, in aqueous solution under a variety of conditions, for example, under various pHs, salt conditions and by the presence of urea and other denaturing conditions.

In that sense one should be able to predict higher level intermediate energy states during folding processes of biomolecules with the help of neural networks, once they are trained on known sets of intermediate energy states, that is, the set of structural data we present to the network in this paper. The great thing about utilizing neural network techniques for the inverse scattering problem of deriving structural information from scattering data is that it goes hand in hand with experiments and DFT calculations in the sense that one of the tools can support the other when it fails. The big endeavor is to produce detailed structural data of proteins from VA and VCD spectra and DFT calculations of small peptides that constitute the whole protein. This can be achieved with the neural networks determining all the dihedral angles in the protein from the VA and VCD data for each subunit.

2. DENSITY FUNCTIONAL ANALYSIS OF HYDRATION EFFECTS ON SMALL PEPTIDES AND AMINO ACIDS

2.1. Introduction

DFT is an *ab initio* method which allows one to calculate the structures and properties of single isolated molecules and various aggregated complexes.[1] It is more complete than Hartree Fock methodology in that it includes electron correlation and is better than other correlated methods because it requires less computational resources (disk, memory and cpu time) than otherwise, for the same computational complexity. DFT formulates the energy of the system in terms of the electron density and electron current density instead of the wave function. The trade off is that one must determine electron density and current density functionals. Various local and nonlocal density functionals and local current density functionals have been determined and tested in their ability to determine structures, relative energies, binding energies, dipole moments, polarizabilities, hyperpolarizabilities, dipole moment derivatives, NMR shielding tensors and various other molecular and aggregate properties.[2] The first Hohenberg-Kohn theorem showed that the electron density determines the energy and hence reformulated the basic equation to solve as one in which one has to determine the electron density rather than the wave function. The energy function can be written as

$$E_v[\rho] = T[\rho] + V_{ne}[\rho] + V_{ee}[\rho] \quad (1)$$

$$= \int \rho(\vec{r})v(\vec{r})d\vec{r} + F_{HK}[\rho] \quad (2)$$

where

$$F_{HK}[\rho] = T[\rho] + V_{ee}[\rho] \quad (3)$$

and $v(\vec{r})$ is the external potential. The second Hohenberg-Kohn theorem provides the energy variational principle which enables one to find the density that minimizes this energy functional. The problem is we do not know the functional $F_{HK}[\rho]$ and many functionals have been developed which try to address this problem.

Recently a hybrid functional, the Becke 3LYP functional,

$$E[\rho(x,y,z), \nabla(\rho(x,y,z))] = E_{RB} + E_{HF} - E_{LYP} \quad (4)$$

has been implemented in Cambridge Analytical Derivatives Package (CADPAC) and Gaussian 94. Becke 3LYP level analytical Hessian and atomic polar tensors (APT) calculations have also been implemented in Gaussian 94 and CADPAC. These Becke 3LYP level force fields have been shown to be more accurate than restricted Hartree fock (RHF) level Hessians which must be scaled to get good agreement with both experimental frequencies and VA and VCD intensities.[3] The nature of the normal modes have been shown to depend very much on the scaling scheme one chooses to scale the Hessian. The advantage of the Becke 3LYP level of theory is that the Hessians appear to be accurate enough to predict the VA and VCD intensities when coupled with accurate APT and distributed origin (DO) gauge atomic axial tensors (AAT) without scaling. The number of molecules for which the Becke 3LYP Hessians have been calculated and then with calculated Becke 3LYP APT and DO gauge RHF AAT the VA and VCD spectra predicted has been quite limited. The good agreement shown to date has included only a small number of functional groups and the comparison has been with measurements of the VA and VCD spectra of molecules in non-polar solvents.

In this work we present some preliminary results on some small peptides (L-alanyl-L-alanine and N-Acetyl-L-alanine N'-methylamide) and an amino acid (L-alanine). L-alanine is a clear example of a biological molecule whose properties and structure in water and gas phase are very different and without the water being explicitly or implicitly accounted for, the gas phase or isolated molecule calculations are of questionable use in analyzing the measurements on this molecule in aqueous solution. The zwitterionic structure of L-alanine, the predominant species in aqueous solution, is not stable in the gas phase according to our 6-31G* Becke 3LYP *ab initio* calculations. Therefore we have explicitly included solvent, in this case water, to stabilize this structure and also to be able to compare to the measurements of the VA and VCD spectra of L-alanine in aqueous solution. The VA and VCD spectra of L-alanine in aqueous solution have been measured by the groups of M. Diem at Hunter College in New York[4] and L. Nafie at Syracuse University in Syracuse, NY.[5] Recently some Raman and ROA measurements and calculations on L-alanine have also been reported by Barron and coworkers.[6] At our *ab initio* 6-31G* Becke 3LYP DFT optimized geometries we have calculated the vibrational frequencies, VA and VCD intensities. The VA and VCD spectra were calculated with the 6-31G* Becke 3LYP force field (Hessian) and APT and the 6-31G** RHF DO gauge AAT.

We also present the relative energies of 18 structures of the neutral species of L-alanyl-L-alanine and three zwitterionic structures of L-alanyl-L-alanine which we were able to stabilize with 4, 6, and 7 water molecules respectively. The zwitterionic structure was not stable relative to proton transfer of one of the protons of the NH_3^+ group to the C=O group of the adjacent amide group while the proton of the N–H group was concurrently transferred to the closest oxygen of the CO_2^- group. The goal here is to determine the structure of the zwitterion of L-alanyl-L-alanine in aqueous solution. As shown by our calculations on the neutral species there are many possible conformers for this molecule and the structure of this molecule in water has not yet been determined. We hope our calculations can help to answer this question.

We also present here optimized structures of NALANMA with four waters starting from our 6-31G* Becke 3LYP optimized structures. The relative energies of these complexes are compared with the isolated molecule values.

The goal here has been to model biomolecules by explicitly adding water molecules to provide calculations which can be use to critically evaluate solvent models and specific models developed for water. The H-bonding properties as exemplified by some of the simple water models are clearing wrong and we feel that conclusions based on these models can be critically evaluated utilizing the better models of water and high level calculations like these where the water molecules have been explicitly included.

Various models have been developed for implicitly and explicitly taking into account water at various levels.[7] At the molecular mechanics level, the force field can be parameterized against experimental data measured on the molecule in the aqueous solution. The force field is then an effective force field in that it is not then useful for doing calculations on the molecule in the gas phase or for other solvents. Another approach to the solvent and/or hydration problem is to use a distance dependent dielectric or some other perturbation to the gas phase potential energy surface and/or interactions to take into account the effect of water without actually adding explicit waters. The goal here is to save computational time and resources because the addition of explicit water molecules adds to the length of the calculation and the multiple minimum questions become exponentially worse when one tries to find the "global minimum" of the molecule solvated by water molecules.

The hydrated structures presented here for L-Alanine, L-alanyl-L-alanine and NA-LANMA can be used to test the various water models before one uses them in expensive molecular dynamic simulations on proteins and nucleic acids. The work is a part of our collaborative work at the German Cancer Research Center and the Technical University of Denmark to model proteins and nucleic acids along with various ligands in the presence of water.

2.2. Methods for Density Functional and Vibrational Calculations

Vibrational absorption and circular dichroism spectra are related to molecular dipole and rotational strengths via,

$$\begin{aligned} \epsilon(\bar{\nu}) &= \frac{8\pi^3 N_A}{3000hc(2.303)} \sum_i \bar{\nu} D_i f_i(\bar{\nu}_i, \bar{\nu}) \\ \Delta\epsilon(\bar{\nu}) &= \frac{32\pi^3 N}{3000hc(2.303)} \sum_i \bar{\nu} R_i f_i(\bar{\nu}_i, \bar{\nu}) \end{aligned} \tag{5}$$

where ϵ and $\Delta\epsilon = \epsilon_L - \epsilon_R$ are molar extinction and differential extinction coefficients respectively, D_i and R_i are the dipole and rotational strengths of the ith transition of wavenumbers $\bar{\nu}_i$ in cm^{-1}, and $f(\bar{\nu}_i, \bar{\nu})$ is a normalized line-shape function and N_A is Avogadro's number. For a fundamental $(0 \rightarrow 1)$ transition involving the ith normal mode within the harmonic approximation

$$\begin{aligned} D_i &= \left(\frac{\hbar}{2\omega_i}\right) \sum_\beta \left\{ \sum_{\lambda\alpha} S_{\lambda\alpha,i} P^\lambda_{\alpha\beta} \right\} \left\{ \sum_{\lambda'\alpha'} S_{\lambda'\alpha',i} P^{\lambda'}_{\alpha'\beta} \right\} \\ R_i &= \hbar^2 \Im \sum_\beta \left\{ \sum_{\lambda\alpha} S_{\lambda\alpha,i} P^\lambda_{\alpha\beta} \right\} \left\{ \sum_{\lambda'\alpha'} S_{\lambda'\alpha',i} M^{\lambda'}_{\alpha'\beta} \right\} \end{aligned} \tag{6}$$

where $\hbar\omega_i$ is the energy of the ith normal mode, the $S_{\lambda\alpha,i}$ matrix interrelates normal coordinates Q_i to Cartesian displacement coordinates $X_{\lambda\alpha}$, where λ specifies a nucleus and $\alpha = x, y$, or z:

$$X_{\lambda\alpha} = \sum_i S_{\lambda\alpha,i} Q_i \tag{7}$$

$P^{\lambda}_{\alpha\beta}$ and $M^{\lambda}_{\alpha\beta}$ $(\alpha, \beta = x, y, z)$ are the APT and AAT of nucleus λ. $P^{\lambda}_{\alpha\beta}$ is defined by

$$P^{\lambda}_{\alpha\beta} = \left\{ \frac{\partial}{\partial X_{\lambda\alpha}} \left\langle \psi_G(\vec{R}) | (\mu_{el})_\beta) | \psi_G(\vec{R}) \right\rangle \right\}_{\vec{R}_o}$$

$$= 2 \left\langle \left(\frac{\partial \psi_G(\vec{R})}{\partial X_{\lambda\alpha}} \right)_{\vec{R}_o} \middle| (\mu^e_{el})_\beta \middle| \psi_G(\vec{R}_o) \right\rangle + Z_\lambda e \delta_{\alpha\beta} \tag{8}$$

where $\psi_G(\vec{R})$ is the electronic wavefunction of the ground state G, $\vec{R}$ specifies nuclear coordinates, $\vec{R}_o$ specifies the equilibrium geometry, $\vec{\mu}_{el}$ is the electric dipole moment operator, $\vec{\mu}^e_{el} = -e \sum_i \vec{r}_i$ is the electronic contribution to $\vec{\mu}_{el}$ and $Z_\lambda e$ is the charge on nucleus λ and $M^{\lambda}_{\alpha\beta}$ is given by

$$M^{\lambda}_{\alpha\beta} = I^{\lambda}_{\alpha\beta} + \frac{i}{4\hbar c} \sum_\gamma \epsilon_{\alpha\beta\gamma} R^o_{\lambda\gamma} (Z_\lambda e)$$

$$= \left(\frac{\partial \vec{\mu}_{mag}(\vec{R}, \dot{\vec{R}})}{\partial \dot{X}_{\lambda\alpha}} \right)_{\dot{\vec{R}}=0, \vec{R}_o}$$

$$I^{\lambda}_{\alpha\beta} = \left\langle \left(\frac{\partial \psi_G(\vec{R})}{\partial X_{\lambda\alpha}} \right)_{\vec{R}_o} \middle| \left(\frac{\partial \psi_G(\vec{R}_o, B_\beta)}{\partial B_\beta} \right)_{B_\beta = 0} \right\rangle \tag{9}$$

where $\psi_G(\vec{R}_o, B_\beta)$ is the ground state electronic wavefunction in the equilibrium structure $\vec{R}_o$ in the presence of the perturbation $-(\mu^e_{mag})_\beta B_\beta$, where $\vec{\mu}^e_{mag} = -[\frac{e}{2mc}] \sum_i (\vec{r}_i \wedge \vec{p}_i)$ is the electronic contribution to the magnetic dipole moment operator, $\vec{\mu}_{mag}$. $M^{\lambda}_{\alpha\beta}$ is origin dependent. Its origin dependence is given by

$$(M^{\lambda}_{\alpha\beta})^0 = (M^{\lambda}_{\alpha\beta})^{0'} + \frac{i}{4\hbar c} \sum_{\gamma\delta} \epsilon_{\beta\gamma\delta} Y^{\lambda}_\gamma P^{\lambda}_{\delta\alpha} \tag{10}$$

where $\vec{Y}^\lambda$ is the vector from O to O′ for the tensor of nucleus λ. Equation (10) permits alternative gauges in the calculation of the set of $(M^{\lambda}_{\alpha\beta})^0$ tensors. If $\vec{Y}^\lambda = 0$, and hence O = O′, for all λ the gauge is termed the Common Origin (CO) gauge. If $\vec{Y}^\lambda = \vec{R}^o_\lambda$, so that in the calculation of $(M^{\lambda}_{\alpha\beta})^0$ O′ is placed at the equilibrium position of nucleus λ, the gauge is termed the DO gauge.[8,9]

The problem with the expressions for the APT and AAT presented so far is that they involve the wavefunction, and wavefunction derivatives, (usually calculated with Coupled Hartree Fock theory). These expressions have been implemented in CADPAC to calculate the APT at the SCF and MP2 levels and the AAT at the SCF level. The advantages of DFT are numerous but the main one for us is to extend rigorous methodology to the calculation of properties of large biological molecules. DFT seems to provide a way to do that. But the expressions must be reformulated as expressions amenable for DFT.

We are currently implementing the calculation of the DFT atomic axial tensors in CADPAC. But for this work the SCF AAT are of sufficient accuracy. The rotational strengths of this molecule with DFT AAT will be presented in a following paper along with the formalism of DFT AATs. Also Frisch and coworkers have reported some preliminary work on their implementation of AAT into a developmental version of G94. As this program has not yet been made available to the public we have been unable to compare our results with theirs. But we look forward to being able to compare our DFT implementation of AAT to theirs in the near future.

Table 1. N-acetyl-L-alanine N'-methylamide conformational energies

Conformer	ϕ^a	ψ^a	Energy[a] kcal/mole	ϕ^b	ψ^b	Energy[b] kcal/mole
C_7^{eq}	−82.	72.	0.000	−94.	128.	0.000
C_5^{ext}	−157.	165.	1.433	−94.	128.	0.000
C_7^{ax}	74.	−60.	2.612	59.	−122.	4.132
β_2	−136.	23.	3.181	−151.	116.	1.886
α_L	68.	25.	5.817	61.	52.	2.754
α_R	−60.	−40.	5.652	−82.	−44.	2.465
α_D	57.	−133.	6.467	67.	−111.	3.715
α'	−169.	−38.	6.853	−153.	−92.	15.140
Cryr	−84.	155.		−98.	112.	5.864

[a]Isolated NALANMA, 6-31G* Becke 3LYP relative energies.
[b]NALANMA with 4 bound waters, 6-31G* Becke 3LYP relative energies.

2.3. Results for the DFT Calculations

In Table 1 we present the relative energies of isolated NALANMA and with 4 bound water molecules. The values of ϕ and ψ (measures of secondary structure in proteins) are also given. The starting structures for the bound water optimizations were the 6-31G* Becke 3LYP optimized geometries. To each of these structures four waters molecules were added by the Insight program (Biosym Technologies, San Diego, CA USA). The details of these calculations and the VA and VCD spectra for this molecule will be presented in a future publication. Here we present just some preliminary results.

As one can see the structures and energetics of this molecule are greatly affected by the water, consistent with large changes in the VCD spectra of this molecule when one changes solvent from carbon tetrachloride to water. Note also that the C_7^{eq} and C_5^{ext} conformers both converge to the same structure, which is the lowest energy structure of NALANMA with 4 bound waters found by us to date.

In Table 2 we present our results for the neutral species of L-alanyl-L-alanine. To date we have found 18 structures. We give their relative energies and backbone torsional angles in Table 2. In Table 3 we present the structures of our three zwitterionic structures stabilized by 4, 6 and 7 waters respectively determined to date. We also present two zwitterionic structures stabilized by the Onsager model implemented in Gaussian 94 with and without explicit waters. The atom numbering in Table 3 for the zwitterionic structures is as in Figure 1. As one can see the backbone angles of the zwitterion structures vary greatly with the number of waters explicitly added and the solvent model gives a backbone structure which is quite different from that found when the explicit waters are present. Clearly the requirement to treat the solvent either explicitly or implicitly has complicated the problem of determining the minimum energy structure of L-alanyl-L-alanine in aqueous solution.

In Tables 4a and 4b we present the vibrational frequencies, VA and VCD intensities for our three zwitterionic structures stabilized by explicit waters without the solvent model in the mid and far IR regions. We present the VA and VCD spectra for all three zwitterions utilizing both the 6-31G* Becke 3LYP APT calculated with G94 with the waters present and the 6-31G** RHF APT calculated with CADPAC without the waters present. Utilizing these APT we then utilize the 6-31G** RHF DO gauge AAT calculated without the waters present with CADPAC to calculate the VCD spectra. For the VCD spectra in in the mid and far IR we see that there are 12, 4 and 3 sign differences between the rotational strengths

Table 2. L-alanyl-L-alanine conformational energies

Conformer	ϕ' HNCC	ψ NCCN	ω CCNC	ϕ CNCC	ψ' NCCO	δ CCOH	Energy[a] kcal/mole
I	87/−156.	15.	−176.	−75.	61.	−3.	0.000
II	158/−84.	−17.	−171.	−76.	60.	−3.	0.071
III	82/−162.	16.	174.	−158.	171.	179.	1.484
IV	87/−156.	16.	171.	67.	−55.	5.	1.633
V	164/−79.	−18.	180.	−157.	171.	178.	1.657
VI	156/−86.	−16.	176.	66.	−54.	5.	1.844
VII	47/−66.	146.	173.	−158.	173.	179.	3.030
VIII	50/−64.	146.	−176.	−76.	62.	−3.	3.333
IX	83/−161.	18.	176.	−149.	−4.	−179.	3.401
X	67/−56.	−9.	176.	−156.	173.	179.	4.332
XI	51/−64.	142.	171.	68.	−56.	5.	5.087
XII	158/−85.	−11.	−165.	−129.	22.	−4.	5.466
XIII	70/−56.	−8.	174.	67.	−56.	5.	6.367
XIV	68/−55.	−8.	178.	−150.	−5.	−179.	6.946
XV	−37/−155.	−147.	180.	−152.	−5.	−179.	7.403
XVI	69/−55.	−5.	−175.	53.	38.	179.	8.594
XVII	71/−53.	−5.	−167.	59.	−159.	−178.	8.834
XVIII	79/−164.	76.	−148.	62.	17.	−8.	9.279

[a]6-31G* Becke 3LYP relative energies.

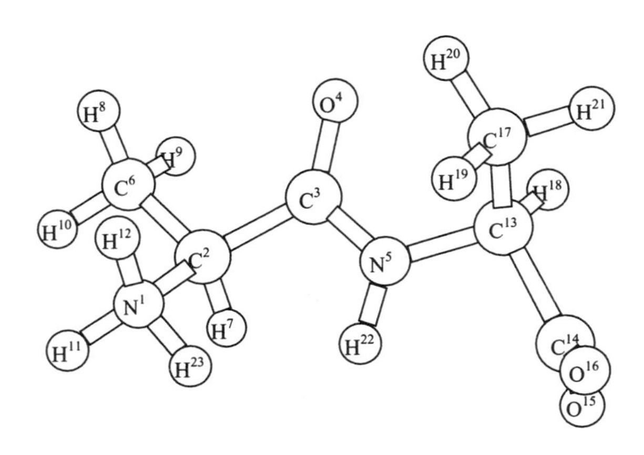

Figure 1. L-alanyl-L-alanine zwitterion atom numbering for Table 3.

calculated with the APT calculated with and without the water present. These preliminary calculations seem to show that one must also include the water molecules in the calculation of the tensors as well as the geometry and force field.

As we have not yet finished our conformational search for the lowest energy zwitterionic structure of L-alanyl-L-alanine the results here only document the feasibility of explicitly adding water to the calculation of the zwitterionic structures of dipeptides and the calculation of the VA and VCD spectra. Our goal here is be able to determine the structure(s) of the zwitterionic state of L-alanyl-L-alanine in aqueous solution. Our preliminary results show that the VCD spectra is very sensitive to the structure of the zwitterion of L-alanyl-L-alanine and one should be able to distinguish which conformer(s) is(are) present if one knew what the VA and VCD spectra for each conformer was. We are pursuing the determination of the other possible structures of the zwitterion of L-alanyl-L-alanine and the VA and VCD spectra thereof and will report the complete results in a future publication.

In Table 5 we present three 6-31G* Becke 3LYP zwitterionic structures for L-Alanine, two in the presence of 4 water molecules and one in the presence of 9 water molecules. We also present the 6-31G** and 6-311++G(2d,2p) optimized geometries for one of the zwitterionic structures. As one can see from the basis set effect, as one adds more polarization functions and diffuse functions the bond length changes are usually less than 0.004 Å. The valence angle changes are usually less than 0.5 degrees.

In Tables 6a and 6b we present the calculated frequencies and VA and VCD intensities for the zwitterionic structures of L-alanine-d_0 We also give the experimental results reported by Diem and coworkers. In Table 6a we also give the calculated vibrational frequencies and VA intensities recently reported by Barron and coworkers.

2.4. Discussion of the DFT Calculations

As seen in Table 1 the explicit addition of water molecules to our structure determination of NALANMA has affected both the relative energies and backbone angles of the gas phase structures. This is consistent with large changes in the measured VCD spectra as one changes the solvent polarity. In Table 2 one sees the relative energies and backbone angles of 18 neutral species of L-alanyl-L-alanine. Even this simple dipeptide gives a clear example of the multiple minimum problem. One is forced with calculating all of the structures of this molecule, the frequencies, dipole strengths and VCD intensities and then comparing to the experimentally measured spectra to determine which structure(s) are present under the experimental conditions that you have made the measurement. When one changes the solvent to water one also has to deal with the problem with different chemical species being present, in the case of L-alanyl-L-alanine and L-alanine, the zwitterionic species rather than the neutral species. In Table 3 one can see that depending on the number of water molecules you have bound to and there to stabilize the zwitterion you get different structures. Also with the Onsager model with and without explicit waters you get different structures. So the ultimate test of the various models is the experimental data, in this case, the vibrational absorption and vibrational circular dichroism spectra. In Table 4 one can see that our predicted vibrational aborption and vibrational circular dichroism spectra for the three zwitterionic structures are different enough that we should be able to distinguish between them in solution. As we have not yet finished all of our zwitterionic structures it would be preliminary to compare with experiment. But here we have documented that the theory should be able to distinguish which of the zwitterionic structure(s) is present under the various experimental conditions. In Table 5 one sees distinguishing features in

Table 3a. L-alanyl-L-alanine zwitterion structures: Bond lengths and dihedral angles

Coordinate	I[a]	I[b]	I[c]	II[d]	III[e]
R(C2N1)	1.5232	1.5090	1.5065	1.5060	1.5086
R(C3C2)	1.5638	1.5532	1.5531	1.5516	1.5460
R(O4C3)	1.2384	1.2348	1.2383	1.2224	1.2311
R(N5C3)	1.3411	1.3437	1.3384	1.3651	1.3524
R(C6C2)	1.5299	1.5216	1.5219	1.5224	1.5288
R(H7C2)	1.0930	1.0955	1.0957	1.0967	1.0917
R(H8C6)	1.0954	1.0929	1.0935	1.0953	1.0944
R(H9C6)	1.0938	1.0928	1.0925	1.0908	1.0943
R(H10C6)	1.0948	1.0950	1.0953	1.0952	1.0964
R(H11N1)	1.0289	1.0359	1.0340	1.0340	1.1524
R(H12N1)	1.0487	1.0566	1.0547	1.0386	1.0227
R(H23N1)	1.0276	1.0381	1.0457	1.0554	1.0462
R(C13N5)	1.4726	1.4754	1.4771	1.4862	1.4731
R(H22N5)	1.0216	1.0191	1.0197	1.0187	1.0168
R(C14C13)	1.5777	1.5590	1.5607	1.5568	1.5590
R(O15C14)	1.2572	1.2773	1.2757	1.2750	1.2814
R(O16C14)	1.2617	1.2465	1.2449	1.2476	1.2457
R(C17C13)	1.5340	1.5265	1.5262	1.5251	1.5284
R(H18C13)	1.0974	1.0951	1.0948	1.0921	1.0898
R(H19C17)	1.0972	1.0936	1.0938	1.0941	1.0955
R(H20C17)	1.0933	1.0947	1.0950	1.0967	1.0950
R(H21C17)	1.0946	1.0944	1.0942	1.0953	1.0947
τ(O4C3C2N1)	−7.35	−123.42	−126.36	−146.63	109.83
ψ(N5C3C2N1)	172.37	57.50	54.50	34.38	−67.77
ω(C13N5C3C2)	178.91	177.48	179.06	162.30	155.23
ϕ(C14C13N5C3)	−169.65	−162.49	−164.92	−135.88	−84.87
ψ(O15C14C13N5)	−2.30	67.31	67.56	85.50	123.81
ψ'(O16C14C13N5)	178.40	−108.61	−107.74	−88.70	−55.18
τ(H8C6C2H7)	178.23	179.03	177.67	175.82	177.19
τ(H9C6C2H7)	−62.17	−61.73	−63.13	−65.66	−62.35
τ(H10C6C2H7)	56.84	57.11	56.08	54.65	56.58
τ(C3C2N1H11)	127.44	−173.41	−174.97	−174.94	−47.16
τ(C3C2N1H12)	8.00	59.80	57.88	57.92	−167.63
τ(C3C2N1H23)	−110.70	−62.10	−63.68	−63.68	73.09
τ(H19C17C13H18)	177.98	−177.54	−176.62	−177.83	180.00
τ(H20C17C13H18)	57.35	60.61	61.26	59.99	58.94
τ(H21C17C13H18)	−63.59	−59.17	−58.66	−59.33	−61.10

[a] Zwitterion stabilized by Onsager solvent model.
[b] Zwitterion stabilized by Onsager solvent model and 7 water molecules.
[c] Zwitterion stabilized by 7 water molecules.
[d] Zwitterion stabilized by 6 water molecules.
[e] Zwitterion stabilized by 4 water molecules.

Table 3b. L-alanyl-L-alanine zwitterion structures: Valence angles

Coordinate	I[a]	I[b]	I[c]	II[d]	III[e]
A(N1C2C3)	104.60	108.46	108.62	111.34	107.58
A(C2C3O4)	117.62	122.01	121.46	119.87	119.28
A(C2C3N5)	118.06	113.69	113.97	115.82	115.75
A(O4C3N5)	124.32	124.30	124.56	124.30	124.92
A(N1C2C6)	110.20	111.07	111.01	110.29	111.26
A(C3C2C6)	113.96	113.57	113.14	111.90	115.40
A(N1C2H7)	106.97	105.17	105.37	105.40	106.62
A(C3C2H7)	110.19	107.92	108.05	106.97	104.87
A(C6C2H7)	110.56	110.27	110.32	110.69	110.57
A(C2C6H8)	111.12	111.80	111.89	110.92	111.81
A(C2C6H9)	109.73	109.25	109.33	109.21	109.84
A(H8C6H9)	108.18	107.80	107.68	107.68	108.46
A(C2C6H10)	111.08	110.38	110.35	110.86	110.70
A(H8H6H10)	108.89	109.32	109.03	108.95	108.11
A(H9H6H10)	107.73	108.18	108.47	109.16	107.80
A(C2N1H11)	113.36	110.71	110.89	110.80	112.75
A(C2N1H12)	104.73	111.06	110.52	110.18	110.16
A(C2N1N23)	113.73	108.64	108.71	109.57	110.58
A(11N1H12)	109.61	113.31	113.79	114.03	107.80
A(11N1H23)	106.50	102.08	101.94	101.56	107.40
A(12N1H23)	108.83	110.64	110.59	110.36	107.98
A(C3N5C13)	123.93	125.32	125.71	122.12	122.25
A(N5C13C14)	109.30	104.48	103.88	103.54	108.17
A(C13C14O15)	116.15	114.42	114.25	115.12	118.40
A(C13C14O16)	115.02	118.54	118.09	116.93	116.18
A(O15C14O16)	128.83	126.90	127.47	127.65	125.41
A(N5C13C17)	111.91	111.91	111.66	111.44	110.49
A(C14C13C17)	110.77	113.70	113.67	113.73	111.73
A(N5C13H18)	108.21	107.24	107.43	106.98	107.05
A(C14C13H18)	108.11	108.86	109.19	109.85	108.40
A(C17C13H18)	108.42	110.31	110.63	110.87	110.84
A(C13C17H19)	110.36	110.28	110.08	110.12	110.64
A(C13C17H20)	111.33	110.40	110.61	110.57	110.92
A(H19C17H20)	108.56	110.07	110.27	110.32	108.97
A(C13C17H21)	108.19	109.90	109.85	110.28	109.84
A(H19C17H21)	108.54	107.52	107.31	107.52	107.81
A(H20C17H21)	108.56	108.61	108.64	107.95	108.58
A(C3N5H22)	124.46	118.18	118.21	117.63	117.85
A(C13N5H22)	111.31	116.38	116.00	114.00	114.98

[a]Zwitterion stabilized by Onsager solvent model.
[b]Zwitterion stabilized by Onsager solvent model and 7 water molecules.
[c]Zwitterion stabilized by 7 water molecules.
[d]Zwitterion stabilized by 6 water molecules.
[e]Zwitterion stabilized by 4 water molecules.

Table 4a. L-alanyl-L-alanine zwitterion VA and VCD spectra

$\nu(cm^{-1})^a$	$A_i{}^b$ III	$R_i{}^c$	$\nu(cm^{-1})^a$	$A_i{}^b$ II	$R_i{}^c$	$\nu(cm^{-1})^a$	$A_i{}^b$ I	$R_i{}^c$
1748.	90.64	−202.35	1793.	107.86	−129.68	1764.	37.59	−100.23
1712.	534.54	−232.77	1769.	282.25	128.79	1742.	441.64	8.05
1701.	197.15	−144.74	1711.	457.65	67.44	1718.	413.77	304.51
1678.	90.92	117.19	1680.	11.09	−35.96	1696.	73.09	−64.26
1625.	1364.87	894.99	1615.	52.23	5.96	1635.	27.49	30.81
1588.	463.42	90.04	1586.	131.26	42.89	1625.	162.49	−67.03
1532.	11.49	7.16	1526.	18.52	11.21	1530.	3.88	12.57
1527.	13.45	−33.48	1524.	6.33	−9.44	1523.	10.48	13.03
1522.	1.59	3.89	1518.	5.85	−2.15	1523.	5.72	−4.90
1520.	7.83	−0.56	1516.	10.56	15.56	1516.	15.82	3.13
1447.	7.38	8.84	1440.	26.35	3.09	1453.	25.63	1.29
1439.	32.36	10.00	1432.	112.75	28.18	1431.	102.61	11.16
1416.	9.56	−49.11	1418.	88.66	42.87	1419.	45.70	51.86
1395.	177.57	−152.64	1414.	46.27	−103.00	1409.	67.20	97.46
1372.	14.14	57.26	1393.	11.95	83.22	1404.	88.05	−201.64
1359.	3.53	0.92	1384.	34.82	−89.21	1372.	12.27	21.19
1329.	25.12	−76.54	1303.	88.62	−140.60	1334.	83.33	−69.71
1309.	21.92	73.74	1275.	10.25	−25.96	1292.	68.84	22.65
1240.	177.71	−73.50	1266.	134.80	131.04	1287.	15.75	18.11
1177.	56.57	−76.81	1203.	30.31	−26.77	1210.	21.24	11.89
1162.	7.82	26.37	1150.	4.71	−34.82	1146.	13.12	1.91
1124.	5.54	−42.03	1137.	13.96	36.27	1141.	8.25	14.75
1110.	33.75	15.87	1112.	16.57	−7.89	1119.	15.40	−9.43
1088.	1.35	−1.63	1070.	5.88	−6.00	1072.	1.98	−9.97
1061.	15.01	−12.32	1049.	24.44	−21.27	1067.	27.22	3.84
1018.	9.14	28.35	1031.	3.59	10.71	1035.	2.63	−11.54
942.	4.94	−17.62	949.	9.11	−38.84	953.	14.15	−28.20
925.	11.94	−24.08	930.	3.20	−5.47	924.	1.78	−8.94
883.	8.12	4.19	883.	8.38	−0.75	883.	8.71	10.84
863.	30.71	−21.61	854.	19.86	6.58	853.	20.83	16.04
800.	40.72	4.39	804.	8.27	27.01	791.	12.67	37.58
797.	8.68	12.85	771.	3.41	16.77	771.	6.78	−28.23
750.	49.21	−9.75	708.	53.83	20.23	726.	39.60	9.79
693.	12.21	10.14	676.	17.62	8.97	665.	18.49	4.71
629.	1.84	−21.16	581.	14.47	−46.19	601.	8.46	−24.13
586.	5.80	−34.30	550.	14.89	−33.20	543.	17.07	3.19
535.	9.53	−29.73	548.	8.96	17.88	531.	7.81	−37.80
505.	16.84	40.15	468.	16.02	39.14	486.	6.19	7.61
394.	37.67	13.16	447.	31.29	51.17	443.	32.06	46.97
385.	62.28	55.27	368.	6.90	13.00	366.	8.57	21.29
342.	7.73	−58.14	361.	6.47	6.92	361.	11.85	14.20
303.	0.36	3.08	325.	7.74	−34.32	311.	4.21	−27.76
276.	25.98	46.00	291.	11.49	49.23	292.	13.35	41.52
253.	3.06	9.40	251.	0.25	−6.17	254.	2.37	−9.71
245.	1.48	−6.81	236.	16.50	24.98	246.	6.49	−11.81
222.	6.41	−34.24	230.	13.88	−58.36	235.	5.07	7.62
193.	24.99	−55.63	189.	13.51	21.87	197.	26.41	0.31
172.	24.98	−16.90	174.	14.19	−8.36	170.	26.67	−53.71
165.	5.27	46.95	134.	4.38	−40.27	157.	6.95	33.04
142.	6.74	21.13	105.	4.36	9.15	124.	5.04	−44.78
125.	2.18	−9.11	83.	3.46	−7.20	109.	0.44	2.06

[a]6-31G* Becke 3LYP frequencies.
[b]6-31G* Becke 3LYP APT.
[c]6-31G* Becke 3LYP APT and 6-31G** RHF DO gauge AAT.

Table 4b. L-alanyl-L-alanine zwitterion VA and VCD spectra

$\nu(cm^{-1})^a$	$A_i{}^b$ III	$R_i{}^c$	$\nu(cm^{-1})^a$	$A_i{}^b$ II	$R_i{}^c$	$\nu(cm^{-1})^a$	$A_i{}^b$ I	$R_i{}^c$
1745.	95.12	−249.30	1793.	172.50	−106.04	1764.	52.15	−39.36
1711.	269.18	182.81	1769.	198.20	116.73	1742.	300.44	3.19
1701.	284.78	−418.71	1711.	641.68	157.82	1718.	599.76	324.44
1678.	187.10	193.42	1680.	67.76	−105.94	1696.	187.86	−109.90
1625.	265.53	388.67	1615.	148.81	10.31	1635.	127.83	65.06
1588.	190.59	94.32	1586.	200.83	70.66	1625.	246.65	−61.06
1532.	2.92	3.80	1526.	19.72	2.27	1530.	1.98	4.76
1527.	23.00	−25.74	1524.	13.84	−15.51	1523.	14.14	6.65
1522.	2.33	−6.22	1518.	7.16	−1.29	1523.	9.94	−11.52
1520.	2.67	3.40	1516.	7.73	6.35	1516.	14.76	3.53
1447.	9.43	15.76	1440.	24.90	2.70	1453.	23.36	2.90
1439.	40.84	11.22	1432.	117.64	24.21	1431.	104.49	10.04
1416.	24.09	−12.95	1418.	113.12	52.53	1419.	44.42	57.15
1395.	232.67	−161.32	1414.	88.42	−132.80	1409.	110.45	78.64
1372.	9.22	−14.27	1393.	11.50	73.31	1404.	101.59	−210.54
1359.	2.81	0.36	1384.	44.91	−109.25	1372.	14.31	21.91
1329.	43.83	−129.21	1303.	107.24	−175.64	1334.	155.88	−122.44
1309.	27.94	74.51	1275.	28.14	−21.82	1292.	60.03	82.10
1240.	15.70	−23.91	1266.	188.86	201.30	1287.	18.46	15.04
1177.	114.51	4.39	1203.	42.40	−69.49	1210.	26.02	−24.61
1162.	15.55	43.66	1150.	4.67	−34.19	1146.	16.95	21.42
1124.	8.32	−40.32	1137.	24.69	49.74	1141.	13.19	17.49
1110.	23.95	−6.83	1112.	26.83	−17.96	1119.	22.80	−18.54
1088.	11.17	11.86	1070.	3.69	−3.71	1072.	2.44	−9.10
1061.	10.17	6.54	1049.	11.13	17.96	1067.	14.73	48.01
1018.	7.54	−9.04	1031.	10.09	−6.41	1035.	10.43	−20.13
942.	8.10	−10.99	949.	14.38	−45.67	953.	18.93	−37.44
925.	8.47	−29.98	930.	7.62	−11.18	924.	9.21	−21.17
883.	8.91	−10.74	883.	13.21	11.28	883.	13.22	38.13
863.	52.91	−33.16	854.	30.93	−5.04	853.	28.92	6.26
800.	49.65	34.95	804.	15.15	30.85	791.	22.82	42.93
797.	14.39	−3.63	771.	6.69	22.72	771.	11.83	−20.59
750.	62.17	−10.17	708.	79.82	21.54	726.	53.53	18.50
693.	14.89	9.25	676.	20.33	0.12	665.	21.44	5.10
629.	1.56	−21.74	581.	12.38	−42.27	601.	16.32	−47.35
586.	3.27	−9.87	550.	3.88	−49.09	543.	19.16	2.85
535.	9.27	−9.73	548.	11.19	19.08	531.	4.79	−48.31
505.	11.96	32.14	468.	11.67	25.32	486.	8.02	19.24
394.	34.00	16.21	447.	9.32	36.02	443.	10.73	31.15
385.	13.66	23.12	368.	4.97	5.29	366.	2.74	8.64
342.	6.62	−49.92	361.	4.56	32.73	361.	12.03	39.00
303.	3.18	13.31	325.	10.25	−37.13	311.	6.43	−33.00
276.	27.88	37.18	291.	19.00	61.69	292.	19.42	46.13
253.	4.52	15.10	251.	0.32	−6.70	254.	3.35	−18.02
245.	1.13	6.52	236.	14.52	25.11	246.	6.28	3.14
222.	10.18	−51.44	230.	14.44	−59.15	235.	7.51	6.94
193.	27.45	−46.99	189.	10.04	1.79	197.	18.45	−6.61
172.	9.89	−15.30	174.	12.99	−8.18	170.	26.80	−42.71
165.	2.71	26.63	134.	4.57	−25.58	157.	8.44	28.75
142.	6.99	18.80	105.	5.51	15.96	124.	3.30	−30.47
125.	2.36	−9.23	83.	6.30	−13.56	109.	0.58	3.01

[a] 6-31G* Becke 3LYP frequencies.
[b] 6-31G** RHF APT.
[c] 6-31G** RHF APT and 6-31G** RHF DO gauge AAT.

Table 5. L-alanine zwitterion structures

Coordinate	I[a] 6-31G*	I[a] 6-31G**	I[a] ++G(2d,2p)	II[a] 6-31G*	III[b] 6-31G*	IV[c] 6-31G*
r(N1–C2)	1.5115	1.5105	1.5129	1.5182	1.5142	1.5123
r(C2–C3)	1.5639	1.5638	1.5593	1.5673	1.5545	1.5516
r(C3–O4)	1.2893	1.2899	1.2854	1.2748	1.2615	1.2647
r(C3–O5)	1.2299	1.2297	1.2211	1.2421	1.2608	1.2601
r(C2–C6)	1.5267	1.5263	1.5229	1.5202	1.5227	1.5220
r(C2–H7)	1.0947	1.0945	1.0889	1.0940	1.0923	1.0925
r(N1–H8)	1.0466	1.0460	1.0344	1.0565	1.0536	1.0484
r(N1–H9)	1.0477	1.0472	1.0346	1.0559	1.0454	1.0464
r(N1–H10)	1.0340	1.0322	1.0250	1.0334	1.0529	1.0583
r(C6–H11)	1.0960	1.0950	1.0907	1.0940	1.0933	1.0938
r(C6–H12)	1.0916	1.0905	1.0866	1.0932	1.0942	1.0940
r(C6–H13)	1.0966	1.0954	1.0913	1.0937	1.0960	1.0955
θ(N1–C2–C3)	111.19	111.28	110.94	105.04	106.82	106.41
θ(C2–N3–O4)	116.14	116.18	115.80	117.31	115.31	115.83
θ(C2–N3–O5)	115.09	115.11	115.72	114.41	117.93	118.10
θ(N1–C2–C6)	109.96	110.05	109.95	110.18	109.95	109.85
θ(C3–C2–C6)	111.60	111.52	112.10	114.17	114.76	114.57
θ(N1–C2–H7)	106.19	106.32	106.13	106.06	105.79	105.78
θ(C3–C2–H7)	107.47	107.25	107.21	109.48	107.73	108.67
θ(C2–N1–H8)	109.05	108.83	109.82	108.64	112.36	111.69
θ(C2–N1–H9)	109.83	109.65	110.22	109.31	108.93	108.99
θ(H8–N1–H9)	106.25	106.27	105.95	106.22	109.40	109.14
θ(C2–N1–H10)	110.63	110.44	111.30	111.72	110.97	110.98
θ(H8–N1–H10)	110.27	110.54	109.88	110.57	109.60	110.36
θ(C2–C6–H11)	110.70	110.62	110.89	110.52	110.93	110.77
θ(C2–C6–H12)	108.27	108.21	108.42	109.44	109.85	109.84
θ(H11–C6–H12)	108.28	108.28	108.21	107.20	108.10	108.01
θ(C2–C6–H13)	111.52	111.52	111.56	111.36	110.91	110.87
θ(H11–C6–H13)	108.81	108.84	108.91	109.69	108.91	109.22
τ(N1–C2–C3–O4)	−0.53	0.08	−3.03	−101.08	94.86	86.84
τ(N1–C2–C3–O5)	−179.57	−179.83	176.59	73.20	−81.69	−89.60
τ(C3–C2–N1–H8)	59.20	58.85	60.96	48.12	80.56	72.44
τ(C3–C2–N1–H9)	−56.85	−57.00	−55.40	−67.39	−40.83	−38.23
τ(C3–C2–N1–H10)	−179.36	−179.64	−177.12	170.36	−156.34	−163.96
τ(H11–C6–C2–H7)	177.17	177.27	177.02	−176.93	178.19	177.51
τ(H12–C6–C2–H7)	−64.29	−64.26	−64.32	−59.09	−62.36	−63.27
τ(H13–C6–C2–H7)	55.87	56.00	55.44	60.89	57.04	56.08

[a]Zwitterions stablized by 4 water molecules.
[b]Zwitterion stabilized by 9 water molecules.
[c]Zwitterion stabilized by 9 water molecules + solvent model.

the 3 different zwitterionic structures of L-Alanine that we here report. As one sees in this table the C–O bond lengths vary by 0.6 Å in one of the zwitterionic structures and in another they are essentially the same. These differences are also manifested in the vibrational frequencies, VA and VCD spectra. Similarly N–H bond lengths vary from structure to structure. Depending on the whether the individual hydrogens are bonded directly or indirectly to an oxygen of the $C–O_2$- group seems to affect the bond length and then also its frequency and VA and VCD intensity. These calculations clearly show that one should be able to distinguish which zwitterionic structure(s) is(are) present, assuming that one has the resolution in the VA and VCD spectra and the mixture is not too complex, that is, not too many structures are present.

In Tables 6a and 6b one sees the differences in the structures as exemplified in the VA and VCD spectra. We also compare to the vibrational frequencies reported by Diem and coworkers for comparison and and the VCD data reported. Again the differences in the VA and VCD spectra combined with the Raman scattering and Raman optical activity should be able to address the question of which of the possible zwitterionic structure(s) of L-alanine are present under the various experimental conditions. This methodology works for smaller systems but it is obvious the number of calculations required for a protein or even a peptide of four or five residues is not tractable. As an alternative to this methodology we have attempted to utilize methods to train on a limited amount of spectroscopic data, that is, vibrational frequencies, VA and VCD intensities for a given secondary structure and then to predict the secondary structure based on a limited set of spectroscopic data. This second part of this article covers this second approach.

3. NEURAL NETWORK ANALYSIS OF SPECTROSCOPIC AND STRUCTURAL CORRELATIONS

3.1. The Inverse Scattering Issue

The inverse scattering problem is concerned with the situation when biomolecules structural properties are to be determined by the information arising from scattered light on the molecules. An inverse problem in an experimental situation is defined by the situation of not having direct structural information about a given object but where this information is provided indirectly by the projections of the object in different planes, e.g., as scattering data in specific directions.

In mathematical terms the inverse scattering problem, given for example in the biomolecular structure measurements mentioned above, can be described by the integral expression:

$$B_i(\vec{r}, t) = \int_V A_j(\vec{r'}, t) \cdot C_{ji}(\vec{r}|\vec{r'}, t) dV' \tag{11}$$

where B_i is the detector signal function localized in r away from a source described by a function A_i localized in r' and integrated over the source volume V. The convolution function C_{ij} is a Green's function matrix. The problem, as it has been formulated here, is mathematically unsolvable and is about determining the source function A from the detector function B. The C matrix contains the detector's projections of the source.[10]

The infrared absorption spectroscopy and optical polarization experiments for determination of the structure of a biomolecule are typical situations of inverse scattering problems.

Table 6a. L-alanine-d_0 zwitterion vibrational frequencies and VA intensities

$\nu(cm^{-1})^a$	$\nu(cm^{-1})^b$ I	$A_i{}^c$	$\nu(cm^{-1})^b$ II	$A_i{}^c$	$\nu(cm^{-1})^b$ III	$A_i{}^c$	$\nu(cm^{-1})^d$	$A_i{}^d$
3080.	3317.	529.5	3327.	497.5	3153.	14.8	3809.	65.3
3060.	3170.	3.6	3155.	12.0	3132.	32.2	3727.	53.9
3020.	3120.	15.9	3132.	19.4	3124.	11.6	3333.	3.7
3003.	3105.	336.6	3099.	10.8	3114.	96.1	3295.	20.7
2993.	3088.	327.4	3057.	29.0	3063.	36.6	3258.	32.2
2962.	3077.	502.0	2956.	683.3	3009.	63.9	3204.	31.9
2949.	3053.	33.9	2940.	691.6	2991.	72.3	2813.	581.0
1645.	1770.	324.3	1777.	77.9	1775.	60.3	1996.	442.0
1625.	1745.	82.9	1745.	181.4	1749.	37.8	1822.	42.1
1607.	1741.	56.8	1719.	221.6	1696.	544.0	1790.	24.6
1498.	1634.	91.1	1647.	78.2	1662.	185.0	1641.	11.9
1459.	1534.	2.2	1525.	4.7	1526.	21.5	1639.	3.9
1459.	1525.	5.2	1513.	6.5	1522.	3.1	1568.	47.0
1410.	1433.	41.3	1444.	44.1	1454.	92.1	1523.	388.0
1375.	1407.	7.0	1415.	5.8	1438.	98.7	1501.	69.9
1351.	1377.	34.6	1405.	63.9	1428.	8.9	1447.	143.0
1301.	1333.	186.1	1342.	157.5	1370.	111.5	1395.	238.0
1220.	1281.	37.5	1297.	14.2	1298.	22.9	1314.	18.0
1145.	1200.	56.0	1199.	20.2	1238.	36.6	1205.	41.8
1110.	1124.	16.9	1115.	13.1	1121.	16.9	1164.	38.7
1001.	1058.	7.9	1075.	7.0	1076.	5.2	1080.	72.9
995.	1026.	7.6	1023.	1.5	1038.	10.1	1060.	22.6
922.	901.	15.3	896.	17.9	921.	17.7	942.	99.2
850.	833.	43.4	840.	28.9	850.	34.5	881.	34.1
775.	771.	9.8	799.	13.4	804.	17.9	838.	16.2
640.	620.	4.2	658.	6.8	672.	8.9	681.	7.9
527.	602.	3.7	639.	1.9	658.	0.7	562.	34.9
477.	521.	16.7	533.	2.3	549.	9.8	413.	3.0
399.	432.	15.7	403.	28.0	400.	11.3	352.	111.0
296.	368.	39.4	352.	45.4	335.	34.9	303.	30.4
283.	287.	8.1	279.	3.7	283.	14.1	275.	17.2
219.	262.	4.2	231.	4.2	232.	3.6	255.	3.3
184.	178.	6.9	170.	3.4	171.	2.4	54.	3.5

[a]M. Diem and coworkers, *J. Am. Chem. Soc.* **104** (1982) 3329.
[b]6-31G* Becke 3LYP frequencies.
[c]6-31G* Becke 3LYP APT.
[d]Reference 7.

Table 6b. L-alanine-d_0 zwitterion vibrational frequencies and VCD intensities

$\nu(cm^{-1})^a$	$10^5(\Delta\epsilon)^a$	$\nu(cm^{-1})^b$ I	$R_i{}^c$	$\nu(cm^{-1})^b$ II	$R_i{}^c$	$\nu(cm^{-1})^b$ III	$R_i{}^c$
3080.		3317.	17.1	3327.	28.0	3153.30	−0.1
3060.		3170.	−1.4	3155.	−2.3	3131.85	−11.5
3020.		3120.	1.7	3132.	−4.1	3123.14	9.2
3003.		3105.	−76.2	3099.	−17.6	3062.90	0.1
2993.		3088.	29.5	3057.	−1.7	2292.70	−1.3
2962.		3078.	15.7	2957.	−55.9	2229.68	0.5
2949.		3053.	−11.6	2940.	21.6	2162.25	1.8
1623.	−1.1	1770.	−53.8	1777.	−77.3	1698.27	42.9
1625.		1745.	−0.6	1746.	162.9	1525.36	−5.4
1607.		1741.	26.4	1719.	−65.2	1522.64	6.4
1498.		1635.	−6.4	1647.	−5.5	1453.01	21.5
1459.		1534.	−4.6	1525.	−2.5	1431.99	−56.5
1459.		1525.	−5.1	1513.	14.9	1406.06	1.3
1410.		1433.	32.3	1444.	26.6	1350.23	32.8
1375.		1407.	19.3	1415.	−35.5	1275.03	8.6
1351.	−27.	1377.	−71.3	1405.	28.6	1252.24	9.6
1301.	16.	1333.	−13.5	1341.	−122.0	1246.32	−55.8
1220.	5.3	1281.	60.1	1298.	18.9	1188.45	−4.1
1145.	−4.2	1200.	−24.8	1198.	37.8	1121.80	−9.8
1110.	−4.7	1124.	−0.7	1115.	12.3	1076.59	10.9
1001.		1058.	0.4	1074.	15.9	940.75	−1.4
995.		1026.	6.6	1022.	−7.1	904.25	−45.6
922.		901.	−23.4	896.	−20.8	900.22	67.0
850.		833.	21.1	839.	16.6	834.87	−25.2
775.		771.	23.6	799.	−5.2	785.71	−8.3
640.		620.	−1.0	658.	−18.2	656.77	0.7
527.		602.	10.1	639.	7.7	539.94	−10.9
477.		521.	−8.5	533.	−7.7	471.04	1.5
399.		432.	14.9	403.	43.7	383.49	16.5
296.		368.	−26.3	352.	51.4	321.03	45.4
283.		287.	39.9	279.	−23.7	280.67	−34.2
219.		262.	−22.9	231.	−25.9	230.78	3.5
184.		179.	−9.3	170.	−35.5	168.87	−14.8

[a]M. Diem, *J. Am. Chem. Soc.* **118** (1988) 6967-6970.
[b]6-31G* Becke 3LYP frequencies.
[c]6-31G* Becke 3LYP APT and 6-31G** RHF DO gauge AAT.

3.2. Methodology

In this section we discuss the application of neural networks to the problem of inverse scattering where the structural information of small biomolecules is predicted from spectroscopic data such as frequency, absorption (dipole strength D) and differential absorption (rotational strength R) data. The structural data are represented in the form of dihedral angles (ϕ and ψ).

In the following we shall first give a description of how to utilize artificial neural network especially with respect to classifying spectroscopic data.

The basic elements of a neural network, the neurons, are processing units that produce output from a characteristic non-linear function of a weighted sum of input data. A neural network is a group of such processing units, the individual members of which can communicate with each other through mutual interconnections between the neurons. The network will gradually acquire a global information processing capacity of classifying data by being exposed (trained) to many pairs of corresponding input and output data such that new output can be generated from new input. If a set of input values is denoted by $\{x_j\}$ and the corresponding output is denoted by $\{y_i\}$ the processing of each neuron i in the net can be described as

$$y_i = f\left(\sum_j W_{ij}x_j + \eta_i\right) \tag{12}$$

where W_{ij} are the weights of the connections leading to the neuron i, η_i and f is the characteristic non-linear function for the neuron. As this equation already tells such type of network can be considered as a non-linear map between the input and output data.

The most straightforward type of neural networks employed for this study were feed forward networks of the multi-layered perceptron type or more complicated recurrent neural networks equivalent to the networks used with real-time recurrent learning (RTRL).[11] The former networks have a unique direction of the data stream such that input are passed through the consecutive layers towards a specific layer of neurons that produce the output while the latter networks have a set of extra feed-back connections. We shall hereafter denote these layers of neurons as, mentioned in the consecutive order, the input layer, the hidden layers and the output layer. The reason for choosing the feed forward network among many other types is due to its renown ability to generalize speech recognition, image processing and molecular biology data[12,13,14,15] and its rather simple structure both with respect to processing of data and training of which the back-propagation error algorithm[16] is the most commonly used and the one we shall employ. The training procedure is performed until a cost function C has reached a local minimum e.g., by a gradient descent. The cost function C is normally written as:

$$C = 1/2\sum_{\alpha,i}(t_i^\alpha - z_i^\alpha)^2 \tag{13}$$

which is simply the squared sum of errors, t_i being the correct target value and z_i the actual value of the output neurons.

It is important when utilizing neural networks to have a few basic facts of common knowledge about the architecture of the network in relation to the training. First of all the network should be dimensioned according to the training set, i.e., the number of adjustable parameters (the synaptic weights and thresholds) should not exceed the number of training examples. There is a heuristic rule that the number of training examples should be around 1.5 times larger than the number of synaptic weights. Basically the ability to learn and recall learned data increases with the size of the hidden layer while the ability to generalize decreases with an increasing number of hidden neurons above a certain limit. This fact can clearly be understood when one considers the network as essentially a curve fitter between points depicting relations between input and output data in the training set. Therefore it is also easy to understand that a network can be over-trained when the training process reaches the point where the spurious data points are memorized. The training process and the construction of the training set is of greatest importance because the predictive power of the network is dependent on how clearly the training set is defined and how many patterns that are exposed. These problems are nicely elucidated in a previous study where neural networks were applied to the task of water binding prediction on proteins.[17]

3.3. Evaluation

In order to evaluate the performance of the network various statistical measures have been proposed. In the case of a dual valued output we shall be using the so-called Mathews coefficient.[18] If we denote the two possible output values by 1 and 2 (e.g., signifying an event or no event) and if p is the number of correctly predicted examples of 1, $\bar{p}$ the number of correctly predicted examples of 0, q the number of examples of 1 incorrectly predicted and $\bar{q}$ is the number of examples of 0 incorrectly predicted then we define the coefficient C_M as:

$$C = \{p\bar{p} - q\bar{q}\}/\{\sqrt{(p+q)(p+\bar{q})(\bar{p}+q)(\bar{p}+\bar{q})}\} \tag{14}$$

For complete coincidence with the correct decisions (ideal performance) the measure is 1 and for complete anti-coincidence C_M is -1. A poor net will give $C = 0$ indicating that it does not capture any correlation in the training set in spite the fact that it might be able to predict several correct values.

3.4. Implementation

The actual neural networks to be used here for the inverse scattering problem of predicting peptide structures can be constructed from a real valued processing neural network system of the feed-forward type. The networks are trained on a large set of corresponding values of spectroscopic and structural data that are produced from extensive density functional calculations of our model peptide system N-Acetyl-L-alanine N'-methylamide.

The input values, the spectroscopic data (ν, D, R), to the network are encoded by real values in the neurons of the input layer. The input numbers are read into a window with 3 numbers (ν, D, R) at a time corresponding to a specific pair of output values (ϕ, ψ). The input values of the frequency will typically range from 40 to 3400 cm^{-1} which will be normalized to the range 1 to 400 and partitioned on 20 neurons so that the first of these 20 input neurons take care of the range 1 to 20, the next neuron of 21 to 40 etc. Values that are just below 20 will cause the first neuron to fire maximally while the other neuron are silent. Beside the 20 neuron for coding the frequencies there will similarly be 40 other input neurons for coding the dipole and rotation values in the same way.

The output values, the structural data (ϕ, ψ) from the network, are encoded into mostly 8 neurons in the output layer each representing one out of 8 sections of the Ramachandran plot that in turn corresponds to a specific range of the dihedral angles. Hence there are 8 possible values of output, 1–8, generated in the output layer and determined by the most active neurons. The actual output value to be read out from the neurons is the position of the neuron closest to a calculated "center of gravity" of a given weighted firing pattern. If, for example, an output firing pattern appears from a symmetric group of neurons around the seventh neuron (containing the maximal signal) it will be assigned the output value 7. A more simple procedure is to classify an unknown pattern by the value corresponding to the largest activation at the output unit is assigned to the pattern. This is the usual winner-takes-all evaluation of the output of a classifier and is obvious to use in the case of binary outputs but not so obvious for a larger set of output units.

In order to facilitate the interpretation of misclassification we can group the light spectral data in larger super classes of structures, such as helical structures, that have a natural one-dimensional order inferred from physical properties of the spectra. It could also simply be yes or no corresponding to a given conformation being present or not.

3.5. Neural Network Results

In this subsection we shall discuss the performance of the network. The calculated set of numbers from the spectra can be randomly divided into a training set and a test set being disjunct from each other. To be sure about the homogeneity of the training/test set one performs a cross validation. A neural network that is trained on the pairs of correlations in the training set can then have the performance monitored by trying to predict the correlations, i.e., the output numbers (structural data), in the test set from the corresponding input values.

The full set of calculated data (480 lines of corresponding input numbers (3) and an output number) was thus divided up into a training set of 384 lines and a test set of 96 lines chosen at random from the full set. When one evaluates the network there will be both a score for how well the network has learned the correlations in the training set (prediction of the training set output values from the input) and the score for how well new correlations can be predicted in the test set.

Below is given a table of the performance results of different configurations (different sizes of input layer, hidden layer and output layer) of the feed forward neural network. The best neural network configuration is apparently the one with 20×3 input neurons, 24 hidden neurons and 8 output neurons. The networks are also much better at super-classification with only 2 output neurons basically classifying stable structures depending on whether the frequency numbers are high or low.

The small network configurations are clearly not able to comprehend any correlation in the data since the corresponding scores are that of random predictability (i.e., 25% for four output neurons). For 8 output neurons a random score is approximately 12% which is far below the actual scores for the larger networks. For the larger networks the performance is improved by increasing the number of training cycles at least up to 2 thousand. In Table 7 we show a typical section of the training set, i.e., the 14 first data-lines in output-class 1 and 8. When testing the networks a predicted output value, varying between 1.0 and 8.0, is considered correct if it differs less than 0.5 from the the correct value.

Due to the limited amount of statistics at this stage it is difficult to perform a detailed sensitivity analysis but it seems nevertheless possible, on the available amount of statistics, to deduce that the neural networks were better in learning the sections in

Table 7. N-acetyl-L-alanine N'-methylamide training-set data

$\nu(\text{cm}^{-1})$	D_i	R_i	$\phi-\psi$-output	$\phi-\psi$-section
3604.80	17.64	10.32	1	α_R
3599.84	17.34	−12.67	1	α_R
3172.63	5.50	1.63	1	α_R
3158.80	9.83	0.59	1	α_R
3148.93	12.96	2.82	1	α_R
3125.12	18.99	−8.34	1	α_R
3118.49	38.93	−0.03	1	α_R
3111.57	41.49	3.28	1	α_R
3073.43	10.77	17.49	1	α_R
3060.08	10.38	1.86	1	α_R
3052.05	26.70	−4.18	1	α_R
3043.91	71.83	−17.13	1	α_R
1798.76	674.23	−321.40	1	α_R
1789.94	240.50	268.28	1	α_R
...	...	...	...	...
3610.52	25.16	1.65	8	C_7^{eq}
3506.23	168.92	10.48	8	C_7^{eq}
3171.22	3.84	−0.70	8	C_7^{eq}
3150.76	19.56	2.45	8	C_7^{eq}
3148.57	20.48	−0.11	8	C_7^{eq}
3142.24	7.36	−2.47	8	C_7^{eq}
3138.64	23.91	−4.24	8	C_7^{eq}
3096.14	46.95	−2.86	8	C_7^{eq}
3085.30	5.46	6.26	8	C_7^{eq}
3068.64	9.02	2.25	8	C_7^{eq}
3066.68	20.25	−4.32	8	C_7^{eq}
3043.20	83.63	13.02	8	C_7^{eq}
1785.84	685.88	24.30	8	C_7^{eq}
1746.35	341.98	−61.38	8	C_7^{eq}
...	...	...	...	...

the (ϕ, ψ)-plane of secondary structure stability, e.g., the α_R region around $(\phi, \psi) \sim (-60, -40)$, than the other sections. Furthermore, for these stability regions, the lower frequency modes seem to be more important for the stability than the high energy modes since they were more accurately learned. This could probably also be due to the dipeptide limitation which means that the high frequency modes do not involve the contribution from the helix H-bonds (from i to $i+4$) and therefore does not contain the most crucial information about α-helix stability. A forthcoming paper will include a sensitivity analysis of molecules comprising helix type H-bond modes.

Table 8 contains the measured scores (in rounded-off percentages) and correlation coefficients of the performances concerning training and testing of various neural network configurations described by the sizes of their neuron layers. The scores are calculated in percentage as the number of correctly predicted output values over the total number of values. A number is correctly predicted if the corresponding neuron has the value ± 0.5 of the correct value being and integer between 1 and 8.

Table 8. Neural network performance results

Network configuration (#in × #hid × #out)	Number of train cycles	Training score %	Test score %	Test correlation coefficient C
(3 × 3 × 4)	100	25	25	0.00
(3 × 3 × 4)	1800	50	30	0.10
(30 × 10 × 8)	1800	60	33	0.21
(60 × 20 × 8)	900	65	40	0.24
(60 × 20 × 8)	1800	74	55	0.41
(60 × 20 × 2)	1800	83	68	0.48
(80 × 40 × 8)	1800	67	51	0.32

4. CONCLUSIONS

The calculation of the VA and VCD spectra of biological molecules in the presence of water is now feasible and these calculations provide benchmark calculations for simpler models for the calculation of VA and VCD spectra of larger biological molecules in aqueous solution. The 6-31G* RHF zwitterionic structure of L-alanine reported recently by Barron, Gargaro, Hecht and Polavarapu[7] did not include water. Their stable zwitterionic structure without water and our reported structures are quite different. Note that at this geometry they also calculated the VA, Raman and Raman optical activity intensities utilizing the 6-31G and 6-31G* tensors and the 6-31G and 6-31G* RHF Hessians. As shown in Table 6 our results are very much different than their results.

The network results show that it is possible to train neural networks on scattering data to predict new correlations fairly successfully. A high performance is obtained when the network is classifying super-class structures such a structures (e.g., helical) limited to one location of the Ramachandran plot. Therefore the networks can be used to predict secondary structures and stability in larger peptides from spectral data.

ACKNOWLEDGMENTS

K. J. Jalkanen would like to thank the German Cancer Research Center for a fellowship making his two year stay at the German Cancer Research Center possible. He would also like to thank the scientists and staff at the DKFZ which made his stay in Heidelberg a wonderful experience, especially A. Retzmann, T. Reber, and D. Ward.

REFERENCES

1. P. Hohenberg and W. Kohn, *Phys. Rev. A* **136** (1964) 864; N. D. Mermin, *Phys. Rev. A* **137** (1965) 1441; W. Kohn and L. J. Sham, *Phys. Rev. A* **140** (1965) 1133.
2. S.-K. Ma and K. A. Brueckner, *Phys. Rev.* **165** (1968) 165; A. K. Rajagopal and J. Callaway, *Phys. Rev. B* **7** (1973) 1912; J. P. Perdew and M. Levy, *Phys. Rev. L.* **51** (1983) 1884; L. J. Sham and M. Schlüter, *Phys. Rev. B* **32** (1985) 3883; M. Lannoo, M. Schlüter and L. J. Sham, *Phys. Rev. B* **32** (1985) 3890; G. Vignale and M. Rasolt, *Phys. Rev. L.* **59** (1987) 2360; G. Vignale, M. Rasolt and D. J. W. Geldart, *Phys. Rev. B* **37** (1988) 2502; G. Vignale and M. Rasolt, *Phys. Rev. B* **37** (1988) 10685; R. W. Godby, M. Schlüter and L. J. Sham, *Phys. Rev. B* **37** (1988) 10159; A. D. Becke, *Phys. Rev. A* **38** (1988) 3098; C. Lee, W. Yang, R. G. Parr, *Phys. Rev. B* **37** (1988) 785; Z. H. Levine and D. C. Allen, *Phys. Rev. L.* **63** (1989) 1719; I. Papai, A. St-Alant, J. Ushio and D. Salahub, *Int. J. Quant. Chem.: Quan. Chem. Symp.*

24 (1990) 29; A. Becke, *J. Chem. Phys.* **96** (1992) 2155; A. Becke, *J. Chem. Phys.* **97** (1992) 9173; J. P. Perdew and Y. Wang, *Phys. Rev. B* **45** (1992) 13244; A. D. Becke, *J. Chem. Phys.* **98** (1993) 1372; A. Komornicki and G. Fitzgerald, *J. Chem. Phys.* **98** (1993) 1398. A. D. Becke, *J. Chem. Phys.* **98** (1993) 5648; S. M. Colwell, C. W. Murray, N. C. Handy and R. D. Amos, *Chem. Phys. L.* **210** (1993) 261; C. Lee and C. Sosa, *J. Chem. Phys.* **100** (1994) 9018; B. G. Johnson and M. Frisch, *J. Chem. Phys.* **100** (1994) 7429; S. M. Colwell and N. C. Handy, *Chem. Phys. L.* **217** (1994) 271; A. M. Lee, S. M. Colwell and N. C. Handy, *Chem. Phys. L.* **229** (1994) 225; G. Schreckenback and T. Ziegler, *J. Phys. Chem.* **99** (1995) 606.

3. P. J. Stephens, F. J. Devlin, C. F. Chabalowski and M. J. Frisch, *J. Phys. Chem.*, **98** (1994) 11623; P. J. Stephens, F. J. Devlin, C. S. Ashvar, C. F. Chabalowski and M. J. Frisch, *Faraday Discuss.* **99** (1995) 103; K. J. Jalkanen and S. Suhai, *Chem. Phys.* in press.
4. M. Diem, *J. Am. Chem. Soc.* **110** (1988) 6967.
5. T. B. Freedman, A. C. Chernovitz, W. M. Zuk, M. G. Paterlini and L. A. Nafie, *J. Am. Chem. Soc.* **110** (1988) 6970.
6. L. D. Barron, A. R. Gargaro, L. Hecht and P. L. Polavarapu, *Spectrochimica Acta* **47A** (1991) 1001.
7. A. Fortunelli and J. Tamasi, *Chem. Phys. L.*, **231** (1994) 34; A. B. Schmidt and R. M. Fine, *Molecular Simulations*, **13** (1994) 347; G. J. Tawa, R. L. Martin, L. R. Pratt and T. V. Russo, *J. Phys. Chem.* **100** (1996) 1515; J. S. Craw, J. M. Guest, M. D. Cooper, N. A. Burton and I. H. Hillier, *J. Phys. Chem.* **100** (1996) 6304; K. Ösapay, W. S. Young, D. Bashford, C. L. Brooks and D. A. Case, *J. Phys. Chem.* **100** (1996) 2698 and more references therein.
8. P. J. Stephens, *J. Phys. Chem.* **89** (1985) 748; ibid. 91 (1987) 1712; R. D. Amos, N. C. Handy, K. J. Jalkanen and P. J. Stephens, Chem. Phys. L. **133** (1987) 21; R. D. Amos, K. J. Jalkanen and P. J. Stephens, *J. Phys. Chem.* **92** (1988) 5571.
9. A. D. Buckingham, P. W. Fowler and P. A. Galwas, *Chem. Phys.* **112** (1987) 1.
10. I. Grabec and W. Sachse, *J. Acoust. Soc. Am.*, **85** (1989) 1226.
11. R. J. Williams and D. Zipser, *Neural Computation*, **1** (1989) 270.
12. Sejnowski, T. J. and Rosenberg C. R. *Complex Systems* **1** (1987) 45.
13. H. Bohr, J. Bohr, S. Brunak, R. M. J. Cotterill, B. Lautrup, L. Nørskov, O. H. Olsen, and S. B. Petersen, *FEBS Lett.* **241** (1988) 223.
14. L. H. Holley, L. H. and M. Karplus, *Proc. Natl. Acad. Sci. USA* **86** (1989) 152.
15. H. Bohr, J. Bohr, S. Brunak, R. M. J. Cotterill, H. Fredholm, B. Lautrup and S. B. Petersen, *FEBS Lett.* **261** (1990) 43.
16. D. E. Rumelhart, J. L. McClelland, *et al.*, Parallel Distributed Processing, (1986) MIT Press.
17. R. C. Wade, H. Bohr and P. G. Wolynes, *J. Am. Chem. Soc.* **114** (1992) 8284.
18. B. W. Mathews, *Biochem. Biophys. Acta* **405** (1975) 442.

20

COMPUTER SIMULATIONS OF PROTEIN–DNA INTERACTIONS

Mats Eriksson and Lennart Nilsson*

Department of Bioscience at NOVUM
Karolinska institutet
S-141 57 Huddinge, Sweden

1. INTRODUCTION

Molecular recognition is an essential component in almost all biomolecular processes, specifically in processes relating to transcription and translation of the genetic material. Much progress has been made in recent years towards characterizing several such systems in structural terms, providing insight into the fundamental issue of the structural basis for sequence dependent interactions and binding; in particular one can identify some principles of recognition and structural organization within the transcription factor families (Pabo & Sauer, 1992). Molecular dynamics (MD) simulation provides a very detailed, structural and dynamic, description of biomolecular systems; this level of detail, which is very difficult to obtain by other means, is very valuable for a thorough understanding of the subtle balance between competing interactions involved in molecular recognition processes. From a comparison (Elofsson *et al.*, 1993) of calculated interaction energies (enthalpies) in substrate:protein complexes, with calculated free energy values as well as with experimental data, it is quite clear that straightforward, intuitive guesses of the outcome of mutation experiments in complicated systems are unreliable. The influences of slight structural changes, interplay with solvent and ions, and entropic effects are very difficult to guess; more precise methods, like free energy perturbation or potential of mean force calculations, therefore are necessary. Although some aspects of these system may also require combined molecular mechanics/quantum mechanics energy calculations, the non-covalent binding processes that are the focus of this report have been studied using classical mechanics and empirical energy functions.

The main aims of the studies are to shed light on the nature of interactions that stabilize the structural elements as well as the complexes, and to understand how small sequence changes can lead to functional differences, either by affecting structural or

* To whom all correspondence should be sent: Lennart.Nilsson@csb.ki.se, Fax Int+46–8–608 9290

Theoretical and Computational Methods in Genome Research, edited by Suhai
Plenum Press, New York, 1997

dynamic properties. Many of these systems undergo conformational changes in conjunction with the functional events, and the dynamics of these transitions are a natural target for investigations by MD simulation. We focus our research on protein-DNA interactions in the DNA-glucocorticoid receptor system.

Since most biomolecular reactions take place in an aqueous environment it is also very important to consider the effects of the solvent, which may be of several kinds: competition for hydrogen bonds, screening of electrostatic interactions or favoring structural arrangements with minimal exposure of hydrophobic groups, to name a few possibilites. There are also cases in which water molecules have been seen to supply crucial and specific contacts between protein and DNA. Water molecules are easily followed and characterized in terms of localization and dynamics in simulation studies, but are rather more elusive in both X-ray and NMR experiments (Levitt & Park, 1993).

Simulation studies of nucleic acids are not as common as of protein systems, mainly for the two reasons that 1/ there are fewer structures available, and 2/ nucleic acids are more difficult to handle due to their highly charged and non-globular nature. With the discovery of the so-called ribozymes it has become clear that there is much more to nucleic acid structure than the canonical DNA duplexes, but still there are very few experimentally determined structures other than some tRNAs and short oligomers of DNA and RNA. Modelling and simulations of nucleic acids have been used in structural refinement over the last decade (Nilsson et al, 1986; Nilges et al, 1987). The early simulations of nucleic acids were usually performed in vacuum (Nilsson & Karplus, 1986) or "pseudo-vacuum" conditions (Nordlund et al, 1989). Today simulations with explicit treatment of solvent dominate (Beveridge & Ravishanker, 1994) and detailed studies of dynamics and thermodynamics of processes like base stacking have been performed on short oligo nucleotides (Norberg & Nilsson, 1995a,b,c).

1.1 Simulations of Protein-DNA Systems

A few dynamics simulations of protein-DNA systems have been reported in the literature. The lac repressor headpiece in complex with its operator was simulated using distance restraints from NMR studies, with a 50ps piece of unrestrained trajectory (de Vlieg et al, 1989). From this simulation the specific contacts between the protein and the DNA were found to be formed mainly by nonpolar contacts, and to a lesser extent through water-mediated hydrogen bonds. The DNA binding domain of the glucocorticoid receptor (GRDBD) in complex with DNA has been the subject of studies by three groups using slightly different approaches. Harris et al (Harris et al, 1994, 1995) have run a 1ns simulation of a model of the GRDBD dimer bound to DNA with a 10Å shell of water surrounding the system (yielding a total of 9625 atoms), and Bishop & Schulten (Bishop & Schulten, 1996) have reported two 90ps simulations of the GRDBD dimer bound to two DNA sequences with different spacings between the recognition half sites, using an ellipsoidal geometry for the surrounding water (ca 13600 atoms in total). Our own simulations of the GRDBD system have used the crystallographic coordinates for the complex (Luisi et al, 1991) in a 42Å radius sphere of water (ca 31000 atoms in total), which was simulated for 200ps (Eriksson et al, 1994; Eriksson et al, 1995). Complexes of GRDBD monomers bound to various halfsite DNA sequences were simulated for 200ps, and also used in free energy perturbation calculations of the relative affinities of the GRDBD for these DNA sequences (Eriksson & Nilsson, 1995).

1.2 Simulations of Other Nucleic Acid Complexes

The thermodyncamics of nucleic acid interactions with proteins and small ligands have also been simulated with results in good agreement with experiment.

1.2.1 DNA-Netropsin Complex. The small dicationic peptide drug netropsin binds preferentially to the minor groove of A-T regions of DNA, which are favored by 4.4 kcal/mol over G-C regions, a difference that has been attributed to the repulsion of the amino group protruding into the minor groove of a G-C pair, but which is absent in an A-T pair. We have applied the thermodynamic cycle-perturbation method to calculate this change in free energy of binding with very encouraging results (Härd & Nilsson, 1992). We obtained a value for $\Delta\Delta G$ in the range from 3.4 to 4.0 kcal/mol with an estimated error of 1–2 kcal/mol. The results do not seem to depend on the size of the system or the length of the simulations within the limits we have tried: 10–20ps runs of either a DNA hexamer duplex in water (1200 atoms) or a DNA octamer duplex in water (2600 atoms).

1.2.2 Ribonuclease T_1. Ribonuclease T_1 is a small enzyme which selectively cleaves single-stranded RNA on the 3′ side of guanines, and for which the 3D structure is known with and without bound substrate. This makes the enzyme ideal for the study of protein-nucleic acid interactions. Our work on this system has so far focussed on changes in the structure of the wild-type enzyme, particularly in its active site, upon binding of the inhibitor 2′′GMP. Initial experimental studies using time-resolved fluorescence showed the presence of a picosecond motion of Trp59 (the only tryptophan in the enzyme) which was altered by the binding of 2′GMP (MacKerell et al, 1987). This was also observed in a series of molecular dynamics simulations of the same systems, in water as well as *in vacuo* (MacKerell et al, 1988a,b,1989).

From a series of free energy simulations comparing the binding of 2-aminopurine versus 3′GMP to the enzyme (Elofsson & Nilsson, 1993; Elofsson et al, 1993) we found that 2-aminopurine has a less favorable interaction (by ~5 kcal/mol) which is reasonable; the effect of the compensatory mutation Glu46-Gln46 seems mainly to be a reduction of the affinity for GMP, with almost no direct effect on 2-aminopurine binding.

1.3 Nuclear Hormone Receptors

The transcription factors which are activacted by hormones such as glucocorticoids, estrogens, and retinoids, constitute the large family of nuclear hormone receptors. These receptors seem to have a common evolutionary origin and to have evolved into three distinct subfamilies, one of which is the steroid hormone receptor group. The receptors are multidomain proteins, with the different functions needed within the signal transduction system performed by different domains, and they bind to the response elements of the DNA in various dimeric combinations. Some of the main domains contain the ligand binding, DNA binding, and transactivation functions.

There are two families of response elements for steroid receptors, the glucocorticoid response elements (GRE) with the halfsite sequence TGTTCT and the estrogen response elements (ERE) with the halfsite sequence TGACCT. Crystallographic structures have been determined for systems of both kinds (Luisi et al, 1991; Schwabe et al, 1993).

2. METHODS

In a molecular dynamics (MD) simulation the motions of a molecular system are computed, using Newton's equations of motion, from an initial structure and knowledge of the potential energy of the system as a function of its atomic coordinates (Brooks et al, 1988). MD simulations were performed using the program CHARMM (Brooks et al, 1983) with all atom energy functions for proteins and nucleic acids (MacKerell et al, 1992, 1995). Relative affinities between slightly different complexes were calculated using free energy perturbation methods (for a review, see Beveridge & DiCapua, 1989), which use the fact that free energy is a thermodynamic state function so that in a closed loop of reactions free energy differences thus can be obtained in a computationally convenient way.

3. PROTEIN-DNA INTERACTIONS

The recently solved structure of the DNA binding domain (GRDBD) of the glucocortioid receptor in complex with DNA shown in Figure 1 (Luisi et al., 1991), together with the related estrogen receptor (ER) DBD bound to DNA (Schwabe et al., 1993), gives us an excellent opportunity to study the complex between DNA and DBD. We have simulated this system in aqueous solution, first as one monomer of GR DBD (Eriksson et al., 1993), and now also as the dimeric complex with DNA (Eriksson et al., 1994, 1995); this is one of the first simulations of a protein-DNA complex, and one of the largest systems to be simulated to date (GRDBD+DNA+water contains 30977 atoms).

Overall the structure of the complex is maintained, a small bend in the DNA becomes slightly more pronounced due to somewhat closer contacts between protein and DNA, and some local distortions in the DBD are also seen. There is one region of DBD where the NMR and X-ray structures differ, and in simulations of the DBD monomer in solution, starting from either of these conformers, the simulated structure remains close to its initial conformation indicating that both conformers are in local minima. Results on backbone dynamics correlate well with NMR relaxation data from ^{15}N-labelled proteins (Eriksson et al, 1993), and in the cases with significant discrepancies an analysis of the actual motions as found in the simulated trajectories revealed that they were generally the result of infrequent processes, e g a backbone N-H hydrogen bond would jump back and forth between different H-bond accepting groups, causing insufficient sampling. An example of the agreement between simulation and experiment is given in Figure 2, where the generalized order parameters S^2 for the backbone N-H vectors are shown. The order parameter characterizes the extent of the mobility of the vector, such that $S^2=1$ means completely immobility and $S^2=0$ means completely unrestricted motion.

In this case the combined use of experiment and simulation provided a more complete description of the backbone dynamics than either method could have given on its own.

Further analysis of the various hydrogen bonding, and water bridging patterns in the complex, shows that the hydrogen bonding networks on either side of the major groove are similar in the GRDBD:GRE and ERDBD:ERE systems; more so in the simulation than when using the crystal structure of GRDBD:GRE because of the increased number of quite stable waters that appear in the protein DNA interface in the simulation.

Free energy calculations are being performed on the relative affinities of the wildtype protein for different DNA response elements (Eriksson & Nilsson, 1995); there is data available from binding and genetic studies on all combinations of changing be-

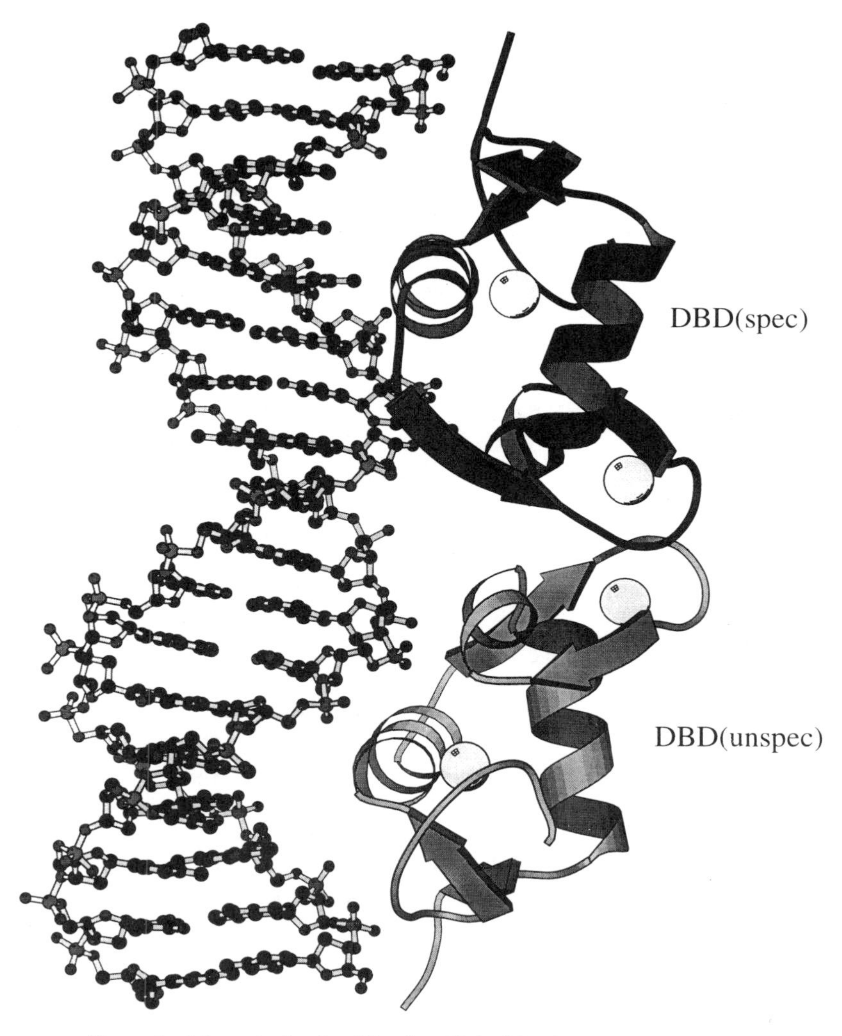

Figure 1. Schematic drawing (Kraulis, 1991) of the GRE:GRDBD complex.

tween the glucocortioid and estrogen response elements (GRE and ERE), as well as for changing the three amino acids determining the specificity of the proteins (Zilliacus et al, 1991).

There is a number of questions, for which a body of experimental genetic and thermodynamic data is building up in this system, relating to DNA-protein interactions (specific and unspecific binding modes), and to protein-protein interactions (allosteric affects) that are amenable to modelling and simulation studies. It has, for instance, been shown (Zilliacus *et al.*, 1992) that in the glucocorticoid receptor single amino acids in the recognition region can form both positive and negative contacts with specific base pairs with the cognate DNA, but only negative contacts with non-cognate DNA; there are also synergistic effects of combinations of amino acids. A first attempt at rationalizing some of these finding has been made with a combined use of genetics and molecular modelling to

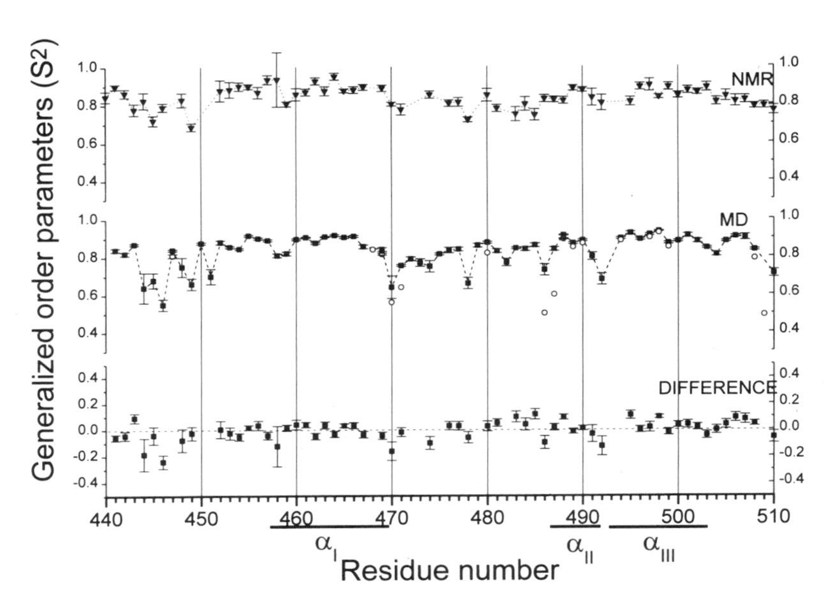

Figure 2. Order parameters vs residue in the GRDBD monomer.

study the activity/"affinity" of all 320 combinations of GR mutants (all 20 amino acids in position 339, the first of the three positions crucial for binding to an ERE or a GRE) with the 16 response elements obtained by replacing the middle two base pairs (which differ betwen ERE and GRE) with all possible base pair combinations (Zilliacus *et al.*, 1995). In addition to the general finding, that amino acids with short side chains seem to fit quite well with all base sequence, the modelling part of the study provided explanations for a few anomalous combintations; eg., the Trp mutant, bulky as this side chain is, binds well to TGGCCT by positioning the side chain perpendicular to the base planes and accepting a hydrogen bond from the C-NH_2 to the Trp imidazole ring, a relation which is only possible for this particular combination. From a computationally more demanding free energy perturbation study of the binding of GRDBD to halfsite response elements corresponding to GRE, ERE and an intermediate (GRE2) with one of the middle basepairs from GRE and the other from ERE, we found relative binding energies very similar to what was seen experimentally, with affinities ordered as GREH-GRDBD > GRE2H-GRDBD > EREH-GRDBD.

4. DISCUSSION

Simulations of protein:DNA systems are feasible on a nanosecond timescale, and yield structural, dynamic and thermodynamic results which agree well with available experimental data. In addition to this we find details concerning the dynamics at the contact interfaces in terms of direct as well as water mediated hydrogen bonds that are not easly seen in experiments. We could also make some suggestions as to the underlying reasons for differences in the stabilities and cooperativity of the complex formation in the GRDBD:GREH and GRDBD:EREH cases (Eriksson & Nilsson, 1995), such as the stiffening of the GRDBD when bound to EREH (Figure 3).

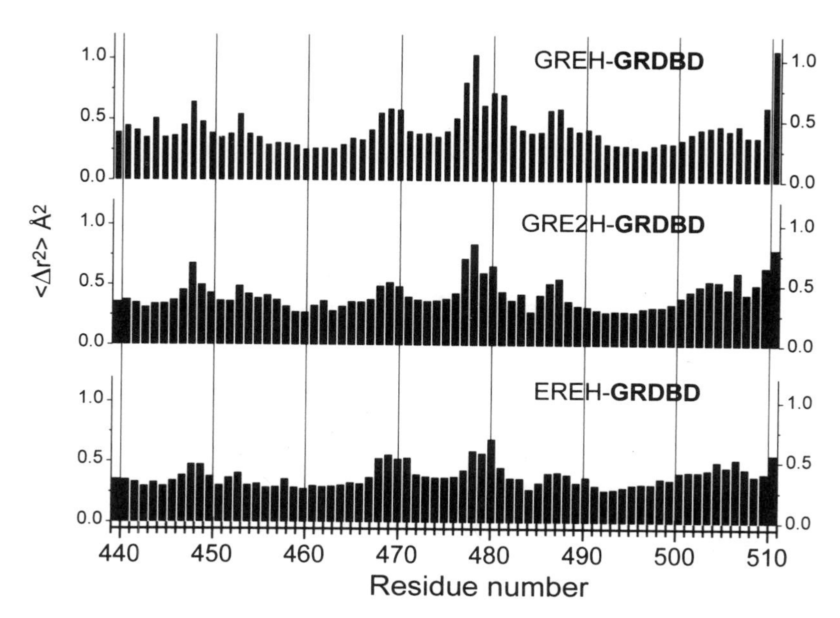

Figure 3. Mean square atomic fluctuations in different complexes of GRDBD and DNA response elements.

The possibilities to change the systems at will, in the computer, should in the future be exploited to make systematic studies of the specific binding, and also of protein-protein interactions that may influence the binding.

5. REFERENCES

Beveridge, D. & Ravishanker, G. (1994) Molecular Dynamics Studies of DNA, Curr. Op. Struct. Biol. **4**, 246–255.

Beveridge, D.L. & DiCapua, F.M. (1994) Free Energy via Molecular Simulation: Applications to Chemical and Biochemical Systems, Ann. Rev. Biophys. Biophys. Chem. **18**, 431–492

Brooks, B.R., Bruccoleri, R.E., Olafson, B.D., States, D.J., Swaminathan, S. & Karplus, M. (1983). CHARMM: A Program for Macromolecular Energy, Minimization and Dynamics Calculations. J. Comp. Chem. **4**, 187–217.

Brooks, C.L, Karplus, M. & Pettitt, B.M. (1988) Proteins: A Theoretical Perspective of Dynamics, Structure and Thermodynamics. Adv. In Chem. Phys. Vol 71, Wiley, New York.

Elofsson, A. & Nilsson, L. (1993) Free Energy Perturbations in Ribonuclease T_1 Substrate Binding. A Study of the Influence of Simulation Length, Internal Degrees of Freedom and Structure in Free Energy Perturbations, Mol. Sim. **10**, 255–276.

Elofsson, A., Kulinski, T., Rigler, R. & Nilsson, L. (1993) Site Specific Point Mutation Changes Specificity. A Molecular Modelling Study by Free Energy Simulations and Enzyme Kinetics of the Thermodynamics in Ribonuclease T_1 Substrate Interactions, PROTEINS **17**, 161–175

Eriksson, M., Berglund, H., Härd, T. & Nilsson, L. (1993) A Comparison of ^{15}N NMR Relaxation Measurements with a Molecular Dynamics Simulation: Backbone Dynamics of the Glucocorticoid Receptor DNA-binding Domain, PROTEINS, **17**, 375–390.

Eriksson, M.A.L., Härd, T. & Nilsson, L. (1994) Molecular Dynamics Simulation of a DNA Binding Protein Free and in Complex with DNA, in "Computational Approaches to Supramolecular Chemistry" (Ed. G. Wipff), NATO ASI Series, Kluwer, Dordrecht, pp 441–456

Eriksson, M.A.L., Härd, T. & Nilsson, L. (1995) Molecular Dynamics Simulations of the DNA -Binding Domain of the Glucocorticoid Receptor as a Dimer in Complex with DNA and as a Monomer in Solution, Biophys. J. **68**, 402–426

Eriksson, M.A.L. & Nilsson, L. (1995) Structure, Thermodynamics and Cooperativity of the Glucocorticoid Receptor DNA-binding Domain in Complex with Different Response Elements, J. Mol. Biol. **253**, 453–472.

Härd, T. & Nilsson, L. (1992) Free Energy Calculations Predict Sequence Specificity in DNA-Drug Complexes, Nucleosides & Nucleotides **11**,167–173.

Kraulis, P.J. (1991) MOLSCRIPT: A Program to Produce Both Detailed and Schematic Plots of Protein Structure, J. Appl. Crystallogr. **24**, 946–950.

Levitt, M. & Park, B.H. (1993) Water: now you see it now you don't, Structure **1**, 223–226.

Luisi B.F., Xu W.X., Otwinowski Z., Freedman L.P., Yamamoto K.R. and Sigler P.B., Nature Vol. 352 (1991) 497–505.

MacKerell, A.D., Rigler, R., Nilsson, L., Hahn, U. & Saenger, W. (1987). Biophysical Chemistry, **26**, 247–261.

MacKerell, A.D., Nilsson,L., Rigler, R. & Saenger, W. (1988a) Biochemistry, **27**, 4547–4556.

MacKerell, A.D., Rigler, R., Nilsson, L., Heinemann, U. & Saenger,W. (1988b). Eur. Biophys. J. **16**, 287–297.

MacKerell, A.D., Nilsson, L., Rigler, R., Heinemann, U. & Saenger, W. (1989). Molecular Dynamics Simulations of Ribonuclease T1: Comparison of the Free Enzyme and the 2'GMP-Enzyme Complex, Proteins **6**, 20–31.

MacKerell Jr., A.D., Bashford, D., Bellott, M., Dunbrack Jr., R.L., Field, M.J., Fischer, S., Gao, J., Guo, H., Ha, S., Joseph, D., Kuchnir, L., Kuczera, K., Lau, F.T.K., Mattos, C., Michnick, S., Ngo, T., Nguyen, D.T., Prodhom, B., Roux, B., Schlenkrich, M., Smith., J.C., Stote, R., Straub, J. Wiorkiewicz-Kuczera, J. and Karplus, M. (1992) Self-consistent parameterization of biomolecules for molecular modeling and condensed phase simulations. FASEB Journal, 6:A143.

MacKerell Jr., A.D., Wiorkiewicz-Kuczera, J. and Karplus, M. (1995) An all-atom empirical energy function for the simulation of nucleic acids, Journal of the American Chemical Society **117**,11946–11975.

Nilges, M., Clore, G.M., Gronenborn, A.M., Brünger, A.T., Karplus, M. & Nilsson, L. (1987) .The Three-Dimensional Solution Structure of the DNA Hexamer 5'd(GCATGC)$_2$. Combined Use of Nuclear Magnetic-Resonance and Restrained Molecular Dynamics. Biochemistry, **26**, 3718–3733.

Nilsson, L. & Karplus, M. (1986). Molecular Dynamics Simulation of the Anticodon Arm of Phenylalanine Transfer RNA. In "Structure and Dynamics of RNA", NATO ASI series A Vol. 110 (Eds. Hilbers, C.W. & van Knippenberg, P.), Plenum Press, New York, .151–159.

Norberg, J. & Nilsson, L. (1995a) Stacking Free Energy Profiles for All 16 Natural Diribonucleoside Monophosphates in Aqueous Solution, J. Am. Chem. Soc. **117**, 10832–10840

Norberg, J. & Nilsson, L. (1995b) Temperature Dependence of the Stacking Propensity of Adenylyl-3',5'-Adenosine, J. Phys. Chem. **99**, 3056–3058.

Norberg, J. & Nilsson, L. (1995c) NMR Relaxation Times, Dynamics and Hydration of a Nucleic Acid Fragment from Molecular Dynamics Simulations, J. Phys. Chem. **99**, 14876–14884.

Nordlund, T.M., Andersson,S., Nilsson, L., Rigler, R., Gräslund, A. & McLaughlin, L.W. (1989). Structure and Dynamics of a Fluorescent DNA Oligomer Containing the *Eco*RI Recognition Sequence: Fluorescence, NMR and Molecular Dynamics Studies. Biochemistry **28**, 9095–9103.

Pabo, C. & Sauer, R.T. (1992) Transcription factors: Structural families and principles of recognition. Ann. Rev. Bioch. **61**, 1053–1095.

Robinson, C.R. & Sligar, S.C. (1993) J. Mol. Biol. **234**, 302–306.

Schwabe, J.W.R., Chapman, L., Finch, J.T. & Rhodes, D. (1993) The crystal structure of the oestrogen receptor DNA-binding domain bound to DNA: How receptors discriminate between their response elements. Cell **75**, 567–578.

Zilliacus, J., Dahlman-Wright, K., Wright, A., Gustafsson, J.-Å. & Carlstedt-Duke, J. (1991) DNA binding specificity of mutant glucocorticoid recptor DNA-binding domains, J. Biol. Chem. **266**, 3101–3106.

Zilliacus, J., Wright, A.P.H., Norinder, U., Gustafsson, J.-Å. & Carlstedt-Duke, J. (1992) Determinants for DNA Binding Site Recognition by the Glucocorticoid Receptor, J. Biol. Chem. **267**, 24941–24947.

Zilliacus, J., Wright, A.P.H., Carlstedt-Duke, J., Nilsson, L. & Gustafsson, J.-Å. (1995) Modulation of DNA Binding Specificity Within the Nuclear Receptor Family by Substitutions at a Single Amino Acid Position, PROTEINS **21**, 57–67.

THE ROLE OF NEUTRAL MUTATIONS IN THE EVOLUTION OF RNA MOLECULES

Peter Schuster*

Institut für Molekulare Biotechnologie
PF 100813, D-07708 Jena, Germany and
Institut für Theoretische Chemie der Universität Wien
Währingerstr.17, A-1090 Wien, Austria

ABSTRACT

Molecular evolution started out from replication *in vitro* and serial transfer experiments with RNA molecules from bacteriophages. Variants of high fitness measured in terms of replication rates were evolved in the test-tube. RNA replication in the Qβ system is well understood in terms of biochemical reaction kinetics. Generation times as short as a few seconds can be achieved in molecular replication and fast adaptations to changes in the environment are readily observed.

The Darwinian view of molecular evolution visualizes optimization processes as adaptive walks on fitness landscapes. Every fittest genotype or master sequence is surrounded by a cloud of mutants called molecular quasi-species which represents the genetic reservoir from which selection chooses structures. In molecular evolution genotype and phenotype are two features of the same molecule, sequence and structure, respectively. In case of RNA secondary structures we have many more sequences than different structures and a high degree of neutrality is the consequence.

Computer based studies on RNA sequence-structure relations revealed two important results: (i) relatively few structures are common in the sense that they are formed by very large numbers of sequences and many structures are rare, and (ii) sequences folding into the same structure need not be related in the sense that they have small Hamming distances. Combining both results yields the principle of shape space covering which implies that only a small fraction of all possible sequences has to be searched in order to find at least one genotype for every common structure, and any arbitrarily chosen sequence is an equally well suited starting point for optimization.

Neutral sequences form extended networks in sequence space along which populations can migrate by a diffusion like mechanism. Neutral networks are essential for the success of

*All correspondence should be sent to: pks@imb-jena.de, Phone: +49 (3641) 65 6444, Fax: +49 (3641) 65 6446

Theoretical and Computational Methods in Genome Research, edited by Suhai
Plenum Press, New York, 1997

evolutionary adaptation because they enable populations to explore vast regions of sequence space without being caught in local fitness optima.

1. EVOLUTION OF RNA MOLECULES

In genome analysis, and likewise in almost every other discipline of life sciences, one can never too often recall Theodosius Dobzhansky's famous statement[8]: "Nothing in biology makes sense except in the light of evolution." In other words, knowledge on genome evolution is indispensible for a comprehensive understanding of genome structure, function and regulation. Methods to study biological evolution are generally facing three tantalizing problems: (i) evolutionary phenomena like adaptation need ten-thousands to millions of generations in order to become observable in populations and this is much too long for any experimental approach, (ii) combinatorial explosion leads to numbers of possible genotypes that exceed all imaginations, and (iii) relations between genotypes and phenotypes are so complex that any realistic modeling is impossible. Even in the case of fast replicating bacteria with generation times of about one hour ten-thousand generations require more than a year and evolution experiments would last several years. (The most patient microbiologists, for example, have just arrived at some ten-thousand generations of *Escherichia coli* under controlled conditions.[37]) The numbers of possible sequences are prohibitive for systematic studies: viroid genomes are about three-hundred nucleotides long and this implies 4×10^{180} possible different RNA sequences of this chain length; we have more different sequences than Avogadro's number already with oligomers of forty nucleotides. The complexity of the relation between genotypes and phenotypes is even more discouraging: to predict the spatial structure of a single biopolymer molecule from its sequence is a very hard (and still unsolved) problem, and the simplest bacteria consist of several thousand different protein and nucleic acids molecules.

Sol Spiegelman[45] did the first optimization experiments on replication of virus specific RNA *in vitro* which demonstrated that evolutionary phenomena are no privilege of cellular life: molecules capable of reproduction and mutation fulfil the prerequisites for Darwin's principle and behave like asexually replicating individuals as far as selection and adaptation to environmental conditions are concerned. About the same time Manfred Eigen[9] developed a theoretical frame for molecular evolution which had its roots in chemical reaction kinetics. *In vitro* evolution of RNA molecules seems to circumvent the three problems indicated above: (i) generation times can be reduced to a few seconds under most favorable conditions and evolutionary phenomena can be observed in a few days, (ii) many genotypes form the same phenotype and thus are selectively neutral in the sense of Motoo Kimura,[34] and (iii) genotype and phenotype are two properties of the same molecule, the sequence and the spatial structure, respectively. Evolution of RNA in the test-tube appears to be the simplest conceivable system to study Darwinian adaptation phenomena.

Since the early days of Spiegelman's pioneering works many different studies on evolutionary phenomena were performed with RNA molecules. RNA replication kinetics has been studied in great detail,[3] serial transfer experiments in the presence of increasing concentration of RNase yielded "resistant" RNA molecules,[46] catalytic RNA molecules of the group I intron family were trained to cleave DNA rather than RNA,[2,36] the SELEX technique[48] was applied to the selection of RNA molecules which bind to predefined targets with high specificity,[27] and ribozymes with novel catalytic functions were derived from libraries of random RNA sequences.[1,4,38] Molecular evolution experiments with RNA molecules provided essentially two important insights into the nature of evolutionary processes: (i) the Darwinian principle of (natural) selection is no privilege of cellular life since it is valid also in evolution in the test-tube and (ii) adaptation to the environment and optimization of molecular properties can be observed in a few days or weeks during the course of a typical laboratory experiment.

In fact, a new discipline called evolutionary biotechnology (or applied molecular evolution) grew out of the *in vitro* evolution studies.[10,28,29,30] In particular, RNA based evolutionary design of biomolecules has become routine already.[13,42] In order to be able to "breed" molecules for predefined purposes one cannot be satisfied with natural selection which is tantamount to searching for the "fittest" or fastest replicating molecular species. In general, the properties that have to be optimized are completely unconnected from fitness. The principles of the evolutionary design of biomolecules are shown schematically in figure 1. Molecular properties are optimized iteratively in a series of selection cycles. The first cycle is often initiated by a sample of random sequences or alternatively by a population created by error prone replication of RNA (or DNA) sequences. Each selection cycle consists of three phases: (i) selection of suitable RNA molecules, (ii) amplification through replication, and (iii) diversification through mutation (with artificially elevated error rates). The first step, the selection of the genotypes that fulfil best the predefined criteria, requires biochemical and biophysical intuition or technological skill in the design of appropriate automata for massively parallel testing. The two strategies are selection of the best suited candidates from a large sample of different genotypes in homogeneous solutions or spatial separation of genotypes and massively parallel screening. Variants are tested and discarded in case they do not fulfil the predefined criteria. The selected genotypes are amplified, diversified by mutation with problem adjusted error rates, and then subjected again to selection. Optimally adapted genotypes are usually obtained after some twenty to fifty selection cycles.

2. RNA PHENOTYPES

Evolution of molecules in the test-tube represents the simplest conceivable case of relations between genotypes and phenotypes since both are features of the same molecule, the sequence being the genotype and the spatial structure the phenotype, respectively. The mappings of genotypes into phenotypes are then reduced to sequence structure relations of RNA molecules (Fig. 2). Three-dimensional structures of RNA molecules are not well known since only very few structures were so far determined by crystallography and nmr spectroscopy. Needless to say, they are also very hard to predict. A course grained version of RNA structure that lists only the Watson–Crick and **GU** base pairs, the so-called secondary structure, is conceptionally much simpler. Secondary structure predictions are more reliable. Some statistical properties of RNA secondary structures were found to be fairly robust and (almost) independent of the choices of algorithms and parameter sets.[47] In addition, secondary structures are sufficiently simple to allow the application of rigorous mathematical analysis and large scale computations to sequence structure relations.

In order to be able to derive global properties of relations between the sequences and the structures of RNA molecules we have to abandon the conventional approach of structural biology which is in essence concerned with the folding of a given sequence into its structure. Instead, we shall consider sequence structure relations as (non-invertible) mappings from sequence space into shape space.[15,41,44] Sequence spaces are well understood objects: each

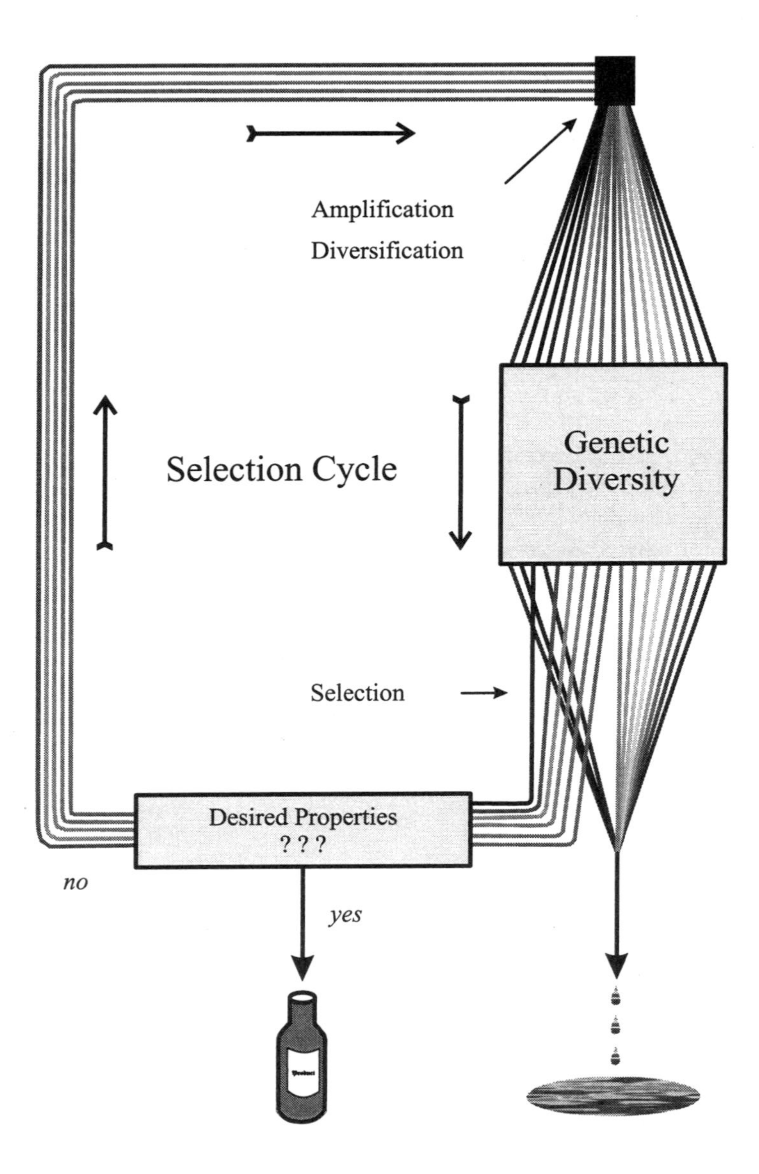

Figure 1. Evolutionary design of biopolymers. Properties and functions of biomolecules are optimized iteratively through selection cycles. Each cycle consists of three different phases: amplification, diversification by replication with high error rates or random synthesis, and selection. Currently successful selection techniques apply one of two strategies: (i) selection in (homogeneous) mixtures using binding to solid phase targets (SELEX) or reactive tags that allow to separate suitable molecules from the rest and (ii) spatial separation of individual molecular genotypes and large scale screening.

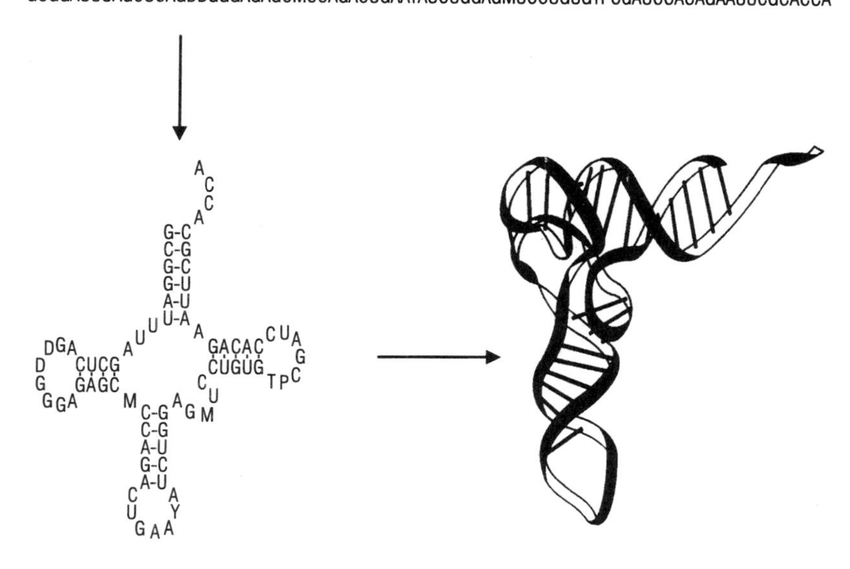

Figure 2. Mapping of RNA genotypes into phenotypes. The sequence being the genotype folds into a three-dimensional structure which represents the phenotype of the molecule. A coarse-grained structure, the so-called secondary structure which, in essence, consists of a list of base pairs, is represented by a planar graph. Secondary structures turned out to be useful for modeling RNA evolution.

sequence is represented by a point in sequence space and distances between sequences are given by the (minimum) number of (elementary) mutations converting two sequences into each other. In case of point mutations this distance is the Hamming distance which is tantamount to the number of positions in which two (properly aligned) sequences differ.[21] The sequence space containing all 2^n binary sequences of chain length n is simply the n-dimensional hypercube. It is not hard to find a formally consistent distance measure in shape space.[15,25,35,40] Such a distance, however, need not appropriate for practical purposes. Indeed, it is much more difficult to define a distance between structures that meets biological intuition and allows to compare structures with respect to function. This problem has not yet been solved satisfactorily.

Molecular properties are understood as functions of molecular structures. The evolutionary relevant aspect of function is fitness which in constant environments can be expressed by a single number. We are then dealing with a combined mapping from sequences into the real numbers:

$$\text{sequences} \Longrightarrow \text{structures} \Longrightarrow \text{fitness parameters}.$$

Such a mapping from sequence space into the real numbers is commonly called a "landscape." Evolution maximizing fitness be visualized as an adaptive walk on a fitness landscape. This suggestive view originally invented by Sewall Wright as a kind of metaphor[49] saw a recent revival in much more precise physical terms.[11,31,44] Landscapes derived from RNA sequences after folding them into secondary structures were characterized and studied extensively in the past.[14,15,18]

RNA secondary structures provide an excellent model system to study relations between genotypes and phenotypes. Application of combinatorics allows to derive an asymptotic expression for the numbers of acceptable structures[24,43]:

$$S_n \approx 1.4848 \times n^{3/2}(1.8488)^n.$$

This expression is based on two assumptions: (i) the minimum stack length is two base pairs ($n_{stack} \geq 2$, i.e., isolated base pairs are excluded) and (ii) the minimal size of hairpin loops is three ($n_{loops} \geq 3$). The numbers of sequences are 4^n for natural RNA molecules, and 2^n for **GC**-only or **AU**-only sequences, respectively. In any case we have many more sequences than structures and this implies neutrality in the sense that several RNA sequences form the same (secondary) structure. Not all acceptable secondary structures are formed as minimum free energy structures of some sequences. The numbers of actually formed structures can be determined only by exhaustive folding of all sequences. This has indeed been carried out in a recent study[19,20] for **GC**-only sequences of chain lengths up to $n = 30$. Some results obtained thereby are summarized in table 1. The numbers of minimum free energy structures are smaller than the numbers of acceptable structures by a factor between two and four which is slightly increasing with the chain length. Secondary structures are properly grouped into two classes, common ones and rare ones. A straightforward definition of common structures was found to be very useful:

$$\mathcal{C} := \text{common} \quad \text{iff} \quad n_{\mathcal{C}} \geq \overline{n_{\mathcal{S}}} = \frac{\kappa^n}{\mathcal{S}_n} = \#(\text{Sequences})/\#(\text{Structures}),$$

wherein κ denotes the size of the alphabet ($\kappa = 2$ for **GC**-only or **AU**-only sequences and $\kappa = 4$ for natural RNA molecules). A structure $\mathcal{C}$ is common if it is formed by more sequences than the average structure.

The results of exhaustive folding suggest two important general properties of the above given definition of common structures[19,20]: (i) the common structures represent only a small fraction of all structures and this fraction decreases with increasing chain length, and (ii) the fraction of sequences folding into the common structures increases with chain length and approaches unity in the limit of long chains. Thus, almost all RNA sequences fold into a small fraction of the secondary structures for sufficiently long chains. The ratio of sequences to structures, in fact, is even larger than originally expected since only the common structures play a role in natural evolution and in evolutionary biotechnology.

3. SHAPE SPACE COVERING

There are many more sequences than secondary structures, and focussing on common structures the numbers of neutral sequences forming one particular structure are very large indeed. Sequences folding into the same structure are (almost) randomly distributed in sequence space. It is straightforward then to compute a spherical environment (around any randomly chosen reference point in sequence space) that contains at least one sequence for every common structure. The radius of such a sphere with a radius, called the covering radius r_{cov}, is much smaller than the radius of sequence space. A covering sphere represents a small subset of all sequences that covers the entire shape space (Fig. 3). The covering radius can be estimated by simple probability arguments[42]:

$$r_{cov} = \min\{h = 1, 2, \ldots, n | B_h \geq \kappa^n / \overline{n_{\mathcal{S}}} = \mathcal{S}_n\},$$

Table 1. Common secondary structures of **GC**-only sequences

	Sequences		Struct.	**GC***		
n	4^n	2^n	S_n	$S_{\mathbf{GC}}$	R_c	n_c
7	16,384	128	6	2	1	120
8	65,536	256	9	3	1	224
9	262,144	512	14	6	1	371
10	1.05×10^6	1,024	22	11	4	859
11	4.19×10^6	2,048	35	20	7	1,648
12	1.68×10^7	4,096	57	31	13	3,502
13	6.71×10^7	8,192	93	48	22	7,384
14	2.68×10^8	16,384	155	73	31	14,657
15	1.07×10^9	32,768	258	116	43	28,935
16	4.29×10^9	65,536	432	195	64	58,886
17	1.72×10^{10}	131,072	730	340	86	115,140
18	6.87×10^{10}	262,144	1,238	582	117	224,713
19	2.75×10^{11}	524,288	2,111	973	183	450,802
20	1.10×10^{12}	1.05×10^6	3,613	1,610	286	902,918
21	4.40×10^{12}	2.10×10^6	6,209	2,615	461	1,826,514
22	1.76×10^{13}	4.19×10^6	10,706	4,258	752	3,716,134
23	7.04×10^{13}	8.39×10^6	18,516	6,936	1,202	7,547,362
24	2.81×10^{14}	1.68×10^7	32,115	11,348	1,866	15,246,819
25	1.13×10^{15}	3.36×10^7	55,848	18,590	2,869	30,745,861
26	4.50×10^{15}	6.71×10^7	97,352	30,501	4,302	61,716,291
27	1.80×10^{16}	1.34×10^8	170,079	49,949	6,372	123,634,231
28	7.21×10^{16}	2.68×10^8	297,748	81,748	9,579	247,907,264
29	2.88×10^{17}	5.37×10^8	522,251	133,782	14,641	497,595,288
30	1.15×10^{18}	1.07×10^9	917,665	218,820	22,718	999,508,805

*The total number of minimum free energy secondary structures formed by **GC**-only sequences is denoted by S_{GC}, R_c is the rank of the least frequent common structure and thus is tantamount to the number of common structures, and n_c is the number of sequences folding into common structures.

with B_h being the number of sequences contained in a ball of radius h. Examples of covering radii are presented in table 2. We consider the case of natural sequences of chain length $n = 100$ to illustrate the role of shape space covering for evolutionary searches: $r_{cov} = 15$ implies that the maximum number of sequences that have to be searched in order to find all common structures is 4×10^{24}. Although 10^{24} is a very large number and exceeds the capacities of all currently available polynucleotide libraries, it is negligibly small compared to the size of the entire sequence space that contains 1.6×10^{60} sequences.

Exhaustive folding allows to test the estimates derived from simple statistics.[20] The agreement for **GC**-only sequences of short chain lengths is as good as it could be in case we use the numbers of minimum free energy structures derived from exhaustive folding. The covering radius increases linearly with chain length with a factor around 1/4. The fraction of sequence space that is required to cover shape space thus decreases exponentially with increasing size of RNA molecules (table 2). Nevertheless, the absolute numbers of sequences contained in the covering fraction increase (also exponentially) with the chain length.

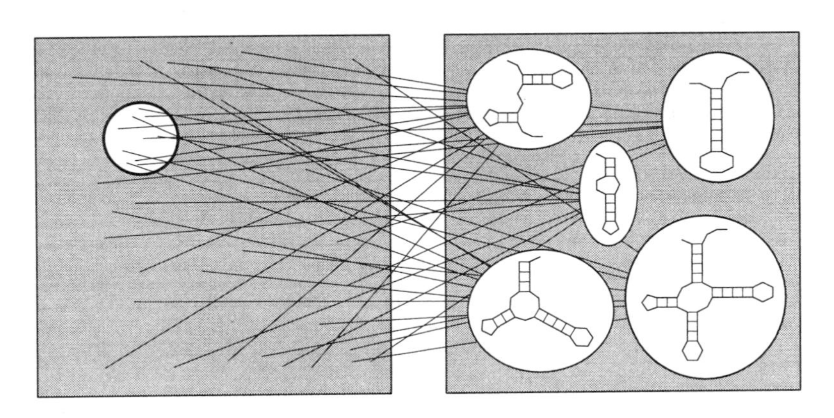

Figure 3. Shape space covering. Only a (relatively small) spherical environment around any arbitrarily chosen reference sequence has to be searched in order to find RNA sequences for every common secondary structure.

Table 2. The shape space covering radius for common secondary structures

n	Covering Radius r_{cov}*				$S_n/4^{\kappa}$‡
	Exhaustive Folding		Asymptotic Value of S_n		
	GC†	**AU**	$\kappa = 2$	$\kappa = 4$	
20	3 (3.4)	2	4	2	3.29×10^{-9}
25	4 (4.7)	2	4	3	4.96×10^{-11}
30	6 (6.1)	3	7	4	7.96×10^{-13}
50			12	6	7.32×10^{-20}
100			26	15	4.52×10^{-37}

*The covering radius is estimated by means of a straightforward statistical estimate based on the assumption that sequences folding into the same structure are randomly distributed in sequence space.

†Exact values derived from exhaustive folding are given in parentheses.

‡Fraction of **AUGC** sequence space that has to be searched on the average in order to find a minimum at least one sequences for every common structure.

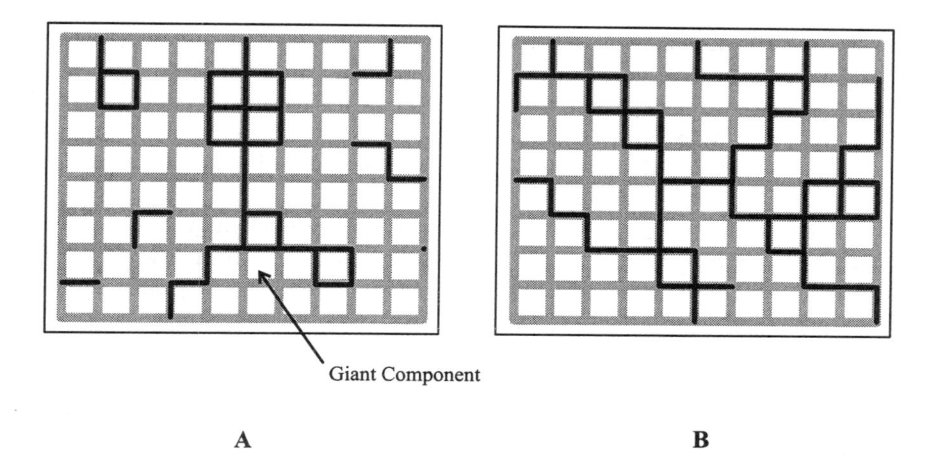

Figure 4. Connectivity of neutral networks. A neutral network consists of many components if the average fraction of neutral neighbors in sequence space (λ) is below a threshold value (λ_{cr}). Random graph theory predicts the existence of one giant component that is much larger than any other component (**A**). If λ exceeds the threshold value (**B**) the network is connected and spans the entire sequence space.

4. NEUTRAL NETWORKS

Since every common structure is formed by a great number of sequences it is important to know how the sequences folding into the same structure are organized in sequence space. The set of sequences forming one structure has been characterized as a neutral network. Two approaches were applied in the study of neutral networks: a mathematical model based on random graph theory[41] and exhaustive folding.[20] The mathematical model assumes that sequences forming the same structure are distributed randomly in the space of compatible sequences (A compatible sequence that can in principle form the structure under consideration, i.e. it has complementary bases in all positions where the structure has a base pair, or, in other words, a compatible sequence forms the structure either as its minimum free energy structure or it contains it in its set of suboptimal foldings.) Neutral networks are modeled by random graphs in sequence space. The statistics of these random graphs is described by a single parameter, λ, measuring the fraction of neighbors in sequence space that are neutral. The structure of neutral networks depends critically on the value λ. The connectivity of networks changes drastically threshold when λ crosses a threshold value:

$$\lambda_{cr}(\kappa) = 1 - \sqrt[\kappa-1]{\frac{1}{\kappa}}.$$

The parameter κ in this equation represents the size of the alphabet, $\kappa = 4$ (**A**, **U**, **G**, **C**) for bases in single stranded regions of RNA molecules and $\kappa = 6$ (**AU**, **UA**, **UG**, **GU**, **GC**, **CG**) for base pairs. Neutral networks consist of a single component that spans whole sequence space if $\lambda > \lambda_{cr}$ and below threshold, $\lambda < \lambda_{cr}$, the network is partitioned into a great number of components, in general, a giant component and many small ones (Fig. 4).

Exhaustive folding allows to check the predictions of random graph theory. The typical sequence of components for neutral networks (either a connected network spanning whole sequence space or a very large component accompanied by many small ones) is indeed found with many common structures. There are, however, also numerous networks with significantly different sequences of components. We find networks with two as well as four equal sized large components, and three components with an approximate size ratio of 1:2:1. Differences between the predictions of random graph theory and the results of exhaustive folding were readily explained in terms of special properties of RNA secondary structures.[20]

The deviations from the ideal network (as predicted by random graph theory) can be identified as structural features that are not accounted for by some simple base pairing logics. All structures that cannot readily form additional base pairs when the sequence requirement is fulfilled behave perfectly normal. There are, however, structures that can form additional base pairs (and will generally do so under the minimum free energy criterion) provided the sequences carry complementary bases at the corresponding positions (Fig. 5). Class II structures, for example, are least likely to be formed when the overall base composition is 50% **G** and 50% **C**, because the probability for forming an additional base pair and folding into another structure is largest then. When there is an excess of **G** ({50+δ}%) it is much more likely that such a structure will actually be formed. The same is true for an excess of **C** and this is precisely reflected by the neutral networks of class II structures with two (major) components: the maximum probabilities for forming class II structures are $\mathbf{G} : \mathbf{C} = (50 + \delta) : (50 - \delta)$ for one component and $\mathbf{G} : \mathbf{C} = (50 + \delta) : (50 - \delta)$ for the second one. By the same token structures of class III have two (independent) possibilities to form an additional base pair and thus they have the highest probability to be formed if the sequences have excess δ and ε. If no additional information is available we can assume $\varepsilon = \delta$. Independent superposition yields then four equal sized components with **G**:**C** compositions of $(50+2\delta) : (50-2\delta)$, $2 \times (50 : 50)$, and $(50-2\delta) : (50+2\delta)$ precisely as it is observed indeed with four component neutral networks. Three component networks are *de facto* four component networks in which the two central (50:50) components have merged to a single one. Neutral networks thus are described well by the random graph model. The assumption that sequences folding into the same structure are randomly distributed in the space of compatible sequences is justified unless special structural features lead to systematic biases.

5. OPTIMIZATION ON COMBINATORY LANDSCAPES

Sewal Wright[49] used hill-climbing as a metaphor for evolutionary optimization. Populations or species are thought to migrate on a fitness landscape. In more recent years this concept has been revived by emphasizing the complex nature of fitness landscapes.[11,17,31,32] Model landscapes[33] as well as those derived from folding RNA molecules into structures and evaluating fitness related properties[14,16,17] were used to study adaptive walks of populations on fitness landscapes. The fundamental concept used in modeling evolutionary optimization is the molecular quasispecies[9,11,12]: the stationary state of a population is characterized by fittest type called master sequence which is surrounded by mutants in sequence space (Fig. 6). The frequency of mutants is determined by their fitness as well as by their Hamming distance from the master. The quasispecies represents the genetic reservoir in asexual reproduction. The concept was originally developed for infinite populations but can be readily extended to finite population sizes.[5,39]

An extension of adaptive evolution to migration of populations through sequence space in absence of fitness differences is straightforward. Neutral evolution has been studied on

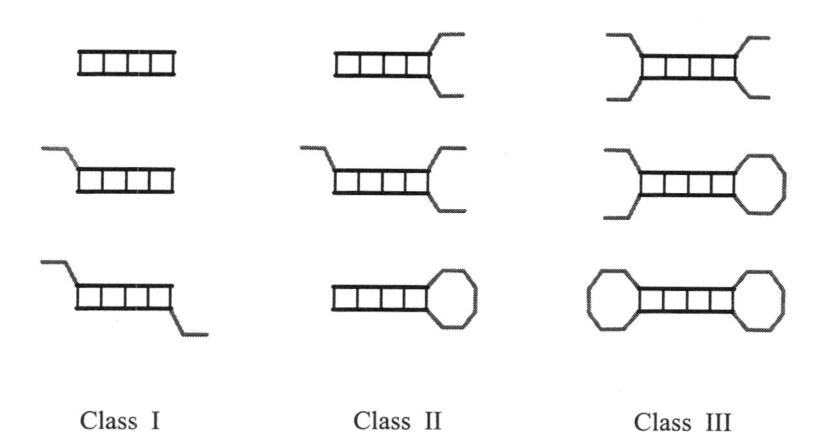

Figure 5. Three classes of RNA secondary structures forming different types of neutral networks. Structures of class I contain no mobile elements (free ends, large loops or joints) or have only mobile elements that cannot form additional base pairs. The mobile elements of structures of class II allow the extension of stacks by additional base pairs at one position. Stacks in class III structures can be extended in two positions. In principle, there are also structures that allow extensions of stacks in more than two ways but they play no role for short chain length ($n < 30$).

model landscapes by analytical approaches[7] derived from the random-energy model[6] as well as by computer simulation.[22,23] More recently the computer simulations were extended to neutral evolution on RNA folding landscapes.[26] In case of selective neutrality populations drift randomly in sequence space by a diffusion-like mechanism. Populations corresponding to large areas in sequence space are partitioned into smaller patches which have some average life time. These feature of population dynamics in neutral evolution was seen in analogy to the formation and evolution of biological species.[23]

In order to visualize the course of adaptive walks on fitness landscapes derived from RNA folding we distinguish single walkers from migration of populations and "non-neutral" landscapes from those built upon extended neutral networks (Fig. 6):

1. Single walkers in the non-neutral case can reach only nearby lying local optima since they are trapped in any local maximum of the fitness landscape.[14] Single walkers are unable to bridge any intermediate value of lower fitness and hence the walk ends when there is no one-error variant of higher fitness.

2. Populations thus have an "smoothing" effect on landscapes. Even in the non-neutral case sufficiently large populations will be able to escape from local optima provided the Hamming distance to the nearest point with a non-smaller fitness value can be spanned by mutation. In computer simulations of populations with about 3000 RNA molecules jumps of Hamming distances up to six were observed.[17]

3. Optimization follows a combined mechanism in the presence of extended neutral networks: adaptive walks leading to minor peaks are supplemented by random drift along networks that enable populations to migrate to areas in sequence space with higher fitness values (Fig. 6). Eventually, the global fitness optimum is reached.

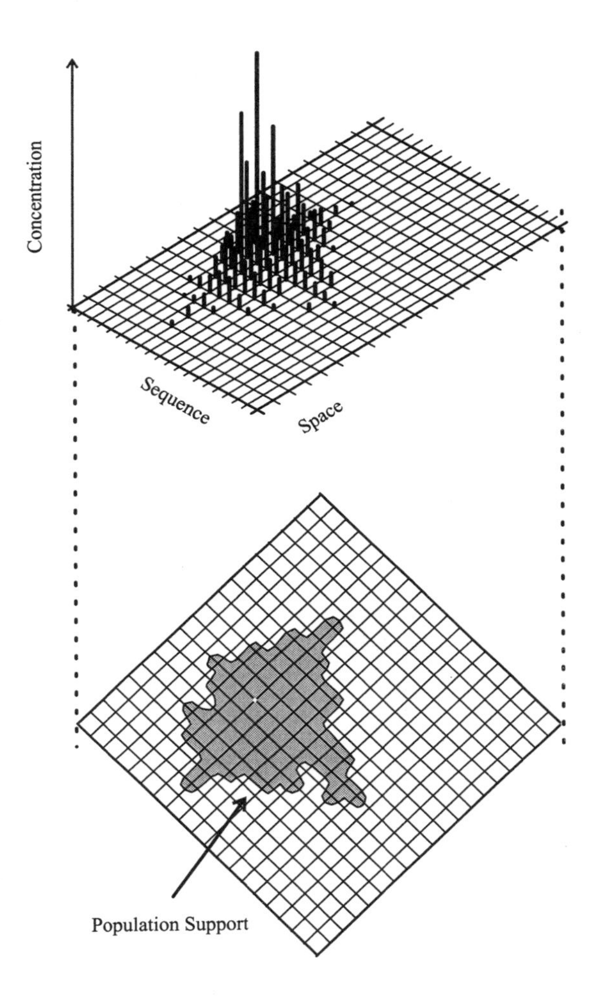

Figure 6. Molecular quasispecies in sequence space. The quasispecies is a stationary mutant distribution surrounding a (fittest and most frequent) master sequence. The frequencies of individual mutants are determined by their fitness values and by their Hamming distances from the master. A quasispecies occupies some region in sequence space called the population support. In the non-stationary case the (population) support migrates through sequence space.

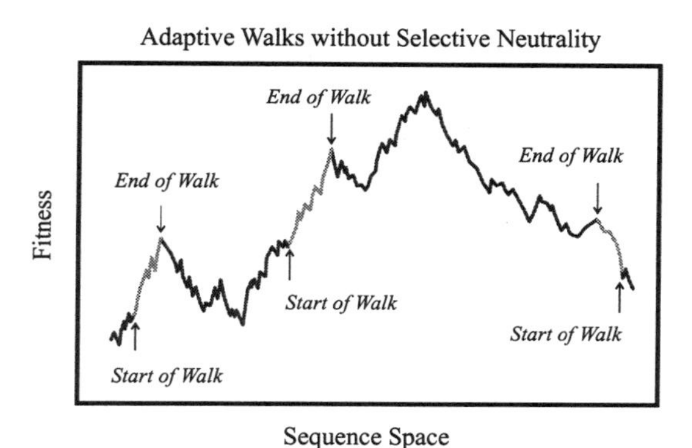

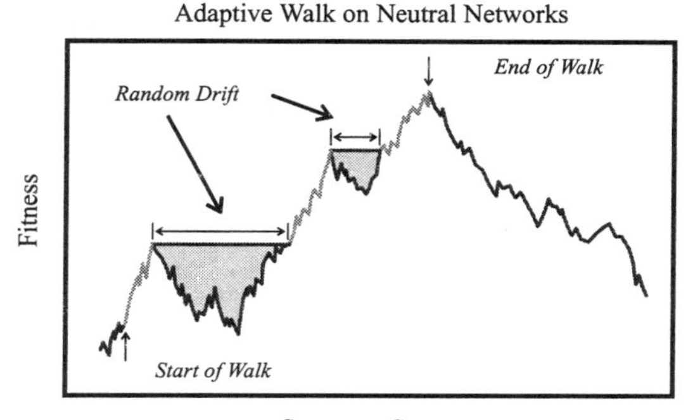

Figure 7. Optimization in sequence space through adaptive walks of populations. Adaptive walks allow to choose the next step arbitrarily from all directions of where fitness is (locally) non-decreasing. Populations can bridge over narrow valleys with widths of a few point mutations. In absence of selective neutrality (upper part) they are, however, unable to span larger Hamming distances and thus will approach only the next major fitness peak. Populations on rugged landscapes with extended neutral networks evolve by a combination of adaptive walks and random drift at constant fitness along the network (lower part). Eventually populations reach the global maximum of the fitness landscape.

6. CONCLUDING REMARKS

Most of the studies discussed here were based on structures evaluated by means of the minimum free energy criterium of RNA folding. The validity of the statistical results, like shape space covering or the existence of extended neutral networks, is, however, not limited to minimum free energy folding since they belong to the (largely) algorithm independent properties of RNA secondary structures.[47] Whether or not our results can be transferred to three-dimensional structures of RNA molecules is an open question. There are strong indications that this will be so although the degree of neutrality is expected to be smaller than with the secondary structures. The answer to this question will be given only by carrying out suitable experiments on screening of RNA structures and properties in sequence space. Corresponding experiments dealing with binding of RNA molecules to predefined targets as the function to be optimized are under way in our group. A whole wealth of data on protein folding and resilience of protein structures against exchanges of amino acid residues seems to confirm the validity of shape space covering and the existence of extended neutral networks for proteins too.

Neutral evolution apparently is not a dispensable addendum to evolutionary optimization. In contrary, neutral networks provide a medium through which evolution can become really efficient. Adaptive walks of populations, usually ending in one of the nearby minor peaks of the fitness landscape, are supplemented by random drift on neutral networks. The phase of neutral diffusion ends when the population reaches an area of higher fitness values. Series of adaptive walks interrupted by (selectively neutral) random drift periods allow to approach the global minimum provided the neutral networks are sufficiently large.

ACKNOWLEDGMENTS

The work reported here has been supported financially by the Austrian *Fonds zur Förderung der wissenschaftlichen Forschung* (Projects No.9942-PHY, No.10578-MAT, and No.11065-CHE), by the Commission of the European Community (CEC Contract Study PSS*0884) and by the Santa Fe Institute. Fruitful discussions with Drs. Walter Fontana, Christian Forst and Peter Stadler are gratefully acknowledged.

REFERENCES

1. D. P. Bartel and J. W. Szostak. Isolation of new ribozymes from a large pool of random sequences. *Science*, 261:1411–1418, 1993.
2. A. A. Beaudry and G. F. Joyce. Directed evolution of an RNA enzyme. *Science*, 257:635–641, 1992.
3. C. K. Biebricher and M. Eigen. Kinetics of RNA replication by Qβ replicase. In E. Domingo, J. J. Holland, and P. Ahlquist, editors, *RNA Genetics. Vol.I: RNA Directed Virus Replication*, pages 1–21. CRC Press, Boca Raton, FL, 1988.
4. K. B. Chapman and J. W. Szostak. *In vitro* selection of catalytic RNAs. *Curr. Opinion in Struct. Biol.*, 4:618–622, 1994.
5. L. Demetrius, P. Schuster, and K. Sigmund. Polynucleotide evolution and branching processes. *Bull. Math. Biol.*, 47:239–262, 1985.
6. B. Derrida. Random-energy model: An exactly solvable model of disordered systems. *Phys. Rev. B*, 24:2613–2626, 1981.
7. B. Derrida and L. Peliti. Evolution in a flat fitness landscape. *Bull. Math. Biol.*, 53:355–382, 1991.
8. T. Dobzhansky, F. J. Ayala, G. L. Stebbins, and J. W. Valentine. *Evolution*. W.H. Freeman & Co., San Francisco, CA, 1977.
9. M. Eigen. Selforganization of matter and the evolution of biological macromolecules. *Naturwissenschaften*, 58:465–523, 1971.
10. M. Eigen and W. C. Gardiner. Evolutionary molecular engineering based on RNA replication. *Pure Appl. Chem.*, 56:967–978, 1984.
11. M. Eigen, J. McCaskill, and P. Schuster. The molecular quasispecies. *Adv. Chem. Phys.*, 75:149 – 263, 1989.
12. M. Eigen and P. Schuster. The hypercycle. A principle of natural self-organization. Part A: Emergence of the hypercycle. *Naturwissenschaften*, 64:541–565, 1977.
13. A. D. Ellington. Aptamers achieve the desired recognition. *Current Biology*, 4:427–429, 1994.
14. W. Fontana, T. Griesmacher, W. Schnabl, P. Stadler, and P. Schuster. Statistics of landscapes based on free energies, replication and degradation rate constants of RNA secondary structures. *Mh. Chem.*, 122:795–819, 1991.
15. W. Fontana, D. A. Konings, P. Stadler, and P. Schuster. Statistics of RNA secondary structures. *Biopolymers*, 33:1389–1404, 1993.
16. W. Fontana, W. Schnabl, and P. Schuster. Physical aspects of evolutionary optimization and adaptation. *Phys. Rev. A*, 40:3301–3321, 1989.
17. W. Fontana and P. Schuster. A computer model of evolutionary optimization. *Biophys. Chem.*, 26:123–147, 1987.

18. W. Fontana, P. Stadler, E. Bornberg-Bauer, T. Griesmacher, I. Hofacker, M. Tacker, P. Tarazona, E. Weinberger, and P. Schuster. RNA folding and combinatory landscapes. *Phys. Rev. E*, 47:2083 – 2099, 1993.
19. W. Grüner, R. Giegerich, D. Strothmann, C. Reidys, J. Weber, I. Hofacker, P. Stadler, and P. Schuster. Analysis of RNA sequence structure maps by exhaustive enumeration. I. Neutral networks. *Mh.Chem.*, 127:355–374, 1996.
20. W. Grüner, R. Giegerich, D. Strothmann, C. Reidys, J. Weber, I. Hofacker, P. Stadler, and P. Schuster. Analysis of RNA sequence structure maps by exhaustive enumeration. II. Structure of neutral networks and shape space covering. *Mh.Chem.*, 127:375–389, 1996.
21. R. W. Hamming. Error detecting and error correcting codes. *Bell Syst. Tech. J.*, 29:147–160, 1950.
22. P. G. Higgs and B. Derrida. Stochastic models for species formation in evolving populations. *J. Physics A*, 24:L985–L991, 1991.
23. P. G. Higgs and B. Derrida. Genetic distance and species formation in evolving populations. *J. Mol. Evol.*, 35:454–465, 1992.
24. I. Hofacker, P. Schuster, and P. Stadler. Combinatorics of RNA secondary structures. *Submitted to SIAM J. Disc. Math.*, 1996.
25. P. Hogeweg and B. Hesper. Energy directed folding of RNA sequences. *Nucleic Acids Research*, 12:67–74, 1984.
26. M. A. Huynen, P. F.Stadler, and W. Fontana. Smoothness within ruggedness: The role of neutrality in adaptation. *Proc. Natl. Acad. Sci. USA*, 93:397–401, 1996.
27. R. D. Jenison, S. C. Gill, A. Pardi, and B. Polisky. High-resolution molecular discrimination by RNA. *Science*, 263:1425–1429, 1994.
28. G. F. Joyce. Directed molecular evolution. *Sci. Am.*, 267(6):48–55, 1992.
29. S. A. Kauffman. Autocatalytic sets of proteins. *J. Theor. Biol.*, 119:1–24, 1986.
30. S. A. Kauffman. Applied molecular evolution. *J. Theor. Biol.*, 157:1–7, 1992.
31. S. A. Kauffman. *The Origins of Order. Self-Organization and Selection in Evolution.* Oxford University Press, Oxford, UK, 1993.
32. S. A. Kauffman and S. Levine. Towards a general theory of adaptive walks on rugged landscapes. *J. Theor. Biol.*, 128:11–45, 1987.
33. S. A. Kauffman and E. D. Weinberger. The n-k model of rugged fitness landscapes and its application to maturation of the immune response. *J. Theor. Biol.*, 141:211–245, 1989.
34. M. Kimura. *The Neutral Theory of Molecular Evolution.* Cambridge University Press, Cambridge, UK, 1983.
35. D. Konings and P. Hogeweg. Pattern analysis of RNA secondary structure. Similarity and consensus of minimal-energy folding. *J. Mol. Biol.*, 207:597–614, 1989.
36. N. Lehman and G. F. Joyce. Evolution *in vitro*: Analysis of a lineage of ribozymes. *Current Biology*, 3:723–734, 1993.
37. R. E. Lenski and M. Travisano. Dynamics of adaptation and diversification: A 10,000-generation experiment with bacterial populations. *Proc. Natl. Acad. Sci. USA*, 91:6808–6814, 1994.
38. J. R. Lorsch and J. W. Szostak. *In vitro* evolution of new ribozymes with polynucleotide kinase activity. *Nature*, 371:31–36, 1994.
39. M. Nowak and P. Schuster. Error thresholds of replication in finite populations. Mutation frequencies and the onset of Muller's ratchet. *J. Theor. Biol.*, 137:375–395, 1989.
40. C. Reidys and P. F. Stadler. Bio-molecular shapes and algebraic structures. *Computers Chem.*, 20:85–94, 1996.
41. C. Reidys, P. F. Stadler, and P. Schuster. Generic properties of combinatory maps - Neutral networks of RNA secondary structures. *Bull. Math. Biol.*, 1995. Submitted. SFI-Preprint Series No. 95-07-058.
42. P. Schuster. How to search for RNA structures. Theoretical concepts in evolutionary biotechnology. *Journal of Biotechnology*, 41:239–257, 1995.
43. P. Schuster, W. Fontana, P. Stadler, and I. Hofacker. From sequences to shapes and back: A case study in RNA secondary structures. *Proc.Roy.Soc.(London)B*, 255:279–284, 1994.
44. P. Schuster and P. Stadler. Landscapes: Complex optimization problems and biopolymer structures. *Computers Chem.*, 18:295–314, 1994.
45. S. Spiegelman. An approach to the experimental analysis of precellular evolution. *Quart. Rev. Biophys.*, 4:213–253, 1971.
46. G. Strunk. *Automatized evolution experiments in vitro and natural selection under controlled conditions by means of the serial transfer technique*. PhD thesis, Universität Braunschweig, 1993.
47. M. Tacker, P. Stadler, E. Bornberg-Bauer, I. Hofacker, and P. Schuster. Algorithm independent properties of RNA secondary structure predictions. Submitted to Eur.Biophys.J., 1996.
48. C. Tuerk and L. Gold. Systematic evolution of ligands by exponential enrichment: RNA ligands to bacteriophage T4 DNA polymerase. *Science*, 249:505–510, 1990.

49. S. Wright. The roles of mutation, inbreeding, crossbreeding and selection in evolution. In D. F. Jones, editor, *Int. Proceedings of the Sixth International Congress on Genetics*, volume 1, pages 356–366, 1932.

22

HOW THREE-FINGERED SNAKE TOXINS RECOGNISE THEIR TARGETS

Acetylcholinesterase-Fasciculin Complex, a Case Study

Kurt Giles,[1,2] Mia L. Raves,[1] Israel Silman,[2] and Joel L. Sussman[1,3]

[1]Department of Structural Biology
[2]Department of Neurobiology
Weizmann Institute of Science
Rehovot, 76100, Israel
[3]Department of Biology
Brookhaven National Laboratory
Upton, New York 11973

1. ABSTRACT

Three-fingered toxins from snake venoms constitute a family of 6–8kDa proteins, which can be divided into a number of groups with widely varying targets. The most studied of these are the α-neurotoxins which have long been used for the purification and characterisation of nicotinic acetylcholine receptors. Another group are the fasciculins, which specifically inhibit another essential synaptic component, the enzyme acetylcholinesterase.

The structure of acetylcholinesterase is described, with emphasis on its unique features of relevance to its interaction with fasciculin. Furthermore, all three-fingered toxins whose structures have been determined are compared at both the sequence and structural level. With the additional use of a multiple sequence alignment of all known three-fingered toxins, residues of fasciculins which were expected to be important in the interaction with acetylcholinesterase were identified.

The recently determined three-dimensional structure of the acetylcholinesterase-fasciculin complex not only validates the use of sequence comparison for identifying important residues, but in addition reveals unexpected interactions between toxin and target that are impossible to predict from sequence alone.

Theoretical and Computational Methods in Genome Research, edited by Suhai
Plenum Press, New York, 1997

2. ACETYLCHOLINESTERASE

Acetylcholinesterase (AChE) is an essential component of cholinergic synapses in both the peripheral and central nervous system, where its main function is the rapid hydrolysis of the neurotransmitter acetylcholine (ACh). Its existence was first proposed by Sir Henry Dale in 1914 [1] and demonstrated experimentally by Loewi and Navratil in 1926 [2]. However, it is only ten years since the first primary sequence of a cholinesterase was determined [3], and just five years since the three-dimensional structure was solved [4].

The active site of AChE was generally considered to consist of two subsites; a negatively charged or 'anionic' site, to which the positively charged quaternary nitrogen of ACh binds, and an esteratic site containing the catalytic residues [5]. A second 'anionic' site, which became known as the peripheral 'anionic' site, around 14Å from the active site, was proposed on the basis of the binding of *bis*-quaternary compounds [6]. The early literature on cholinesterases has been extensively reviewed in a book by Ann Silver [7], and a number of reviews cover the more recent work [8–13].

The structure determination of AChE from the electric ray *Torpedo californica* (Fig. 1) [4] revealed the active site to contain a catalytic triad. However, rather than the Ser-His-Asp triad usually found in serine hydrolases [14], it was found to be Ser-His-Glu (S200, H440, E327). Mutagenesis of the glutamic acid of the catalytic triad (E327) to

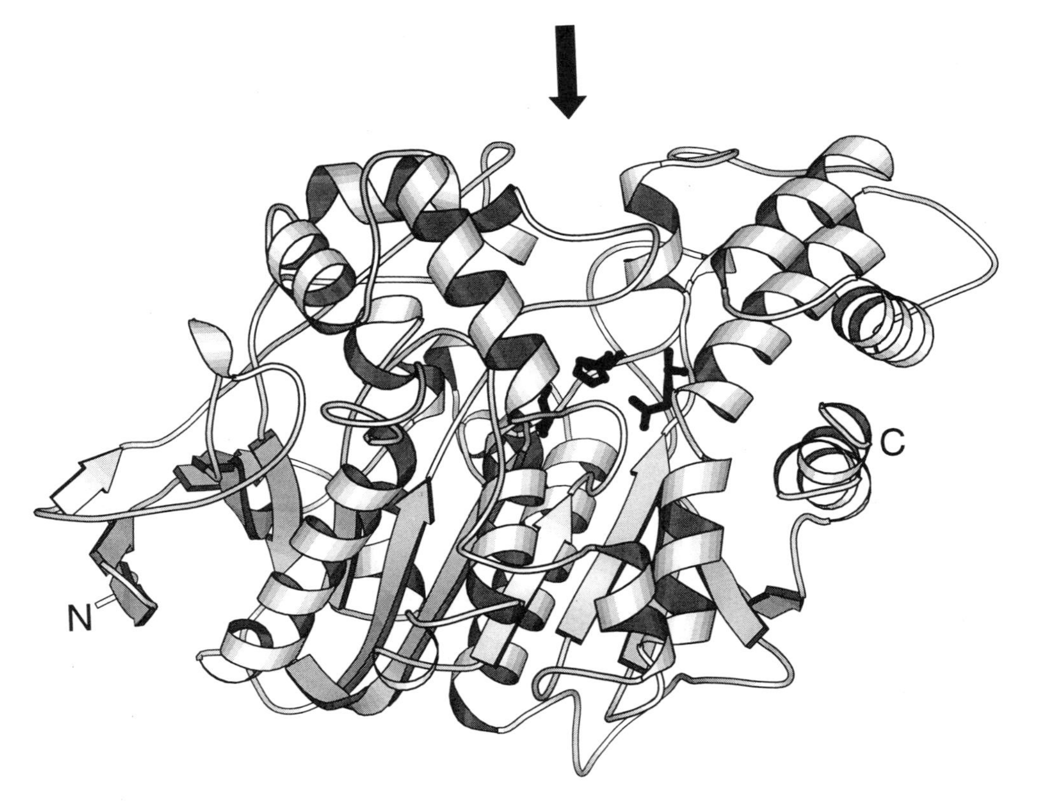

Figure 1. Three-dimensional structure of *T. californica* AChE (PDB code 1ACE). The residues of the catalytic triad are depicted in bold type as balls and sticks. The arrow delineates the entrance to the active site gorge, and the assumed direction of substrate entry.

aspartic acid rendered the enzyme inactive [15]. At the time of the structure determination only one other enzyme was known with a Ser-His-Glu catalytic triad: a structurally related lipase from the fungus *Geotrichum candidum* [16]. Other examples of the Ser-His-Glu catalytic triad have since been found in the lipase [17] and cholesterol esterase [18] of *Candida rugosa*, which both belong to the same structural family as AChE and *G. candidum* lipase, that of the α/β hydrolase fold [19].

The 'anionic' site, to which the positively charged quaternary amino group of the ACh molecule binds, was unexpectedly found not be composed of negatively charged residues, but to contain a tryptophan residue (W84), conserved in cholinesterases of all species. The quaternary group binds primarily via interaction with the π-electrons of the tryptophan indole ring; such 'cation-π' interactions have recently been the subject of considerable interest [reviewed in 20]. Furthermore, the active site was found to lie near the bottom of a narrow gorge that reaches deep into the protein. A substantial portion of the surface of the active site gorge is lined by fourteen aromatic residues, most of which are conserved in all known AChEs [21].

The *bis*-quaternary compound decamethonium was shown to span the gorge, binding to both the 'anionic' subsite of the active site and the peripheral 'anionic' site at the top of the gorge (Fig. 2) [22], thus identifying a tryptophan residue (W279) as part of the peripheral 'anionic' site.

T. californica AChE has an overall net charge of -11e and an unequal spatial distribution of charged residues, which give rise to a large dipole moment, approximately aligned with the active site gorge, of nearly 1700 Debye at physiological ionic strength and pH [23]. Experimental determination of the dipole moment for an AChE purified from the venom of the snake *Bungarus fasciatus*, yielded a value of 1000 Debye [24]. This has led to the suggestion of an electrostatic mechanism for the guidance of the positively charged substrate, ACh, to the active site [25, 26]. In order to test this hypothesis, a multiple mutant of human AChE was made in which seven negatively charged residues from around the top of the gorge were replaced by neutral ones [27]. Although mutagenesis greatly decreased the negative isopotential surface over the entrance to the active site gorge of AChE, the mutant was found to retain around half the catalytic activity of the wild-type enzyme [27]. This may be due to the necessity to take into account the contribution of charged residues within the gorge and on the positively charged hemisphere surrounding the bottom of the gorge [23, 28].

AChE has amazing catalytic power, possessing one of the fastest turnover numbers reported for enzyme catalysis [11]. The diffusion of ACh to the active site is probably rate determining [29]. Although ACh would be electrostatically attracted to the active site of AChE, one of the products, choline, retains a net positive charge, and its expulsion from the active site gorge is, therefore, opposed by the electric field [25]. In addition, crystals of *T. californica* AChE allow large ligands containing quaternary ammonium groups to diffuse into the active site [22], even though the top of the active site gorge is tightly blocked by a symmetry-related molecule in the crystal. Furthermore, the width of the gorge at its narrowest point is smaller than the quaternary ammonium group that is supposed to pass through it (Fig. 2) [30]. All these factors raise cogent questions with respect to traffic of substrate, products and solute to and from the active site. This apparent paradox led to the proposal of an alternative route to the active site termed the 'back door' [30, 31]. Residues C67-C94 of AChE form the so called Ω-loop, whose analogous region in structurally related lipases [19] forms a 'lid' over the active site which must open for substrate entry and product exit [32, 33]. A molecular dynamics simulation of AChE showed the transient opening of a putative 'back door' around residue W84 [31]; however, muta-

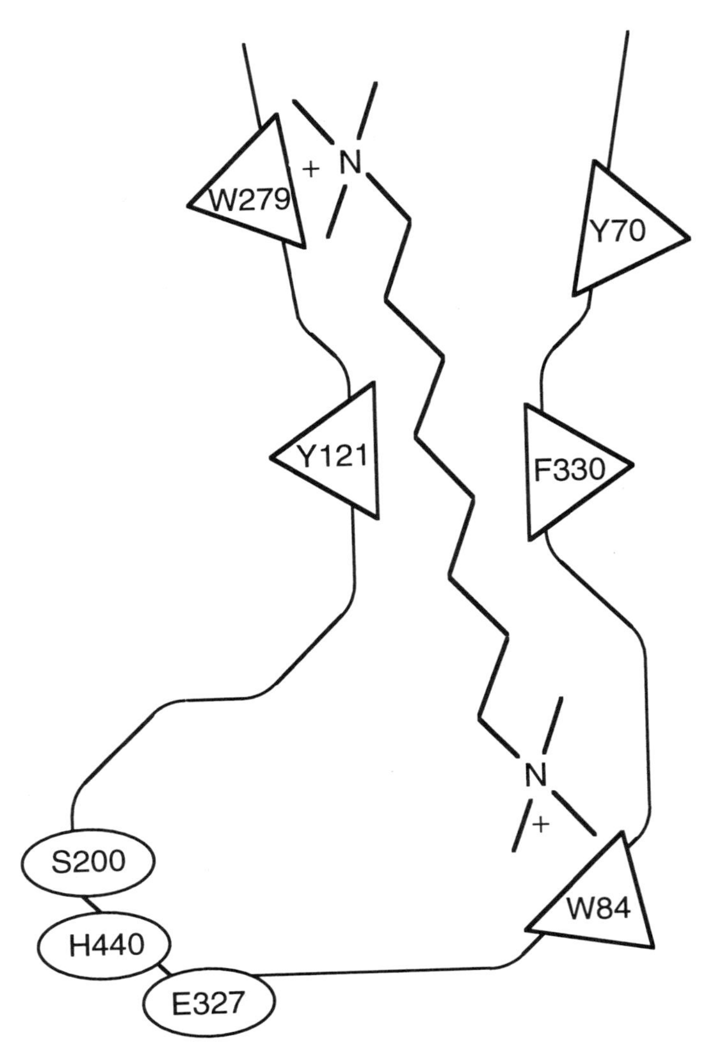

Figure 2. Schematic representation of the active site gorge of AChE with a molecule of decamethonium bound. Triangles represent conserved aromatic residues which line the active site gorge, ovals represent residues of the catalytic triad; residue numbers refer to *T. californica* AChE.

tion studies in which the 'back door' might be closed by a salt bridge [34] or, more substantially, by a disulphide bond [35], both resulted in enzyme with activity similar to that of the wild-type enzyme.

3. THREE-FINGERED SNAKE TOXINS

The most potent reversible inhibitors of AChE are the fasciculins, a set of toxins isolated from mamba (*Dendroaspis*) snake venoms, which have inhibition constants in the region of 10^{-8}–10^{-11} M for AChEs from various species [36]. Fasciculins belong to a family of 6–8kDa proteins isolated from the venoms of snakes of the *Elapidae* and *Hydrophida* families. Nearly 200 such proteins have been found which display significant sequence homology, suggesting that they are evolutionarily and structurally related. However, despite this conservation, they vary considerably in their pharmacological actions [37]. In addition to the AChE-specific fasciculins, some such toxins are specific for ACh receptors; muscle-type nicotinic (α-neurotoxins) [38], neuronal-type nicotinic (κ-neurotoxins) [39], and muscarinic [40]; others inhibit calcium channels [41], and the cytotoxins are involved in cell depolarisation and lysis by an unknown mechanism. There are also a number of related toxins which have little or no direct toxicity but possibly work synergistically with other venom components [37].

Both X-ray crystallography and nuclear magnetic resonance (NMR) have been used for the determination of the three-dimensional structures of snake toxins, and the structures of 20 different three-fingered toxins are now available from the Brookhaven Protein Data Bank [42]. All structures have essentially the same secondary structural elements; two β-sheets, one consisting of three antiparallel β-strands, and the other of two short antiparallel β-strands near the amino terminus. The conserved secondary-structural elements can be seen in a multiple sequence alignment of these 20 toxins (Fig. 3a). The tertiary

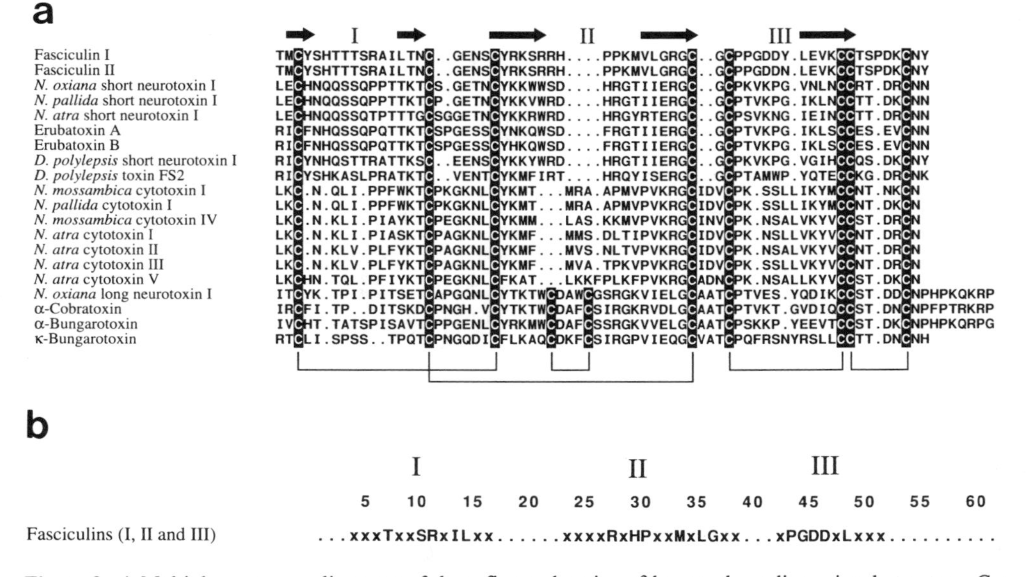

Figure 3. a) Multiple sequence alignment of three-fingered toxins of known three-dimensional structure. Conserved regions of secondary structure (β-strands) depicted by arrows above the alignment, and disulphide bond connectivity at the bottom. b) Predicted important residues in the finger regions of fasciculins.

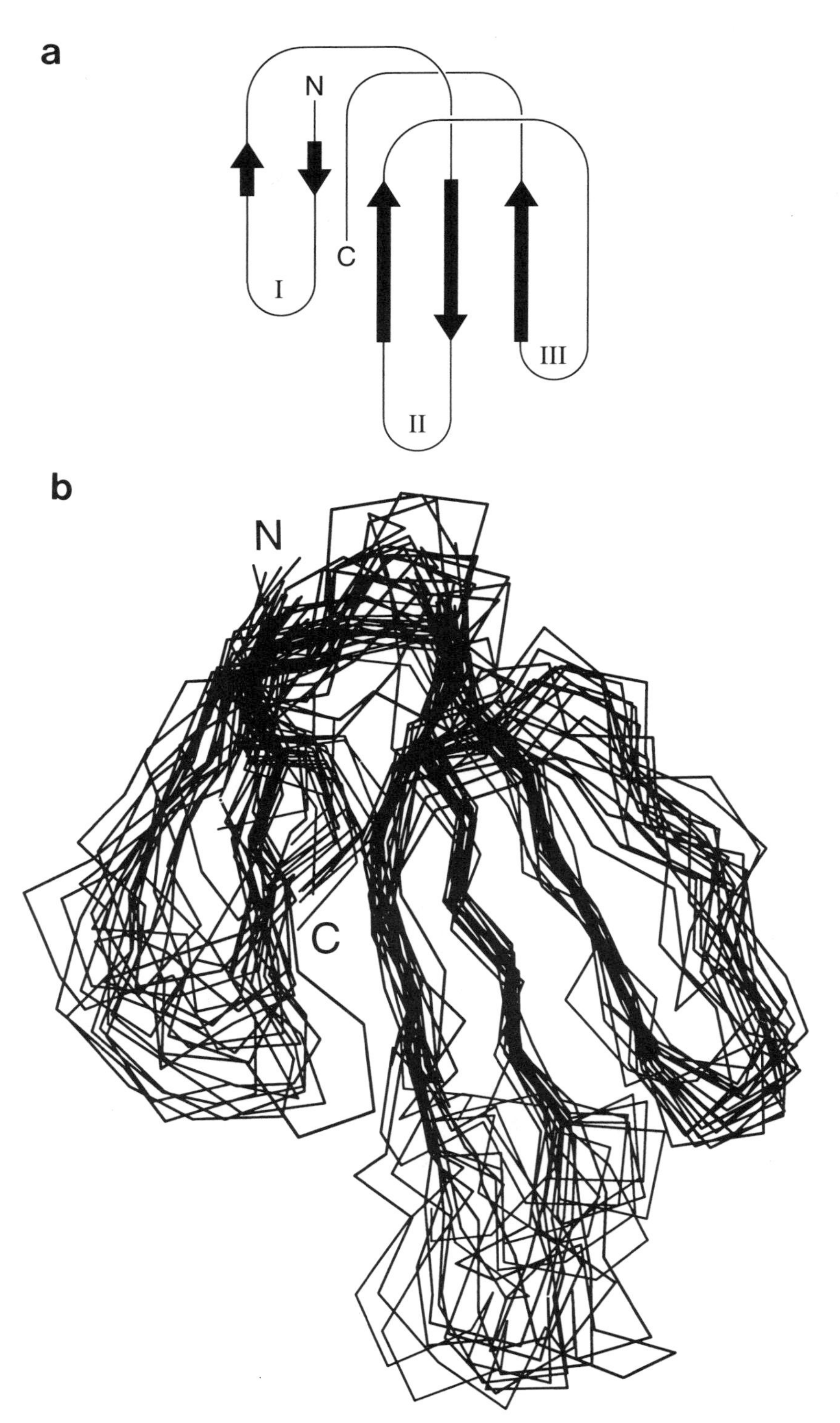

Figure 4. Tertiary structure of three-fingered snake toxins. a) Schematic representation showing the topologically conserved fold; β-strands depicted by arrows. b) Overlay of C_α atoms of 20 different three-fingered toxins, obtained by aligning the conserved β-strand residues of each toxin with those of fasciculin I (PDB code 1FAS).

structure of these toxins contains a core region, consisting of a series of short loops and four disulphide bonds, from which three longer loops, containing the β-strands, protrude like the central fingers of a hand. The tips of the 'fingers' define three short loops; loop I between the first two β-strands, loop II between the third and fourth β-strands, and loop III comprising the four residues leading to the beginning of the fifth β-strand (Fig. 4a). Alignment of these toxins by overlap of the β-strands of each toxin (Fig. 4b), shows most of the structure, including the disulphide bonds, to be highly conserved. There is, however, significant structural deviation at the tips of the fingers, especially in loops I and II. This flexibility is also revealed by an ensemble of NMR structures for a single toxin, which gives an idea of the conformational freedom of a toxin in solution. For example, in *D. polylepsis* short neurotoxin I (PDB code 1NTX) the position of the C_α atom of a threonine residue (T10) at the tip of loop I varies 3.4Å between ensemble members, and the position of the C_α atom of a histidine residue (H32) at the tip of loop II varies 6.8Å between ensemble members [38].

The variations in primary sequence and tertiary structure of the three loops are likely to account for the various pharmacological specificities. As can be seen from Fig. 3a, and

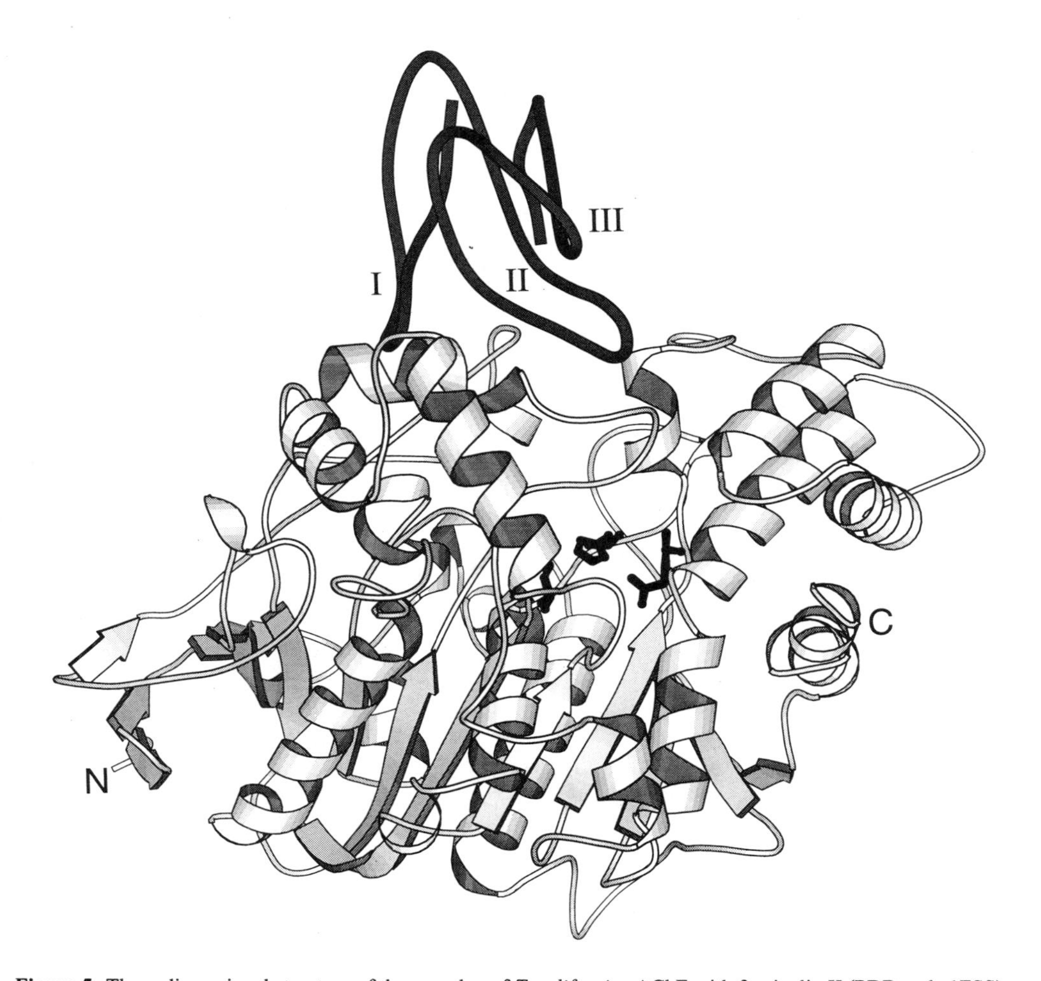

Figure 5. Three-dimensional structure of the complex of *T. californica* AChE with fasciculin II (PDB code 1FSS).

as is confirmed by a much larger alignment of all known three-fingered toxin sequences (KG, unpublished results), certain residues in the loops are specific to particular classes of toxins, for example, the cytotoxins or short neurotoxins. Since loop II protrudes further than loop I or III (Fig. 4), it is probable that it bears the major 'specificity-determining' sequence, and that the residues of loops I and III modulate bonding to the target.

It is likely that residues common to a certain class of toxins (for example the fasciculins) but absent or rare at the same locus in other toxins, determine their pharmacological specificity. These residues were, therefore, identified for the fasciculins (Fig. 3b).

4. ACHE-FASCICULIN COMPLEX

The structure of the complex of AChE and fasciculin II (FAS-II) has recently been solved for both *T. californica* [43] and mouse [44] AChE, representing the first examples of a three-fingered snake toxin in complex with its target molecule. In the complex, FAS-II is located at the peripheral site of AChE, blocking the entrance to the active site gorge (Fig. 5).

The differences in the conformation of both AChE and FAS-II in the complex are very small compared to their respective uncomplexed structures. Most differences for residues of FAS-II can be attributed to the binding to AChE, whereas most of the changes in the AChE structure seem unrelated to binding and may be a result of the different packing interactions between molecules in the crystals of the complex versus crystals of uncomplexed AChE. Although fasciculin I (FAS-I) and FAS-II differ in only a single residue (T47→N), the positions of their C_α atoms differ by up to 10Å for residues of loop I, although the positions of loops II and III are almost identical [45, 46]. It was proposed that this was simply due to the presence of detergent in the crystallisation medium of FAS-II [46].

The difference in the position of loop I in FAS-I and FAS-II explains the failure of an attempt to model the structure of the AChE-FAS complex by docking the rigid structure of FAS-I with *T. californica* AChE [47]. If the FAS-I structure is overlayed onto the FAS-II structure in the AChE-FAS-II complex, loop I clashes with residues 72–77 of AChE.

The affinity of FAS-I and FAS-II for AChE is very similar [48], which implies that FAS-I must undergo a considerable conformational change in loop I, on binding to AChE, from the conformation defined in the uncomplexed crystal structure. The idea that a toxin can have a very different conformation when bound to its target compared to its uncomplexed structure has major implications for studies on the interactions of other three-fingered toxins with their targets. For example, inferences about the quaternary structure of the nicotinic ACh receptor, reached on the basis of labelling studies with chemically modified toxins, were made on the assumption of a fairly rigid toxin structure [49].

There is very large surface complementarity between FAS-II and AChE in the complex; the recess between loops I and II of FAS-II is complementary to a ridge in the AChE structure. Approximately 2000Å^2 of residue surface are buried on binding [43], one of the largest contact areas so far reported. All interactions between AChE and FAS-II in the complex have previously been detailed [43]. Loops I and II of FAS-II account for nearly all the interactions with AChE, loop III does not make any contact with AChE, and the only other interaction is a hydrogen bond between the C-terminal residue of FAS-II (OH of Y61) and the NH_3 group of K341 in AChE.

Loop I lies on the outside of AChE, and interacts with two regions of the AChE molecule (Fig. 6a). An arginine residue (R11) of FAS-II forms hydrogen bonds with the main-chain oxygen atoms of E82 and N85, which restricts the movement of the Ω-loop, and hence may hold the putative 'back door' of AChE closed.

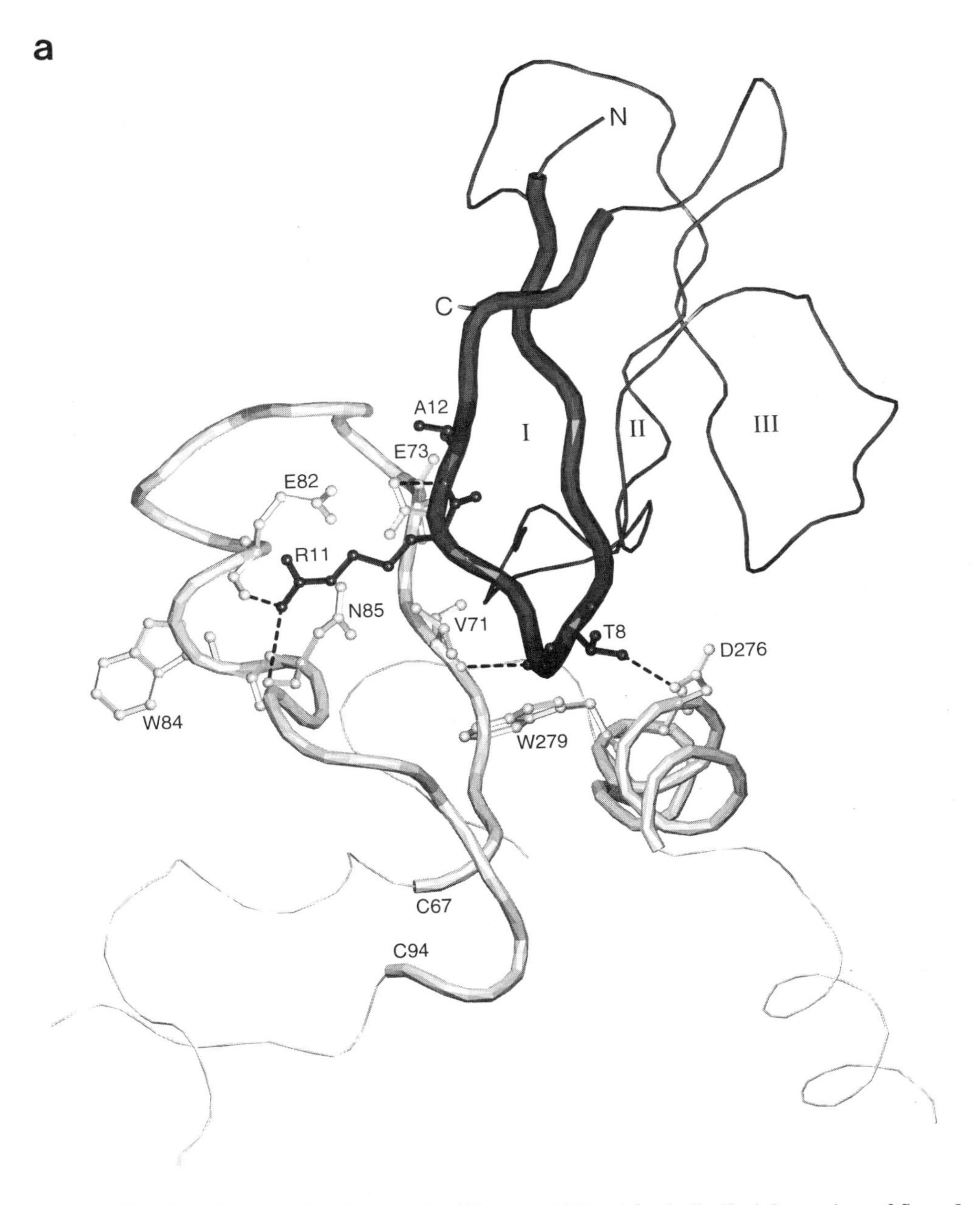

Figure 6. Details of the interaction between *T. californica* AChE and fasciculin II. a) Interactions of finger I of FAS-II. b) Interactions of finger II of FAS-II.

Loop II (Fig. 6b) completely blocks the top of the active site gorge. In the crystalline state, the complementarity between AChE and FAS-II is so good that not even a water molecule can enter the gorge. There are a large number of interactions between loop II and AChE (detailed in ref. [43]). An unusual and interesting interaction was seen between a methionine residue of FAS-II (M33) and a tryptophan residue of AChE (W279) [43, 44]. The S-CH_3 moiety of M33 lies approximately parallel to the plane of the W279 ring, at a

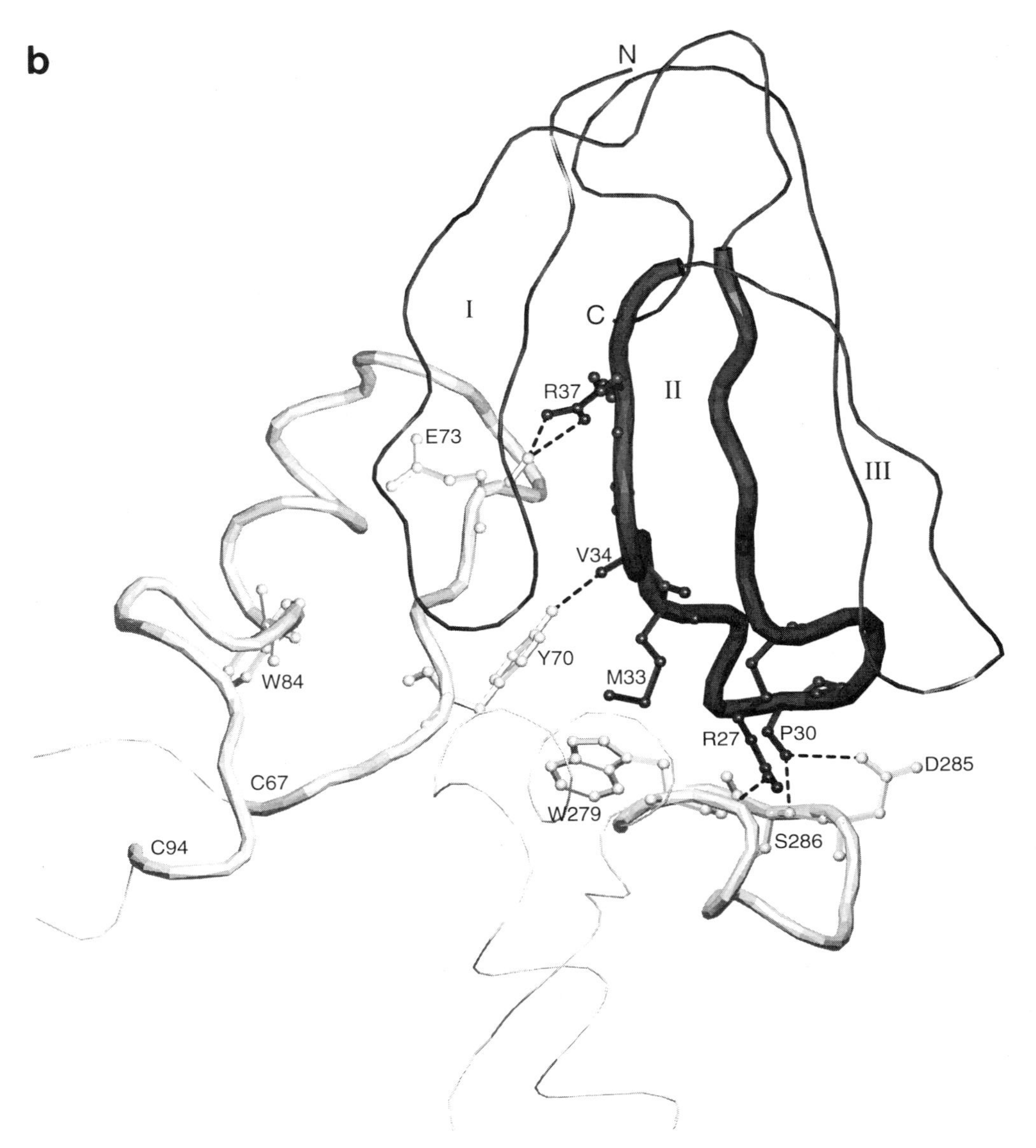

Figure 6. *(Continued)*

distance of around 3.5Å. Such an interaction has been seen in a few other three-dimensional structures, ost notably in AChE itself, between W84 of the 'anionic' site and M83 [43].

It is interesting to note that, like AChE, FAS-II has a large dipole for its size. This is a consequence of a concentration of positive charges in the second finger, and negative charges in the C-terminal region; a characteristic of all fasciculins, but not found in other three-fingered toxins [50]. At physiological ionic strength and pH, FAS-II has a dipole moment of 185 Debye approximately aligned with the second finger [43]. In the complex of AChE and FAS-II, the dipoles are roughly parallel, and may therefore be important in the attraction of fasciculin to AChE [43, 44].

5. DISCUSSION

From a multiple sequence alignment of all known three-fingered toxins (KG, unpublished results), and a comparison of their known three-dimensional structures, it is possible to identify residues that may be important in determining the specificity of these toxins for their targets. The recent solution of the structure of one such toxin in complex with its target [43, 44], suggests that models of interaction between a toxin and its target based on the uncomplexed toxin structure may not be sufficiently accurate, since the toxin may undergo a conformational change on binding to its target.

A number of residues common to fasciculins but rare in all other three-fingered toxins, identified from the multiple sequence alignment, are important in the AChE-FAS-II interaction. However, multiple sequence analysis alone is not sufficient to predict all interactions. For example, the unusual interaction between M33 of fasciculin and the peripheral site tryptophan residue of AChE (W279 in *T. californica*) could not have been predicted.

The complementarity of loop II of FAS-II with AChE in the crystalline state is so good that there is not even space for entry of a water molecule into the gorge; in addition, the putative 'back door' appears to be held closed by loop I. It is, therefore, very surprising to find that the AChE-FAS complex in solution displays both residual activity [51], and the ability to bind ligands at the active site [48]. The rationale for this is presently obscure.

6. ACKNOWLEDGMENTS

This work was supported by the US Army Medical Research and Development Command under Contract DAMD17–93-C-3070, EC-Israel research grant ISC*CT940105, and the Kimmelman Center for Biomolecular Structure ans Assembly, Rehovot, Israel.

7. REFERENCES

1. Dale, H. H. (1914) The action of certain esters and ethers of choline, and their relation to muscarine. *J Pharm Exptl Therap*, **6**, 147–90.
2. Loewi, O. and Navratil, E. (1926) Über humorale Übertragbarkeit der Herznervenwirkung. *Pflügers Archiv*, **214**, 689–96.
3. Schumacher, M., Camp, S., Maulet, Y., Newton, M., MacPhee-Quigley, K., Taylor, S. S., Friedmann, T. and Taylor, P. (1986) Primary structure of *Torpedo californica* acetylcholinesterase deduced from its cDNA sequence. *Nature*, **319**, 407–9.
4. Sussman, J. L., Harel, M., Frolow, F., Oefner, C., Goldman, A., Toker, L. and Silman, I. (1991) Atomic structure of acetylcholinesterase from *Torpedo californica:* a prototypic acetylcholine-binding protein. *Science*, **253**, 872–9.
5. Nachmansohn, D. and Wilson, I. B. (1951) The enzymic hydrolysis and synthesis of acetylcholine. *Adv Enzymol*, **12**, 259–339.
6. Bergmann, F., Wilson, I. B. and Nachmansohn, D. (1950) The inhibitory effect of stilbamidine, curare and related compounds and its relationship to the active groups of acetylcholine esterase. *Biochim Biophys Acta*, **6**, 217–24.
7. Silver, A. (1974), *The biology of cholinesterases*, North Holland Publishing Company, Amsterdam.
8. Rosenberry, T. L. (1975) Acetylcholinesterase. *Adv Enzymol*, **43**, 103–218.
9. Massoulié, J. and Bon, S. (1982) The molecular forms of cholinesterase and acetylcholinesterase in vertebrates. *Ann Rev Neurosci*, **5**, 57–106.

10. Silman, I. and Futerman, A. H. (1987) Modes of attachment of acetylcholinesterase to the surface membrane. *Eur J Biochem*, **170**, 11–22.
11. Quinn, D. M. (1987) Acetylcholinesterase: enzyme structure, reaction dynamics, and virtual transition states. *Chem Rev*, **87**, 955–79.
12. Massoulié, J., Pezzementi, L., Bon, S., Krejci, E. and Vallette, F.-M. (1993) Molecular and cellular biology of cholinesterases. *Prog Neurobiol*, **41**, 31–91.
13. Taylor, P. and Radic, Z. (1994) The cholinesterases: from genes to proteins. *Ann Rev Pharmacol Toxicol*, **34**, 281–320.
14. Blow, D. M. and Steitz, T. A. (1970) X-ray diffraction studies of enzymes. *Ann Rev Biochem*, **39**, 63–100.
15. Duval, N., Bon, S., Silman, I., Sussman, J. and Massoulié, J. (1992) Site-directed mutagenesis of active-site-related residues in *Torpedo* acetylcholinesterase. Presence of a glutamic acid in the catalytic triad. *FEBS Lett*, **309**, 421–3.
16. Schrag, J. D., Li, Y., Wu, S. and Cygler, M. (1991) Ser-His-Glu triad forms the catalytic site of the lipase from *Geotrichum candidum*. *Nature*, **351**, 761–4.
17. Grochulski, P., Li, Y., Schrag, J. D., Bouthillier, F., Smith, P., Harrison, D., Rubin, B. and Cygler, M. (1993) Insights into interfacial activation from an open structure of *Candida rugosa* lipase. *J Biol Chem*, **268**, 12843–7.
18. Ghosh, D., Wawrzak, Z., Pletnev, V. Z., Li, N., Kaiser, R., Pangborn, W., Jornvall, H., Erman, M. and Duax, W. L. (1995) Structure of uncomplexed and linoleate-bound *Candida cylindracea* cholesterol esterase. *Structure*, **3**, 279–88.
19. Ollis, D. L., Cheah, E., Cygler, M., Dijkstra, B., Frolow, F., Franken, S. M., Harel, M., Remington, S. J., Silman, I., Schrag, J., Sussman, J. L., Verschueren, K. H. G. and Goldman, A. (1992) The α/β hydrolase fold. *Protein Eng*, **5**, 197–211.
20. Dougherty, D. A. (1996) Cation-π interactions in chemistry and biology: a new view of benzene, Phe, Tyr and Trp. *Science*, **271**, 163–8.
21. Cygler, M., Schrag, J. D., Sussman, J. L., Harel, M., Silman, I., Gentry, M. K. and Doctor, B. P. (1993) Relationship between sequence conservation and three-dimensional structure in a large family of esterases, lipases, and related proteins. *Protein Sci*, **2**, 366–82.
22. Harel, M., Schalk, I., Ehret-Sabatier, L., Bouet, F., Goeldner, M., Hirth, C., Axelsen, P. H., Silman, I. and Sussman, J. L. (1993) Quaternary ligand binding to aromatic residues in the active-site gorge of acetylcholinesterase. *Proc Natl Acad Sci U S A*, **90**, 9031–5.
23. Antosiewicz, J., McCammon, J. A., Wlodek, S. T. and Gilson, M. K. (1995) Simulation of charge-mutant acetylcholinesterases. *Biochemistry*, **34**, 4211–9.
24. Porschke, D., Créminon, C., Cousin, X., Bon, C., Sussman, J. L. and Silman, I. (1996) Electrooptical measurements demonstrate a large permanent dipole moment associated with acetylcholinesterase. *Biophys J*, **70**, in press.
25. Ripoll, D. R., Faerman, C. H., Axelsen, P. H., Silman, I. and Sussman, J. L. (1993) An electrostatic mechanism for substrate guidance down the aromatic gorge of acetylcholinesterase. *Proc Natl Acad Sci U S A*, **90**, 5128–32.
26. Tan, R. C., Truong, T. N., McCammon, J. A. and Sussman, J. L. (1993) Acetylcholinesterase: electrostatic steering increases the rate of ligand binding. *Biochemistry*, **32**, 401–3.
27. Shafferman, A., Ordentlich, A., Barak, D., Kronman, C., Ber, R., Bino, T., Ariel, N., Osman, R. and Velan, B. (1994) Electrostatic attraction by surface charge does not contribute to the catalytic efficiency of acetylcholinesterase. *EMBO J*, **13**, 3448–55.
28. Ripoll, D. R., Faerman, C. H., Gillilan, R., Silman, I. and Sussman, J. L. (1995) Electrostatic properties of human acetylcholinesterase, in *Enzymes of the cholinesterase family*, D. M. Quinn, A. S. Balasubramanian, B. P. Doctor, P. Taylor, Eds. Plenum Press, New York, pp. 67–70.
29. Bazelyansky, M., Robey, E. and Kirsch, J. F. (1986) Fractional diffusion-limited component of reactions catalyzed by acetylcholinesterase. *Biochemistry*, **25**, 125–30.
30. Axelsen, P. H., Harel, M., Silman, I. and Sussman, J. L. (1994) Structure and dynamics of the active site gorge of acetylcholinesterase: synergistic use of molecular dynamics simulation and X-ray crystallography. *Protein Sci*, **3**, 188–97.
31. Gilson, M. K., Straatsma, T. P., McCammon, J. A., Ripoll, D. R., Faerman, C. H., Axelsen, P. H., Silman, I. and Sussman, J. L. (1994) Open "back door" in a molecular dynamics simulation of acetylcholinesterase. *Science*, **263**, 1276–8.
32. Lawson, D. M., Brzozowski, A. M. and Dodson, G. G. (1992) Lifting the lid off lipases. *Curr Biol*, **2**, 473–5.

33. van Tilbeurgh, H., Egloff, M. P., Martinez, C., Rugani, N., Verger, R. and Cambillau, C. (1993) Interfacial activation of the lipase-procolipase complex by mixed micelles revealed by X-ray crystallography. *Nature*, **362**, 814–20.
34. Kronman, C., Ordentlich, A., Barak, D., Velan, B. and Shafferman, A. (1994) The "back door" hypothesis for product clearance in acetylcholinesterase challenged by site-directed mutagenesis. *J Biol Chem*, **269**, 27819–22.
35. Faerman, C., Ripoll, D., Bon, S., Le Feuvre, Y., Morel, N., Massoulié, J., Sussman, J. and Silman, I. (1996) Site-directed mutants designed to test back-door hypothesis of acetylcholinesterase function. *FEBS Lett*, in press.
36. Cervenansky, C., Dajas, F., Harvey, A. L. and Karlsson, E. (1991) Fasciculins, anticholinesterase toxins from mamba venoms: biochemistry and pharmacology, in *Snake Toxins*, A. L. Harvey, Ed. Pergamon Press, New York, pp. 303–21.
37. Dufton, M. J. and Harvey, A. L. (1989) The long and the short of snake toxins. *Trends Pharmacol Sci*, **10**, 258–9.
38. Brown, L. R. and Wüthrich, K. (1992) Nuclear magnetic resonance solution structure of the α-neurotoxin from the black mamba (*Dendroaspis polylepis polylepis*). *J Mol Biol*, **227**, 1118–35.
39. Sutcliffe, M. J., Dobson, C. M. and Oswald, R. E. (1992) Solution structure of neuronal bungarotoxin determined by two-dimensional NMR spectroscopy: calculation of tertiary structure using systematic homologous model building, dynamical simulated annealing, and restrained molecular dynamics. *Biochemistry*, **31**, 2962–70.
40. Segalas, I., Roumestand, C., Zinn-Justin, S., Gilquin, B., Menez, R., Menez, A. and Toma, F. (1995) Solution structure of a green mamba toxin that activates muscarinic acetylcholine receptors, as studied by nuclear magnetic resonance and molecular modeling. *Biochemistry*, **34**, 1248–60.
41. Albrand, J. P., Blackledge, M. J., Pascaud, F., Hollecker, M. and Marion, D. (1995) NMR and restrained molecular dynamics study of the three-dimensional solution structure of toxin FS2, a specific blocker of the L-type calcium channel, isolated from black mamba venom. *Biochemistry*, **34**, 5923–37.
42. Stampf, D. R., Felder, C. E. and Sussman, J. L. (1995) PDBBrowse - a graphics interface to the Brookhaven Protein Data Bank. *Nature*, **374**, 572–4.
43. Harel, M., Kleywegt, G. J., Ravelli, R. B. G., Silman, I. and Sussman, J. L. (1995) Crystal structure of an acetylcholinesterase-fasciculin complex: Interaction of a three-fingered toxin from snake venom with its target. *Structure*, **3**, 1355–66.
44. Bourne, Y., Taylor, P. and Marchot, P. (1995) Acetylcholinesterase inhibition by fasciculin: crystal structure of the complex. *Cell*, **83**, 503–12.
45. le Du, M.-H., Marchot, P., Bougis, P. E. and Fontecilla-Camps, J. C. (1992) 1.9-Å resolution structure of fasciculin 1, an anti-acetylcholinesterase toxin from green mamba snake venom. *J Biol Chem*, **267**, 22122–30.
46. le Du, M.-H., Housset, D., Marchot, P., Bougis, P. E., Navaza, J. and Fontecilla-Camps, J. C. (1996) Structure of fasciculin 2 from green mamba snake venom: evidence for unusual loop flexibility. *Acta Cryst*, **D52**, 87–92.
47. van den Born, H. K. L., Radic, Z., Marchot, P., Taylor, P. and Tsigelny, I. (1995) Theoretical analysis of the structure of the peptide fasciculin and its docking to acetylcholinesterase. *Protein Sci*, **4**, 703–15.
48. Marchot, P., Khelif, A., Ji, Y. H., Mansuelle, P. and Bougis, P. E. (1993) Binding of ^{125}I-fasciculin to rat brain acetylcholinesterase. The complex still binds diisopropyl fluorophosphate. *J Biol Chem*, **268**, 12458–67.
49. Machold, J., Weise, C., Utkin, Y., Tsetlin, V. and Hucho, F. (1995) The handedness of the subunit arrangement of the nicotinic acetylcholine receptor from *Torpedo californica*. *Eur J Biochem*, **234**, 427–30.
50. Karlsson, E., Mbugua, P. M. and Rodriguez-Ithurralde, D. (1984) Fasciculins, anticholinesterase toxins from the venom of the green mamba Dendroaspis angusticeps. *J Physiol (Paris)*, **79**, 232–40.
51. Eastman, J., Wilson, E. J., Cervenansky, C. and Rosenberry, T. L. (1995) Fasciculin 2 binds to the peripheral site on acetylcholinesterase and inhibits substrate hydrolysis by slowing a step involving proton transfer during enzyme acylation. *J Biol Chem*, **270**, 19694–701.

23

PROTEIN SEQUENCE AND STRUCTURE COMPARISON USING ITERATIVE DOUBLE DYNAMIC PROGRAMMING

William R. Taylor

Division Mathematical Biology
National Medical Research
Ridgeway, Hill
London 1AA, UK

1. INTRODUCTION

One of the central aims of molecular bioinformatics is to develop new algorithms and software to facilitate the structural, functional and evolutionary interpretation of sequence data. For protein sequence data, an important aspect of this is to produce greater synergism between the sequence and structural databanks: extending the structural information of the latter to illuminate as many sequences as possible while also using the (aligned) sequence data to help understand the evolutionary pressures that maintain structure. Through this ongoing pursuit, a greater general understanding of protein structure can be gained as well as insight into a specific sequence or family which was previously unconnected with any known structure or function.

To date, the most successful approach is empirical — meaning that the structural context of sequences in known structures is analyzed to find recurring patterns or *motifs* which can then be used predictively. The analysis side of this approach requires computational tools that can compare (align) sequences and structures. Similarly, the synthetic aspect of the problem requires tools that can take sequence/structure relationships and construct a plausible molecular model*. In pursuit of the central aim, these tools must be continually improved to encompass increasingly remotely related proteins.

In the current paper, methods will be considered that can be used to establish an alignment between sequences, sequences and structure and between structures. The conversion of these alignments into detailed molecular models will not be considered.

*'*Model*', can be used both to mean a mathematical model (a set of equations) or a molecular model (a set of coordinates). If unqualified, the latter sense should be assumed.

Theoretical and Computational Methods in Genome Research, edited by Suhai
Plenum Press, New York, 1997

1.1. Comparison Methods

The importance of comparison methods. The construction of a molecular protein model for a sequence of unknown structure is typically achieved through sequence alignment. When the two proteins share good sequence similarity this is sufficient, but when they are more highly divergent it is necessary to use constraints from multiple sequences and from the known structure itself. On the side of the known structure (referred to below as the *structural side*) additional data can be recruited in the form of proteins either with or without a known structure. Similarly, on the side to be modelled (referred to below as the *sequence side*), additional sequences can also be recruited. (If any of these have a known structure then the problem becomes simply structure/structure comparison). With this central importance for comparison methods on both the sequence and structural side of the modelling problem, in addition to the key comparison between both sides, it is important to have accurate and robust methods (that do not take too long to run).

Methods can be classified by the type of data (sequence or structural) on which they operate thus giving three broad classes: **sequence/sequence** comparison (or sequence alignment) methods, **sequence/structure** comparison (or threading) methods, and **structure/structure** comparison methods. Those that will be considered below have been developed in the Division of Mathematical Biology at the National Institute for Medical Research (MRC, London) around a common algorithm, but before considering some novel developments, the basic methods will be outlined.

1.2. Sequence/Sequence Comparison

At an early stage, it was important to develop a robust multiple sequence alignment program. This was achieved by aligning the most similar pairs of sequences first and using these as a core on which to add their more remote relatives.[1] With each 'condensation' of sequences, a consensus (or average) can be generated to further improve alignment quality[2] (see Ref. 3,for review).

This approach, which is implemented in the program `MULTAL` remains our standard alignment method. More recently, it has been extended through the use of a very fast peptide-based pre-processing stage to aid in the comparison of very large numbers of sequences. This method can identify only clear similarity (better than 40% identity) but is able to compare 10,000 sequences/second (on an Iris 'Challenge' computer) allowing a sequence data-bank of 65,000 sequences to be compared with itself in a few days (18.5 Giga-comparisons). Such analyses (on a smaller databanks) have resulted in new mutational exchange matrices both for globular[4] and trans-membrane[5] proteins.

1.3. Structure/Structure Comparison

The conventional approach to protein structure comparison was to superpose one structure onto another with both behaving as rigid bodies. For remotely related structures and recurring fragments (motifs) this approach is unreliable and difficult to automate. A new approach was developed based on the comparison of the environments of individual residues.[6,7] However, the need to incorporate insertions and deletions into this approach introduced a difficult computational problem since the correct equivalence of residues between the two protein is needed to compare the local environments. This circularity was overcome by calculating a best local residue equivalence based on each pair of residues using the basic Dynamic Programming algorithm[8] commonly used in simple sequence alignment. With a score for the comparison of each pair of residues, the same algorithm was then reapplied at a higher level to extract the best overall alignment. The algorithm has come to be referred to as the Double Dynamic Programming (DDP) algorithm and will be discussed further below.

Through the application of sensible heuristics[9,10] significant increases in computation speed were achieved, making it possible to compare the protein structure data-bank to itself.[10,11] Using specific alignment methods for sub-fragments[12] novel local motifs have been found.[13] Most recently, the method has been integrated with the multiple sequence alignment program (MULTAL) to produce a true multiple structure alignment program based on an internal structural consensus representation.[14]

1.4. Sequence/Structure Comparison

Given a protein structure and the hypothesis that a new sequence might adopt the same fold, a molecular model can be constructed and evaluated. Using conventional methods, this would require that a tentative alignment is established which can then be refined in the light of the environments occupied by each residue in the model. 'By-hand' (using interactive graphics), this is a laborious process and to avoid it we have automated the modelling process to generate an optimal fit — or threading — of the sequence over the structural backbone of a protein of known structure.[15]

This method required the solution of two problems: firstly, the definition of a rapid evaluation function for any given threading; and secondly, a method to generate only sensible threadings out of the 'astronomic' number of possibilities when sequence gaps are permitted. The difficulty in the first problem was that to specify side-chain positions would make the method too slow. This, however, was overcome using empirical pairwise potentials based only on main-chain and β-carbon atoms,[16] together with average solvation preferences. The second problem, of finding the optimal threading, was solved using using a modification of the Double Dynamic Programming algorithm (as used in structure comparison).

Local and non-local matching. Structural constraints can be evaluated and imposed at two levels: the simpler requires the consideration only of features that occur locally in the sequence and can take the form of a bias to match known secondary structure with predicted secondary structure of like type, or to match regions of sequence variation with regions known to be exposed to solvent in the structure. These simple constraints can all be imposed directly during the alignment of the two sequences (or sub-families). The more complex constraints involve interactions that are non-local in the sequence — such as the preference to have two cysteins sufficiently close to form a disulfide bridge, or (more generally) two hydrophobic residues packed in the core. Such interactions seemingly can only be evaluated after the construction of a model and are therefore computationally more difficult to incorporate.

Following the work of Refs. 17 and 18, the former (simple) constraints will be referred to as "1D/3D comparison" while the establishment of an alignment using true spatial interaction will be referred to as 'threading' — following the usage introduced by Ref. 15.

Double Dynamic Programming in threading. Given a sequence of amino acids $A = \{a_1, a_2, \ldots a_M\}$ (of length M) and a structure $B = \{b_1, b_2, \ldots b_N\}$ (of length N) consisting of one spatial position (b) per residue, a simple dynamic programming algorithm can find the optimal alignment of A and B given a score relating all pairs of positions. If these scores reside in an $M \times N$ matrix (**S**), then the alignment is the highest scoring path from one end of the matrix to the other (top or left edge to right or bottom edge). The required scoring scheme is a measure of how well, say, residue a_i fits in location b_j. This quantity, however, cannot be obtained until the full environment around position b_j is determined and this requires residues to be assigned to the surrounding positions. The problem is apparently circular since the final alignment (or a good part of it) must be known in order to take the first step towards calculating it.

The problem can be overcome by inverting the approach and asking not "how 'happy' is a_i on b_j?" but "if a_i is placed on b_j, then how 'happy' can it be made?". The answer to this reformulated question can be found by applying the dynamic programming algorithm to the matrix (${}^{ij}\mathbf{R}$) in which each element (${}^{ij}R_{mn}$) is a measure of the interaction of a_i (on b_j) with a_m on b_n. The best path through this matrix has an associated score, which provides a measure by which the validity of the original assumption (of placing a_i on b_j) can be assessed. These scores themselves can be taken to form the higher-level matrix (**S**), from which a set of equivalences can be extracted again by a dynamic programming calculation (hence '*double*'), finding the highest scoring path compatible with a sequence alignment. However, following the use of the same algorithm in protein structure comparison,[6] the individual scores (${}^{ij}R_{mn}$) along the optimal path traced in each low-level matrix (${}^{ij}\mathbf{R}$) were summed to produce the high-level matrix. This can be represented as:

$$\mathbf{S} = \sum_{i=1}^{N} \sum_{j=1}^{M} \mathcal{Z}({}^{ij}\mathbf{R}) \tag{1}$$

where the function $\mathcal{Z}$ sets all matrix entries to zero that do not lie on the optimal path.

2. NOVEL DEVELOPMENTS

2.1. Iterative Double Dynamic Programming

An iterative DDP algorithm has been described before (applied to protein structure comparison), however, its reformulation described below has some novel advantages. Previously, a large number of pairs were selected for the initial comparison and after this only 20 were taken.[9] In the reformulated algorithm, this trend is reversed and an initially small selection (typically 20–30) pairs are selected and gradually increased with each iteration. Stability through the early sparse cycles is maintained by using the initial rough similarity score matrix as a base for incremental revision. As the cycles progress, the selection of pairs becomes increasingly determined by the dominant alignment, approaching (or attaining) by the final cycle, a self-consistent state in which the alignment has been calculated predominantly (or completely) from pairs of residues that lie on the alignment.

The selection of positions for comparison saves considerable computation time but if taken too far will lead to a deterioration in the quality of results, as informative comparisons will have been neglected. This approach was investigated for the comparison of protein structures where repeated cycles of selection and comparison were found to be both fast and effective.[9]

2.2. Selection and Iteration

Initial selection. The initial selection of pairs to compare was made on the basis of sequentially local measures, including exposure and secondary structure. In addition, a sequence component was included based on the mean pairwise similarity of two sequence positions as calculated in the multiple sequence alignment program MULTAL. These three components were summed for each pair of positions giving a matrix **Q**, which will be referred to as the '*bias-matrix*' and were scaled to give a roughly equal contribution.

Iteration. The elements of **Q** were then sorted and the highest scoring were taken to initiate the calculation. For each cycle of the iteration the number of pairs selected was determined by a function of the size of the two proteins (M and N) and the iteration number (J). After each comparison, the scores along the best path in the high-level matrix (**S**) were combined with the current bias-matrix (${}^{J}\mathbf{Q}$) as:

$$\mathbf{Q}' = \frac{{}^{J}\mathbf{Q}}{2} + \frac{\log(1 + \mathcal{Z}(\mathbf{S}))}{10} \qquad (2)$$

The scores in the high-level matrix are generally large (often up to 1000) and the damping applied above reduces these into a range with an effective maximum of 1 which is more commensurate with the range of values (maximum 3) found in the bias-matrix (**Q**). (As above, the function $\mathcal{Z}$ sets all matrix elements that do not lie on the optimal path to zero). To further ensure that the **Q** matrix does not become dominated by extreme values, its elements were normalized on each cycle to have a root-mean-square deviation from zero of 1 with no element over 3.

As the cycles proceed, the selected pairs (which are initially scattered in the matrix) tend to become increasingly selected along the dominant alignment while by the final cycle, if the iteration has converged, all selected pairs lie on the alignment. To encourage the attainment of this ideal state, the low-level matrices (**R**) were increasingly substituted by the bias-matrix as a function of the iteration cycle number. On the initial cycle the bias-matrix has no effect and by the tenth cycle there is effectively no contribution from the low-level matrix.

3. MULTIPLE SEQUENCE THREADING

A wide variety of packing potentials have been used to assess the alignment of a sequence on a structure. These range from the very complex multi-atomic, multi-faceted potentials of mean-force[16] through complex contact potentials[19] to simple hydrophobic packing scores.[20] It is difficult to generalize the more complex methods to use multiple sequence data effectively and in the method outlined below, a simple empirical contact preference was developed following earlier work on tertiary structure modelling.[21,22,23,24]

This threading algorithm uses three sources of data for assessing similarity: solvent exposure, secondary structure state and pairwise hydrophobic packing. In other methods these sources are sometimes explicit and sometimes intimately bound together. For example, in the threading method of Ref. 4, exposure is treated explicitly but the pairwise 'interaction' preferences of Ref. 16 used in that approach, also contain implicit secondary structure information in their non-carbon atom distance distributions combined with the separate distributions for different residue spacings along the sequence. By contrast the 3D/1D matching method of Ref. 25 (which does not contain pairwise information) has explicit secondary structure and exposure components, while the method of Ref. 26 contains all components together.

Interacting residues. Following the work of Ref. 2728, the method outlined here focuses on the protein core. Pairs of residues in the structure that are distant and, more importantly, shielded from interacting by other residues should not contribute to the assessment of the pairwise interactions. Previously these have excluded by a simple distance cutoff (say, 10Å).[29] In the current method an equivalent cutoff was applied, but more importantly, residue pairs were assessed for any other residue shielding their interaction. Shielding was calculated from the perpendicular distance of the atom being tested to the line connecting the pair of pseudo-β-carbons of the pair being assessed. A simple Gaussian function was then applied to transform the distance d into a shielding factor f.

As this shielding factor (f) ranges from 1 for no shielding to 0 for co-linear intervention, the product of f values over all the atoms was taken as the overall shielding (F) for the residue pair. The solvent accessibility (e) of the residues was estimated by the fast conic algorithm of Ref. 22 and used to further select buried pairs. Those remaining are referred to as the '*deep-packed*' pairs. The importance of a residue in the network of deep-packed pairs is reflected in the number of residues packed with it. This contact count (C) was used to emphasize the key interactions giving a score r for each pair.

Conserved hydrophobic positions were identified[21] and when these positions occur in a deep-packed pair, a score was calculated from both the degree of interaction of the pair (r) and the degree of conserved hydrophobicity (c) of the two sequence positions. As all these components are required to be present for a good interaction, their product was taken.

$$^{ij}t_{mn} = r_{jn}(c_i + 1)(c_m + 1) \tag{3}$$

3.1. Incorporating Local Preferences

Solvent exposure. Solvent exposure (e) was estimated in the known structure and the sequence the measure of hydrophobic conservation (c) was taken as a prediction of burial. The product of these values, both of which lie in the range $\{-1 \rightarrow +1\}$ (exposed$\rightarrow$buried), is positive when conserved-hydrophobic positions are buried or variable-hydrophilic positions are exposed. This simple score was then modified by the addition of the packing (or contact) number, C, to the exposure measure, giving:

$$u_{ij} = c_i(e_j + C_j) \tag{4}$$

(for a position i in a sequence alignment and j in a structure).

Secondary structure. Secondary structure was predicted from the multiple sequence alignment using the GOR method Ref. 30, this simple approach has recently be shown by Ref. 31 to give remarkably good results (approaching 70%, without any post-processing). The GOR method also allows the percentage of predicted structure to be varied by shifting the predicted propensity curves for each structure relative to each other. This allowed the percentage of predicted structure to match the observed composition. It is important in the current application to identify regions where there is a clear prediction with a good chance of being correct. As shown previously[32] this corresponds with the excess of the highest prediction score over those of other structure types. The excess value was then calculated as:

$$^{z}P_i = \min_{\forall y \neq z} ({}^{z}G'_i - {}^{y}G'_i) \tag{5}$$

and the resulting P values were normalized into the range $0 \rightarrow 1$ and used directly as a score (v) when the predicted and observed secondary structures matched.

3.2. Weighted Combination

Each of the three components — pairwise interaction, exposure and secondary structure were combined into a weighted sum in the low-level matrix (${}^{ij}\mathbf{R}$), as:

$$^{ij}R_{mn} = {}^{t}w({}^{ij}t_{mn}) + {}^{u}w(u_{mn}) + {}^{v}w(v_{mn}). \tag{6}$$

4. NON-TRIVIAL ALIGNMENT TEST DATA

4.1. Protein Family Selection

Many methods described in the current literature, such as 3D/1D alignment, threading methods and hidden Markov models, result in a sequence alignment. To test these methods, and justify their additional complexity over simple sequence comparison, requires alignment problems for which the correct alignment cannot easily be attained by sequence alignment, and importantly, for which the correct alignment is known. As the correct alignment is best defined by structural comparison, both families must contain a protein of known structure that can be taken to produce an unambiguous structural alignment.

A guide to the selection of suitable data can be found in the SCOP classification (`http://scop.mrc-lmb.cam.ac.uk/scop/`)[33] An entry at the 'fold' classification level in SCOP that is divided at the 'super-family' level usually identifies suitable protein groups that are difficult for multiple alignment methods to align across the super family level, but, in general, proteins below the 'super-family' level can be aligned given reasonable multiple sequence data. Accurate structural alignment, however, is also difficult across the 'super-family' level, but these comparisons are usually unambiguous in the core — allowing a set of core motifs to be defined.

4.2. Alignment Scoring

The correctness of alignments is often assessed at the level of individual residues, by comparison to some predetermined reference (such as an alignment derived from the comparison of two structures). However, the detailed alignment of residues in the less structured parts of the protein (loops) are often of little interest or poorly determined (even by structure comparison) relative to those in the more conserved core. This has led to the assessment of alignments by the correspondence of short motif regions which are, typically, associated with core secondary structures.[7,34,35]

Using the core motifs defined by structure comparison it would be possible to score their alignment as either correct or incorrect, however, in very difficult alignments it is often not too serious an error to misalign a helix by a turn or a β-strand by two positions. To introduce this flexibility into the scoring of alignments, motifs have been defined using a number of *key* positions that reflect the secondary structure in which they occur.

Motifs in α-helices typically have key positions corresponding to buried positions, giving rise to the characteristic pattern: `AAaaAAaaAAaaAA` (where lower-case letters indicate key positions). In exposed β-strands the alternating pattern: `EeEeEeE`, was used; while in buried strands a continuous string of key positions was used: `BBBbbbbBBB`. Using the number of matched key positions as a score, the preceding patterns are tolerant of minor structural errors: such as, misaligning a helix by one position (even a turn) or a strand by two residues: both preserving the hydrophobic phasing of their respective structures. However, matches in (hydrophobic) antiphase will be costly, even if the matched structures still overlap substantially.

5. CURRENT PROBLEMS AND SUGGESTIONS

5.1. Parameter Minimisation versus Parameter-Space Analysis

The different components that contribute to a sequence/structure match can be parameterized by weights for predicted/observed exposure matching, predicted/observed secondary structure matching and pairwise interaction. Together with the gap-penalty this gives four parameters for which optimal values might be obtained. This situation, in which a number of parameters can be optimized, occurs frequently in alignment problems: for example: even in the relatively simple application of sequence/sequence comparison, there can be a number of components to the gap-penalty combined with different scoring parameters.[36] From experience with sequence comparison, it has been found to lead to greater understanding of the problem if the parameters are not blindly (stochastically) optimized, but instead, visualized in their joint parameter-space. However, as it is difficult to visualize any parameter-space of more than three dimensions directly, and as projections give a restricted view, a technique used previously to assess multiple protein sequence alignments is recommended.[37] In this approach, the area of the correct (or nearly correct) alignment measured on the space of two parameters was plotted for each combination of the other two.

An important aspect of visualizing the parameter-space is that when good results are obtained, their stability (under parameter variation) can be visualized and evaluated by the area over which the correct (or good) result extends. However, even with only four parameters, characterizing the performance of a method can pose a difficult and computationally intensive task.

The problem of memorization. Many of the current threading methods are based on empirical potentials or functions derived from a widespread analysis of the known structures. When these become encoded in great detail, for example; in large neural nets,[25] Hamiltonians[38] or detailed potentials,[16,27,26] the resulting methods acquire the capacity to recognize a sequence/structure match from specific, rather than general, features. To counteract this potential for memorizing a structure, careful statistical techniques have been employed during the evaluation of the methods — however, many believe that the only true test for such methods is to recognize a novel fold that has been elucidated after the empirical potentials were compiled (see Ref. 39, and following papers).

An alternate strategy is to use only very general properties of proteins and to avoid any capacity in the method to memorize a structure. For example, in the simple threading method outlined above, the only components determining the match are the general principles that hydrophobic residues should pack to form a conserved hydrophobic core and that predicted and observed secondary structures should match. These match properties were drawn from simple tables of hydrophobicity and secondary structure preference, which, when taken together with the simple structural representation, gives no scope to encode specific features.

The relative importance of match contributions. While various threading methods are based on different structural aspects, (like exposure or secondary structure) few have attempted to assess the relative importance of these components within a single method. In early studies it was demonstrated that matching predicted and observed exposure alone was very effective.[17] Other successful early attempts also used predicted secondary structure in combination with residue conservation.[40,41,42] More recently, explicit pairwise interactions have been widely employed in determining a sequence/structure alignment along with simpler methods that do not consider pairwise spatial interactions (3D/1D alignment). The most direct approach can be found in the method of Ref. 26 (which contains all components together). Although these were originally weighted to give equal contributions to the score (without considering their relative importance in determining recognition or alignment) an indirect assessment was made by randomizing the sequence fixed in its correct register and monitoring how the total energy varied. This suggested that the pairwise interaction term was the least important component. This conclusion has received support from the recent carefully controlled 'blind' trials,[39] from which it has further become apparent that the role of pairwise interaction may be less important than originally thought as both the 1D/3D and true threading methods generate equivalent results. This raises the question of whether the great additional computational complexity required for (non-local) pairwise interactions is justified.

To answer this question is not simple as the optimal balance of components will be a direct function the quality of the basic methods that are combined. For example a random secondary structure prediction should be relatively unimportant when combined with a good prediction of exposure. While absolute answers cannot be given, the relative balance of the components within the multiple sequence threading method outlined above can be taken as a simple model. With no pairwise interaction contribution, the method obtained results that were comparable to THREADER (one of the 'state-of-the-art' threading methods applied in blind tests[39,29]). (This result was obtained with an equal balance on the secondary structure matching and exposure matching). Comparing this result to those generated with the pairwise interaction weight revealed a slight increase for the all-α proteins, no increase for the β/α proteins and a 10% increase for the all-β proteins.

This limited and uneven effect may simply be a reflection of the simple model adopted for pairwise interaction which was based only on conserved hydrophobicity and was therefore bound to overlap to a considerable extent with the exposure matching component. The two components are, nonetheless, distinct as might be appreciated from the following hypothetical example. Consider a β-sheet of three strands composed of hydrophobic residues: if the central strand is deleted, then a third of the exposure score is lost but all the inter-strand pairwise interactions will be lost. Such an effect may explain the persistent contribution of the pairwise interaction score.

The possible conclusions from the above result are: either that the simple potential employed has scope for further improvement, or that there is little to be gained from explicit pairwise interactions that cannot be captured by (sequentially) local match-scores. While further improvement of the current method may be possible through elaboration of the pairwise potential, the quality of the alignments suggest that the second option is probably the more realistic conclusion.

5.2. Recognition versus Alignment

As argued above, it is necessary to find data that could not be aligned easily using multiple sequence alignment but could be aligned unambiguously by structure comparison. This difficult problem of checking the alignment quality for remotely similar proteins has been avoided by most threading (and related) methods which instead concentrate on the score of a match and whether high scoring proteins are members of the same superfamily. While this is a valid and useful approach, recent studies have suggested that many methods can recognize a correct match without, necessarily, having identified the correct alignment.[39,29] Recognition might be based on fragments either in the correct phase or out-of-phase with the correct alignment. Taking a hypothetical example, it can be seen how this might easily happen: for example; most of the alternating character of an eight-fold alternating β/α (TIM) barrel could be recognized but in a register one β/α unit out of phase. Similarly, with methods that rely on secondary structure matching, phase problems could be tolerated while still recognizing the approximate number and type of the secondary structures.

Reliance on score (rather than alignment quality), while necessary in 'real' applications, can become especially fraught with difficulties when comparisons are made between different proteins. With pure sequence comparison suitably realistic random trials can be constructed to provide some guide to the significance of a given score, however, the equivalent randomization for protein structures is not such a simple task. A potential solution can be found in the iterative DDP in which it is possible to start the iteration with a fully, or partially, randomized bias-matrix. Repeated many times, this results in a spread of scores giving a measure of confidence to the threading (or structure comparison). This can then be combined with the simple molecular representation (only α-carbons) which facilitates the construction of random or pseudo-random models[22] (respectively). However, a 'random' model that can usefully be applied in this context can be generated simply by reversing the chain direction of one of the proteins being compared — either the sequence on one side or the structure on the other. (The latter is made possible by using just the α-carbons). This test, which was used previously in a pure sequence context to assess the specificity of pattern matching,[42] has the great advantage that it preserves not only the length and composition of the protein but also the symmetric local patterns in the sequence due to secondary structure. When combined with the randomization element in the iterative DDP algorithm, two distributions 'normal' and 'reversed' can be generated and the separation of their means tested for significance.

REFERENCES

1. Taylor, W. R. (1987). Multiple sequence alignment by a pairwise algorithm. *CABIOS*, 3:81–87.
2. Taylor, W. R. (1988). A flexible method to align large numbers of biological sequences. *J. Molec. Evol.*, 28:161–169.
3. Taylor, W. R. (1990). Hierarchical method to align large numbers of biological sequences. In Doolittle, R. F., editor, *Molecular Evolution: computer analysis of protein and nucleic acid sequences*, volume 183 of *Meth. Enzymol.*, chapter 29, pages 456–474. Academic Press, San Diego, CA, USA.

4. Jones, D. T., Taylor, W. R., and Thornton, J. M. (1992b). The rapid generation of mutation data matrices from protein sequences. *CABIOS*, 8:275–282.
5. Jones, D. T., Taylor, W. R., and Thornton, J. M. (1994). A mutation data matrix for transmembrane proteins. *FEBS Lett.*, pages 269–275.
6. Taylor, W. R. and Orengo, C. A. (1989b). Protein structure alignment. *J. Molec. Biol.*, 208:1–22.
7. Taylor, W. R. and Orengo, C. A. (1989a). A holistic approach to protein structure comparison. *Prot. Eng.*, 2:505–519.
8. Needleman, S. B. and Wunsch, C. D. (1970). A general method applicable to the search for similarities in the amino acid sequence of two proteins. *J. Mol. Biol.*, 48:443–453.
9. Orengo, C. A. and Taylor, W. R. (1990). A rapid method for protein structure alignment. *J. Theor. Biol.*, 147:517–551.
10. Orengo, C. A., Brown, N. P., and Taylor, W. R. (1992). Fast protein structure comparison for databank searching. *Prot. Struct. Funct. Genet.*, 14:139–167.
11. Orengo, C. A., Flores, T. P., Taylor, W. R., and Thornton, J. M. (1993b). Identification and classification of protein fold families. *Prot. Engng.*, 6:485–500.
12. Orengo, C. A. and Taylor, W. R. (1993). A local alignment method for protein structure motifs. *J. Mol. Biol.*, 233:488–497.
13. Orengo, C. A., Flores, T. P., Jones, D. T., Taylor, W. R., and Thornton, J. M. (1993a). Recurring structural motifs in proteins with different functions. *Current Biology*, 3:131–139.
14. Taylor, W. R., Flores, T. P., and Orengo, C. A. (1994). Multiple protein structure alignment. *Prot. Sci.*, 3:1858–1870.
15. Jones, D. T., Taylor, W. R., and Thornton, J. M. (1992a). A new approach to protein fold recognition. *Nature*, 358:86–89.
16. Sippl, M. J. (1990). Calculation of conformational ensembles from potentials of mean force. an approach to the knowledge-based prediction of local structures in globular proteins. *J. Mol. Biol.*, 213:859–883.
17. Bowie, J. U., Clarke, N. D., Pabo, C. O., and Sauer, R. T. (1990). Identification of protein folds: matching hydrophobicity patterns of sequence sets with solvent accessibility patterns of known structures. *Proteins*, 7:257–264.
18. Lüthy, R., Mclachlan, A. D., and Eisenberg, D. (1991). Secondary structure-based profiles - use of structure-conserving scoring tables in searching protein-sequence databases for structural similarities. *Proteins-structure Function Genetics*, 10:229–239.
19. Crippen, G. M. (1991). Prediction of protein folding from amino-acid-sequence over discrete conformation spaces. *Biochemistry*, 30:4232–4237.
20. Taylor, W. R. (1991). Towards protein tertiary fold prediction using distance and motif constraints. *Prot. Engng.*, 4:853–870.
21. Taylor, W. R. (1993). Protein fold refinement: building models from idealised folds using motif constraints and multiple sequence data. *Prot. Engng.*, 6:593–604.
22. Aszódi, A. and Taylor, W. R. (1994). Secondary structure formation in model polypeptide chains. *Prot. Engng.*, 7:633–644.
23. Dandekar, T. and Argos, P. (1992). Potential of genetic algorithms in protein folding and protein engineering simulations. *Protein Engineering*, 5:637–645.
24. Hänggi, G. and Braun, W. (1994). Pattern recognition and self-correcting distance geometry calculations applied to myohemerythrin. *FEBS Lett.*, 344:147–153.
25. Rost, B. and Sander, C. (1995). Progress of 1D protein structure prediction at last. *Prot. Struct. Funct. Genet.*, 23:295–300.
26. Nishikawa, K. and Matsuo, Y. (1993). Development fo pseudoenergy potentials for assessing protein 3D-1D compatability and detecting weak homologies. *Protein Engineering*, 6:811–820.
27. Bryant, S. H. and Lawrence, C. E. (1993). An empirical energy function for threading protein-sequence through the folding motif. *Proteins-structure Function Genetics*, 16:92–112.
28. Madej, T., Gibrat, J.-F., and Bryant, S. H. (1995). Threading a database of protein cores. *Proteins-structure Function Genetics*, 23:356–369.
29. Jones, D. T., Miller, R. T., and Thornton, J. M. (1995). Successful protein fold recognitionby optimal sequence threading validated by rigorous blind testing. *Prot. Struct. Funct. Genet.*, 23:387–397.
30. Zvelebil, M. J., Barton, G. J., Taylor, W. R., and Sternberg, M. J. E. (1987). Prediction of protein secondary structure and active sites using the alignment of homologous sequences. *J. Mol. Biol.*, 195:957–961.
31. Levine, J. M., Pascarella, S., Argos, P., and Garnier, J. (1993). Quantification of secondary structure prediction improvement using multiple alignments. *Prot. Engng.*, 6:849–854.
32. Thornton, J. M. and Taylor, W. R. (1989). Protein structure prediction. In Findlay, J. B. C. and Geisow, M. J., editors, *Protein Sequencing: A Practical Approach*, chapter 7, pages 147–190. IRL Press, Oxford.
33. Murzin, A. G., Brenner, S. E., Hubbard, T., and Chothia, C. (1995). SCOP: a structural classification of proteins database for the investigation of sequences and structures. *J. Molec. Biol.*, 247:536–540.

34. McClure, M. A., Vasi, T. K., and Fitch, W. M. (1994). Comparative analysis of multiple protein-sequence alignment methods. *Mol. Biol. Evol.*, 11:571–592.
35. Taylor, W. R. (1995b). Sequence alignment of proteins and nucleic acids. In Meyers, R. A., editor, *Encyclopedia of Molecular Biology and Biotechnology*, pages 856–859. VCH, New York, USA.
36. Taylor, W. R. (1996). Multiple protein sequence alignment: algorithms for gap insertion. In Doolittle, R. F., editor, *Computer methods for macromolecular sequence analysis*, Meth. Enzymol. Academic Press, Orlando, FA, USA. in press.
37. Taylor, W. R. (1995a). An investigation of conservation-biased gap-penalties for multiple protein sequence alignment. *Gene*, 165:GC27–35. Internet journal Gene Combis: `http://www.elsevier.nl/locate/genecombis`.
38. Friedrichs, M. S., Goldstein, R. A., and Wolynes, P. G. (1991). Generalised protein tertiary structure recognition using associative memory Hamiltonians. *J. Mol. Biol.*, 222:1013–1034.
39. C. M-R. Lemer, Rooman, M. J. and Wodak, S. J. (1995). Protein structure prediction by threading methods: evaluation of current techniques. *Prot. Struct. Funct. Genet.*, 23:337–355.
40. Crawford, I. P., Niermann, T., and Kirschner, K. (1987). Prediction of secondary structure by evolutionary comparison: application to the α-subunit of tryptophan synthase. *Prot. Struct. Funct. Genet.*, 1:118–129.
41. Niermann, T. and Kirschner, K. (1990). Improving the prediction of secondary structure of 'TIM-barrel' enzymes. *Protein Eng.,*, 4:137–147.
42. Taylor, W. R. (1986). Identification of protein sequence homology by consensus template alignment. *J. Mol. Biol.*, 188:233–258.

INDEX